Advanced Nanomaterials and Their Applications

This book covers various facets of nanomaterials and their applications including low-dimensional materials along with discussions on in vitro cell imaging, bioanalyses, UV laser applications of scheelite-type nanomaterials, and nanosized cyanobridged metal-organic frameworks, including high spin transition metal ions. It explains transition metal dichalcogenides and magnetic tunnel junction devices as an alternative to complementary metal-oxide semiconductors. One of the main aims of this book is to grow interest in the atomistic simulation process and characterization of these nanoscale devices.

- Details the recent advances in the application of nanomaterials for nanoelectronics devices, sensors, and memories
- Describes the first-principles approach to ultrasensitive electrically doped biosensors
- Discusses the application of nanomaterials in spintronic devices, specifically magnetic tunnel junction devices with new architectures
- Covers nanomaterials in water purification and conducting polymer nanocomposites in electrochemical supercapacitors
- Presents the theoretical background of next-generation MRI contrast agents with nanosized cyanobridged metal-organic frameworks including high spin transition metal ions

This book is aimed at researchers and graduate students of materials engineering and nanoelectronics.

Advanced Nanomaterials and Their Applications

Edited by
Bikash Sharma and Chandan Kumar Sarkar

CRC Press
Taylor & Francis Group
Boca Raton London New York

CRC Press is an imprint of the
Taylor & Francis Group, an **informa** business

Designed cover image: © Shutterstock

First edition published 2024
by CRC Press
2385 NW Executive Center Drive, Suite 320, Boca Raton FL 33431

and by CRC Press
4 Park Square, Milton Park, Abingdon, Oxon, OX14 4RN

CRC Press is an imprint of Taylor & Francis Group, LLC

© 2024 selection and editorial matter, Bikash Sharma and Chandan Kumar Sarkar; individual chapters, the contributors

ISBN: 9781032347226 (hbk)
ISBN: 9781032347240 (pbk)
ISBN: 9781003323518 (ebk)

DOI: 10.1201/9781003323518

Typeset in Times
by Newgen Publishing UK

Contents

PART I Nanomaterials and Electronic Application

PART II Advanced Nanomaterials: Bio-Medical Applications

Preface

Nanomaterials have been receiving tremendous attention in recent years due to their unique features and promising uses in a wide range of sectors. Controlling the size, shape, and composition of nanoparticles has opened up new avenues for the development of materials with specific characteristics and functionality. Nanomaterials have been employed in a variety of applications, including electronics and energy, as well as healthcare and environmental science. As such, there is a rising demand for a comprehensive reference that includes the latest breakthroughs in nanomaterials and their applications.

Advanced Nanomaterials and Their Applications aims to provide a comprehensive overview of the recent advancements in the field of nanomaterials. The book covers a wide range of topics, including theoretical modeling, synthesis, characterization, properties, and applications of various nanomaterials. The book is intended to be a useful reference for scientists, engineers, and researchers working in the area of nanomaterials and their applications.

The book consists of two major sections. The first part of this book embarks on a journey into the world of nonvolatile memories, with a specific focus on emerging dielectric materials for resistive random-access memory (RRAM) and ferroelectric memory. These materials hold promise for revolutionizing data storage and retrieval, and their fundamental properties and applications are thoroughly examined. Continuing our exploration, we delve into the fascinating realm of group III–V materials, investigating their potential as transistors. These materials possess unique properties that make them attractive for high-performance electronic devices, and their synthesis, characterization, and applications are discussed in detail.

Next, we shift our attention to transition metal dichalcogenides (TMDs), which have gained considerable attention in recent years. The chapters dedicated to TMDs shed light on their application as gas sensors, particularly in monitoring environmental pollution. Their exceptional sensing properties and synthesis strategies are examined, showcasing the potential for creating efficient and sensitive gas detection systems. Supercapacitors, with their rapid charge-discharge capabilities and long cycle life, have emerged as essential components for energy storage systems. Nanomaterials play a crucial role in enhancing supercapacitor performance, and this book encompasses the latest research on conducting polymers such as polyaniline (PANI), polypyrrole (PPy), and polythiophene (PTh). Binary and ternary nanocomposite electrode materials are extensively discussed, providing valuable insights into the advancements in supercapacitor technology. The book explores the incredible potential of graphene-based materials in designing cost-efficient and modern supercapacitors. Graphene's exceptional electrical, mechanical, and chemical properties have propelled it to the forefront of material research, and its application in energy storage devices is an area of immense interest.

The second section of the book focuses on the use of advanced nanomaterials in biomedical applications. This section delves into the fascinating realm of first-principles approaches based on density functional theory (DFT) and nonequilibrium Green's function (NEGF). These theoretical frameworks offer powerful tools for designing electrically doped nanodevices, circumventing the challenges associated with conventional doping processes. By leveraging the unique properties of nanomaterials, researchers are pushing the boundaries of device engineering, paving the way for innovative solutions in nanoelectronics.

One of the exciting applications of nanomaterials in the biomedical field is the use of nanoparticles as contrast agents in magnetic resonance imaging (MRI). The chapters in this section provide insights into the development and future perspectives of MRI contrast agents based on nanoparticles. The utilization of paramagnetic nanoparticles, including FDA-approved PB and PBA, with their high contrasting efficiency, holds immense potential as both MRI contrast agents and therapeutic agents. These advancements open up new possibilities for enhanced diagnostics and targeted therapy in the field of medical imaging. However, as with any emerging technology, it is crucial to address the potential risks associated with the use of nanomaterials. This part of the book also sheds light on nanoparticle toxicology and the long-term consequences of their cellular mechanisms. Understanding the interactions between nanomaterials and biological systems is paramount to ensuring their safe and responsible use in biomedical applications. By examining the current state of nanoparticle toxicology, researchers can make informed decisions and develop guidelines for the ethical and sustainable implementation of nanomaterials in healthcare.

The authors of this book are leading experts in the field of nanomaterials, with extensive experience in the synthesis, characterization, and application of these materials. Their contributions are a great resource for academic and scientific fraternity who want to stay current on the newest achievements in this fast-expanding subject. We express our deepest gratitude to the esteemed authors who have contributed their expertise and insights to this collection. Their dedication to advancing the field of nanotechnology has made this book a valuable resource for readers across various disciplines.

We think that *Advanced Nanomaterials and Their Applications* will be a useful resource for researchers, scientists, engineers, and students interested in nanomaterials and their applications. The book presents a detailed summary of the field's most recent advances and emphasizes the promise of nanomaterials for future advancements. We believe that this book will spur additional research and development in the field of nanomaterials and their applications, and we look forward to watching the impact that these materials will have on numerous industries.

Contributors

Writam Banerjee
Center for Single Atom-based Semiconductor Device, Department of Material Science and Engineering, Pohang University of Science and Technology (POSTECH), Pohang, Republic of Korea

Nayan Kamal Bhattacharyya
Department of Chemistry, Sikkim Manipal Institute of Technology, Sikkim, India

Rabina Bhujel
Alpine University, Kamrang, Sikkim, India

Joydeep Biswas
Department of Chemistry, Sikkim Manipal Institute of Technology, Sikkim, India

Anand Shankar Chakraborty
Berlin Wireless Research Centre, Berkeley, CA, USA

Bibek Chettri
Department of Physics, Sikkim Manipal Institute of Technology Sikkim, India

Pronita Chettri
Department of Physics, Sikkim Manipal Institute of Technology, Sikkim, India

Dipesh Choudhury
School of Materials Science and Nanotechnology, Jadavpur University, Jadavpur, Kolkata, India

Sanat Kr. Das
Department of Physics, Sikkim Manipal Institute of Technology, Sikkim, India

Debashis De
Department of CSE, Maulana Abul Kalam Azad University of Technology, West Bengal, India
Department of Physics, University of Western Australia, Crawley, Perth, WA, Australia

Chandan Kumar Ghosh
School of Materials Science and Nanotechnology, Jadavpur University, Jadavpur, Kolkata, India

Bhanita Goswami
Department of Chemistry, Guwahati University, Assam, India

Nibedita Haldar
School of Materials Science and Nanotechnology, Jadavpur University, Jadavpur, Kolkata, India

Prasanna Karki
Department of Physics, Sikkim Manipal Institute of Technology, Sikkim, India

Sudip Kundu
School of Materials Science and Nanotechnology, Jadavpur University, Kolkata, India

Debajyoti Mahanta
Department of Chemistry, Guwahati University, Assam, India

Snehasis Mishra
School of Materials Science and Nanotechnology, Jadavpur University, Kolkata, India

Tanmoy Mondal
School of Materials Science and Nanotechnology, Jadavpur University, Jadavpur, Kolkata, India

Abhishek Mukherjee
Agricultural and Ecological Research
    Unit, Biological Science Division,
    Indian Statistical Institute, Giridih,
    Jharkhand, India

Sadhna Rai
Department of CMSNT, Sikkim
    Manipal Institute of Technology,
    Sikkim, India

Debarati Dey Roy
B. P. Poddar Institute of Management &
    Technology, India

Pradipta Roy
B. P. Poddar Institute of Management &
    Technology, India

Panchanan Sahoo
Indian Statistical Institute, Giridih,
    Jharkhand, India

Bikash Sharma
Department of ECE, Sikkim Manipal
    Institute of Technology, Sikkim, India

Suveksha Tamang
Department of Chemistry, Sikkim
    Manipal Institute of Technology,
    Sikkim, India

About the Editors

Bikash Sharma (*SMIEEE (USA), FIE (I), FIETE, MIAAM (Sweden), MInstP (UK), MSEE (South Asia), LMMRSI, LMIPS, LMSSI, LMISTE*) is currently serving as Associate Professor in the Department of Electronics and Communication Engineering, Sikkim Manipal Institute of Technology (SMIT), Sikkim, India, since 2009. He has served as System Analyst at M/s Envision Pvt. Ltd., Sikkim, from 2004 to 2005 and Assistant Professor at SMIT from 2006 to 2009. He is currently the Head of the Department of Electronics and Communication Engineering, SMIT, and the Entrepreneurship Cell and Nodal Officer of the Intellectual Property Rights Cell (supported by the Department of Science and Technology, Government of Sikkim), SMIT. He is also the Vice President of the Institution's Innovation Council, SMIT, under the aegis of the Ministry of Education's Innovation Cell, Government of India and has been its convener in the past. He is a recipient of All India Council for Technical Education (AICTE), Quality Improvement Programme (Engineering) National Fellowship (2014–2017) for Ph.D. at Jadavpur University, Kolkata.

Currently he is Secretary of the IEEE Electron Devices Society (EDS) Kolkata Chapter, India and Secretary of the Atomic and Molecular Interest Group, Institute of Physics, UK. He is also the past Treasurer of IEEE EDS Kolkata Chapter, India, from 2017 to 2019. He is Fellow of the Institute of Engineers (IE(I)) and Fellow of the Institution of Electronics and Telecommunication Engineers (IETE).

His area of research includes nanoelectronics, spintronics, advanced nanomaterials and low-dimensional materials and its application in non-volatile memories and nanoelectronics devices and sensors. His work is majorly focused on advancement of low-dimensional nanomaterials, especially two-dimensional nanomaterials for magnetic tunnel junction (MTJ) memory devices and biosensors. He has published more than 60 peer-reviewed papers in international journals and conferences.

He is a recipient of Dr. Sarvepalli Radhakrishnan Teacher's Excellence Award 2021 by AICTE, Government of India and Shikshak Kalyan Foundation. He is also recipient of the 2017–2018 Albert Nelson Marquis Lifetime Achievement Award, USA. His biographical profile was listed in The Editors of Marquis *Who's Who in Science and Engineering* in 2007–2008 and 2009–2010 editions of *Who's Who in Science and Engineering*, USA; IBC Top 100 Engineers in 2009 by the International Biographical Centre, Cambridge, England; and *Great Minds of the 21st Century* in 2007–2008, a major reference directory by the American Biographical Institute.

Chandan Kumar Sarkar is ex-Head of Department of Electronics & Telecommunication Engineering, Jadavpur University. He completed his M.A. Status from Oxford University in 1984, D. Phil from Clarendon Laboratory, Oxford University in 1983, and Experimental Physics Ph.D. from Radio Physics & Electronics, Calcutta University in 1979.

He has a total of 49 years of research experience. His areas of interest and research are Electron Transport in Semiconductors and Alloys, Quantum Transport in Low-Dimensional Systems, MOS Device Physics, Thin Film and Related Devices,

Microwave and MM Wave Devices and Systems, Nanocrystal Embedded Flash Memory Design, III–V Heterostructure Devices, and Nanotechnology and Gas Sensors.

He has been INSA visiting fellow at IIT Kharagpur (1992) and TIFR (1997), U.G.C. Career Award (1994), 1851 Exhibition Research Fellow at Oxford (U.K.) (1983–1985), 1851 Exhibition Science Scholar at Oxford University (U.K.) (1980–1982), Junior Research Fellow at Wolfson College, Oxford (1983–1985), Graduate award at Wolfson College, Oxford (1982–1983), INSA – Royal Society Visiting Fellowship (1993), Koirala Foundation Fellow to Visit Kathmandu, TWAS Associate Program at Shanghai Institute of Metallurgy, China, CBPF, Brazil, Distinguished Lecturer of IEEE ED Society, EDS Chair, Kolkata Chapter, Third World Academy of Science, Trieste at Italy selected as a TWASUNESCO ASSOCIATE at Center of Excellence CBPF, Rio de Zenireo, Brazil.

Part I

Nanomaterials and Electronic Application

1 Materials in Emerging Nonvolatile Memory Devices[1]

Writam Banerjee

1.1 INTRODUCTION

In the era of advanced technology, electronic memory is an essential element to boost new applications. In general, memory devices are divided into two broad groups based on the requirement of power to memorize the stored information. One needs constant power to remember the state, referred to as volatile memory (VM). In contrast, another is capable of remembering the data without the cost of power, referred to as nonvolatile memory (NVM). So far, the need for temporary and permanent data storage is fulfilled by the complementary metal-oxide-semiconductor-based memories, i.e., VM-type dynamic random-access memory (DRAM) and static random-access memory (SRAM) and NVM-type flash memory. The recent progress has experienced the "memory wall," i.e., the speed gap between logic and memory. To overcome the critical system performance bottleneck and fundamental limitations associated with shrinking device size and increased process complexity, emerging NVM with exciting architectures have been proposed. In semiconductor technology innovation, high-performance computing is the driving tool. However, in the era of the internet of things (IoT), consumer electronics is moving toward data-centric applications, with new requirements such as ultralow power operation, low-cost design, high density, highly reliable, longer data storage capability, etc. (Banerjee, 2020). Over the last few decades, the computing capabilities have enhanced with the miniaturization of transistor size. Flash memory is the basic NVM available in the semiconductor market and is dominating to date. However, over the years, this technology has adopted several high-κ oxide materials because of superior scalability of equivalent oxide thickness (EOT) and other advantages as compared to SiO_2 (Banerjee, Kashir, et al., 2022; Banerjee, Liu, et al., 2020; Banerjee, Liu, Long, et al., 2017; Jeon, 2020; Kol & Oral, 2019; Nikam et al., 2021; Rajendran et al., 2021; Ray et al., 2013). Among several adopted high-κ materials, substoichiometric hafnium oxide (HfO_x) or stoichiometric hafnium oxide (HfO_2) is an attractive material with a high dielectric constant of 20–25 and bandgap of 5.3–5.7 eV (Jeon, 2020). In the flash memory domain, HfO_x has been used in several forms such as gate oxide, charge trapping layer, doped oxide, formation of a nanolaminate layer with other high-κ material, and so on. However, the structural limitations of flash memory impose a multitude of challenges, including device scaling and data retention (Banerjee,

DOI: 10.1201/9781003323518-2

Maikap, Tien, et al., 2011; Goda, 2021; Hamzah et al., 2019; Lan et al., 2013; W. C. Li et al., 2011; W.-C. Li et al., 2010a, 2010b; Maikap et al., 2011; Parat & Dennison, 2015; Shirota, 2019). To overcome those problems, a new device needs to be introduced in which, instead of using the concepts of charge trapping, different mechanisms such as phase change, filament formation/dissolution, and change of dipoles have been adopted. The phase-change memory (PCM) is a current prototype memory in which the memory is defined with the change of crystalline phase to amorphous phase. In this type of device, HfO_2 mostly plays a supporting role (T. C. Chang et al., 2011; J. S. Lee, 2011; Y. Lu, Song, Song, Ren, Liu, et al., 2011; Y. Lu, Song, Song, Ren, Peng, et al., 2011; Monzio Compagnoni & Spinelli, 2019). Several emerging NVMs have garnered massive interest from researchers for prospective applications in high-density memory, storage class memory, neuromorphic computing, hardware security, etc. Among emerging NVMs, the resistive random-access memory (RRAM) and ferroelectric memory are the best-emerging recourses for all kinds of applications (A. Chen, 2016; Lanza et al., 2019; Suri et al., 2013; Wong & Salahuddin, 2015; Zidan et al., 2018). Figure 1.1(a) shows the schematics of the basic structure of RRAM, switching mechanism, and typical current-voltage (I-V)

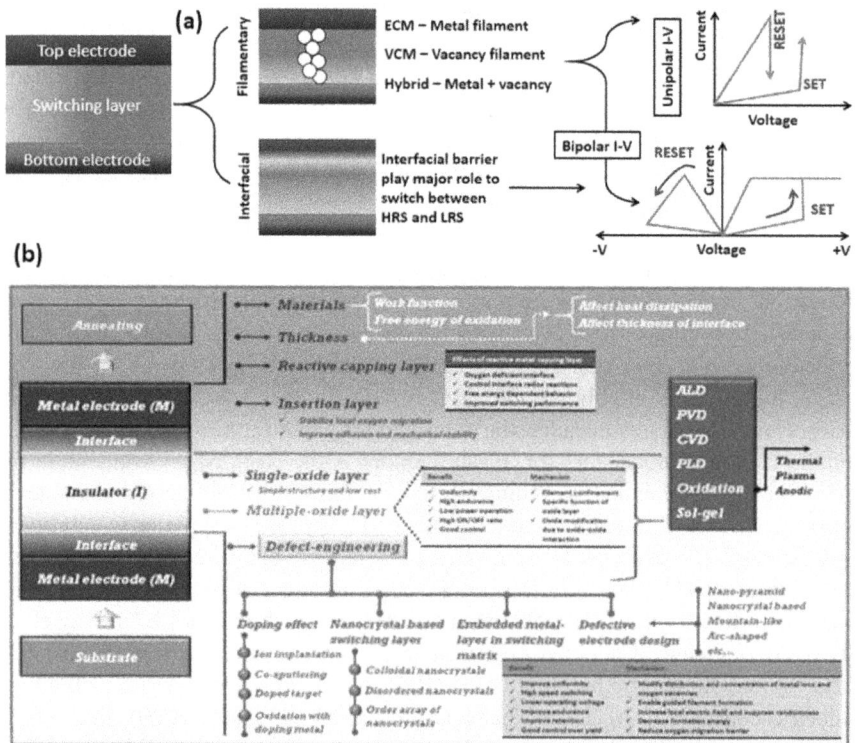

FIGURE 1.1 The schematic representation of the device architecture, mechanism schematic, and typical I-V curve for (a) RRAM and (b) its fabrication. (a) Reproduced with permission (Banerjee, Kashir, et al., 2022). Copyright 2022, Wiley-VCH. (b) License under CC by 3.0 (Banerjee, 2020).

performance. Details about the device engineering of RRAM are shown in Figure 1.1(b). Switching can be filamentary or interfacial type. The filamentary RRAM can be of subcategories such as metal filament–based electrochemical metallization memory (ECM), vacancy filament–based valence change memory (VCM), and a mixed (metal + vacancy) filament–based hybrid memory. Depending on the design, the filamentary RRAM can show both the unipolar or bipolar I-V. In contrast, interfacial RRAM can only show bipolar I-V. In different types of RRAM devices, HfO_2 is one of the most researched switching oxide materials (Azzaz et al., 2016; Banerjee, Cai, Zhao, Liu, et al., 2017; Banerjee, Karpov, et al., 2020a, 2020b; Banerjee, Kim, Lee, Lee, & Hwang, 2021; Banerjee, Kim, Lee, Lee, Lee, et al., 2021; Banerjee, Liu, Lv, et al., 2017; Banerjee, Lu, et al., 2018; Banerjee, Wu, et al., 2018; Banerjee, Xu, Lv, Liu, et al., 2017a, 2017b; Banerjee, Zhang, et al., 2018; Banerjee & Hwang, 2019, 2020a, 2020b; Böttger et al., 2020; C. Y. Chen et al., 2015; Y. Y. Chen et al., 2013; Ding et al., 2020; Fantini et al., 2013, 2014; S. Fujii et al., 2019; Giri et al., 2021; Goux et al., 2012; Govoreanu et al., 2011, 2013, 2016; Grisafe et al., 2019; Guy et al., 2013; Hang et al., 2019; Hua et al., 2019; Hui et al., 2017; Jameson et al., 2016, 2018; Jameson & Kamalanathan, 2016; Jiang et al., 2017; D. Lee et al., 2019; H. Y. Lee et al., 2008, 2010; J. S. Lee et al., 2015; M. J. Lee et al., 2011; S. Lee et al., 2015, 2021; Y. Li, Long, et al., 2015; Lin et al., 2021; N. Lu et al., 2015; Luo et al., 2016, 2018; Lv, Xu, Liu, et al., 2015; Lv, Xu, Sun, et al., 2015; Maikap & Banerjee, 2020; Mallol et al., 2017; Menzel et al., 2019; Milo et al., 2019; Molas et al., 2015, 2018; Nikam et al., 2019; Niu et al., 2016; H. W. Park et al., 2021; J. Park et al., 2019; Pebay-Peyroula et al., 2020; Pi et al., 2019; Prakash et al., 2013, 2015; Raghavan, Degraeve, et al., 2013; Raghavan, Fantini, et al., 2013; Roy et al., 2020; Saadi et al., 2016; Sassine et al., 2016, 2019; Shi et al., 2018; Shin et al., 2016; Shubhakar et al., 2015; Shukla et al., 2017; Stathopoulos et al., 2017; H. Tian et al., 2013, 2014; Traore et al., 2015; Traoré et al., 2015; Tsai et al., 2016; C. Wang et al., 2018; Q. Wang et al., 2019; W. Wang et al., 2019; Waser et al., 2009; Q. Wu, Banerjee, et al., 2018; Q. Wu, Wang, et al., 2018; Yong et al., 2021; Yuan et al., 2017; Zhao et al., 2017; Zhirnov et al., 2011). In RRAM devices, the actual resistive switching process is through filament formation or by controlling the interface. The HfO_2 is one of the established materials in CMOS domain along with SiO_2, Al_2O_3, etc. (Banerjee et al., 2015; Banerjee, Maikap, Chen, et al., 2012; Banerjee, Maikap, Lai, et al., 2012; Banerjee, Maikap, Rahaman, et al., 2011; Banerjee, Rahaman, et al., 2011, 2012; Bozano et al., 2004; Burr et al., 2017; T. Chang et al., 2011; A. Chen, 2017; Cho et al., 2013; B. J. Choi et al., 2016; S. J. Choi et al., 2011; Du et al., 2015; T. Fujii et al., 2011; Gao et al., 2013; Gogurla et al., 2013; Grossi et al., 2018; Gubicza et al., 2016; Guo et al., 2007; Hong et al., 2018; Hu et al., 2014; Indiveri et al., 2013; Kim et al., 2011; Kozicki & Mitkova, 2006; M. K. F. Lee et al., 2019; Y. Li et al., 2013; Y. Li, Xu, et al., 2015; Q. Liu et al., 2012; S. Liu et al., 2016; Luo, Xu, Liu, Lv, Gong, Long, Liu, Sun, Banerjee, Li, Gao, et al., 2015; Luo, Xu, Liu, Lv, Gong, Long, Liu, Sun, Banerjee, Li, Lu, et al., 2015; Merolla et al., 2014; S. Park et al., 2013; Pearson et al., 2013; C. N. Peng et al., 2012; P. Peng et al., 2012; Pi et al., 2019; Seok et al., 2014; Sheridan et al., 2017; Sun et al., 2014; Sung et al., 2018; Theodore M Wong, 2012; X. Tian et al., 2014; C. Wang et al., 2019; W. Wang et al., 2016; Z. Wang et al., 2007; Z. Wang, Joshi, et al., 2017; Z. Q. Wang et al., 2012; Waser & Aono, 2007; Z. Wu et al., 2019; Xia & Yang, 2019; Xu et al., 2016; X. B. Yan et al., 2014;

Z. B. Yan & Liu, 2013; J. J. Yang et al., 2013; Y. C. Yang et al., 2009; L. Q. Zhu et al., 2014; X. Zhu et al., 2012; Zidan et al., 2018). However as compared to the conventional CMOS compatible materials like SiO_2, Al_2O_3, etc., HfO_2 has moderately lower band gap. This can produce relative higher leakage current in HfO_2 and enable a lower forming voltage in HfO_2-based RRAM devices (T. C. Chang et al., 2011; Padovani & Larcher, 2018; I.-S. Park et al., 2013; Petzold et al., 2019; Sokolov et al., 2017). Previously, several groups modeled the switching properties and predicted similar behavior. Both theoretical and experimental findings equip this technology for emerging applications (Banerjee, Nikam, et al., 2022; W. Choi et al., 2020; Fumarola et al., 2018; Hong et al., 2018; Islam et al., 2019; Moon et al., 2018; Z. Wang et al., 2019; Z. Wang, Joshi, et al., 2018; Z. Wang, Kang, et al., 2017; Z. Wang, Rao, et al., 2018). Depending on the structural design and materials, the switching in RRAM devices can be further classified as volatile threshold switch (TS) and nonvolatile memory switch (MS). Generally, TS devices are applicable for selectors, and MS devices are useful as memory. Interestingly, with proper tuning of the structure, HfO_2 is useful to design all possible variations. This chapter describes materials for NVM technology, especially HfO_2, its properties, how it is useful for baseline NVMs and emerging NVMs, and its prospects and concomitant challenges.

1.2   PROPERTIES OF HAFNIUM OXIDE

Details about the properties of HfO_2 were reported by Banerjee et al. (Banerjee, Kashir, et al., 2022). HfO_2 belongs to the transition metal binary oxides, which crystallize into a cubic structure at ~3000 K. In bulk form, it has three stable polymorphs, which appear consecutively on cooling, i.e., cubic (c, $Fm\overline{3}m$), tetragonal (t, $P4_2/nmc$) (at 2803 K), and monoclinic (m, $P2_1/s$) (at 2050 K) structures (Figure 1.2(a–c)). Therefore, at the ambient condition, the formation of an m-HfO_2 is expected. With a wide band gap (~5.5 eV), high dielectric constant ϵ_r, extreme thinness, good reliability, and high binding energy between the oxygen and transition metal ions, and the compatibility with the current complementary CMOS technology, HfO_2 has attracted the attention of theoretical and experimental scientists as one of the most promising materials as a new high-κ material candidate in silicon technology. Currently, the use of HfO_2 in the semiconductor fabrication process is in the mature stage. Its advantages, especially its compatibility with CMOS technology, have motivated various electronic enterprises to invest a considerable proportion of their budget to implement the HfO_2-based film as a dielectric replacement for SiO_2 in DRAMs, which can practically resolve the scaling issues related to SiO_2 ultrathin films. Indeed, a relatively large electronic band gap E_G combined with a high ϵ_r gives elbow room for creative researchers toward the fabrication of ideal high-κ devices (Figure 1.2(d)). Ferroelectric properties are another important part of HfO_2. Unlike the conventional ferroelectric materials such as PZT and BTO, which are strongly affected by scaling down film thickness, the HfO_2-based materials exhibit superlative ferroelectric properties at the nanoscale, which is in favor of the miniaturization of electronic devices. In fact, strong ferroelectricity at the nanoscale enables the fabrication of excellent sub-10 nm emergent ferroelectric-based devices. Moreover, a

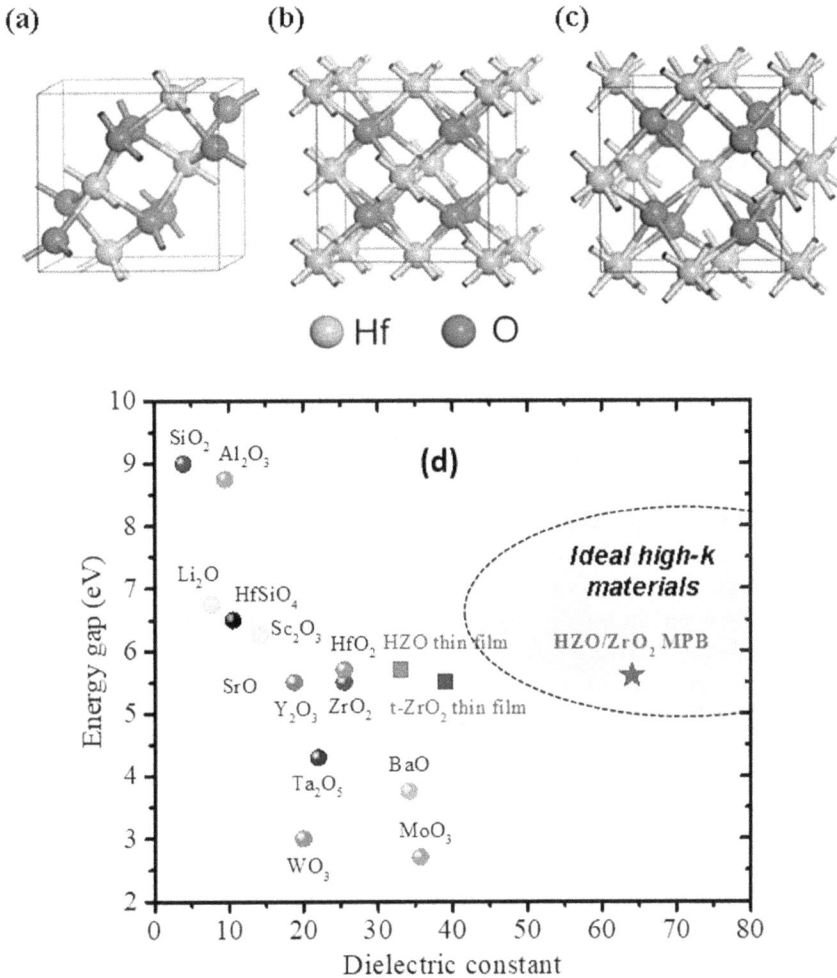

FIGURE 1.2 The structure of HfO_2 in (a) monoclinic; (b) cubic; and (c) tetragonal phases. All these polymorphs are centrosymmetric, showing linear dielectric response. (d) Dielectric features of oxides. Reproduced with permission (Banerjee, Kashir, et al., 2022). Copyright 2022, Wiley-VCH.

relatively low crystallization temperature has made it a popular choice for industrial purposes.

1.3 USE OF DIELECTRIC PROPERTIES OF HAFNIUM OXIDE FOR MEMORY APPLICATIONS

The fundamental concept of traditional nonvolatile flash memory depends on the tunneling of charge from the silicon substrate and storing the same in the trapping layer (Banerjee, Maikap, Tien, et al., 2011; T. C. Chang et al., 2011; Goda, 2021; Hamzah et al., 2019; Lan et al., 2013; W. C. Li et al., 2011; W.-C. Li et al., 2010a, 2010b; Maikap et al., 2011; Monzio Compagnoni & Spinelli, 2019; Parat & Dennison, 2015; Shirota, 2019). Because of their silicon compatibility, SiO_2 and Si_3N_4 have been extensively used in flash devices. The bandgap/dielectric constant of SiO_2 and Si_3N_4 are 8.4–11 eV/3.6 and 5.3 eV/9.5, respectively. Equation 1.1 evidences the fact that higher dielectric constant material is a critical consideration for reducing EOT.

$$EOT = Thickness_{high-\kappa} \left(\frac{\text{Dielectric constant}_{SiO_2}}{\text{Dielectric constant}_{high-\kappa}} \right) \tag{1.1}$$

HfO_2 is one of the most widely used high-κ materials in the flash memory arena. The basic construction of the flash memory is based on three layers on the top of the silicon substrate, namely the tunnel layer, trapping layer, and blocking layer. The development process of flash memory has experienced several structural evolutions, from charge-trapping memory to nanocrystal memory. Several companies have reported flash memory devices (Banerjee, Liu, Long, et al., 2017; Ray et al., 2013). As a mature high-κ material, HfO_2 has outstanding properties. As a gate oxide material, a 25-nm-thick HfO_x layer was reported by Park et al. (Hamzah et al., 2019). The presence of oxygen vacancy (V_O) can increase gate leakage as they can be charged. Tse et al. (Banerjee, Maikap, Tien, et al., 2011) reported good passivation of V_O by fluorine. Choi et al. (Maikap et al., 2011) reported that an optimized postannealing process at 750°C can noticeably decrease gate leakage and increase the capacitance as compared to the as-deposited HfO_x. However, the higher postannealing temperatures can increase the gate leakage because of the formation of grain boundaries. W. C. Li et al. (2011) reported a charge-trapping flash memory based on n-Si/SiO_2/HfO_x/Al structure. The thickness of the HfO_x layer is a crucial factor in this case. As compared to a single SiO_2 layer, the SiO_2/HfO_x can improve the charge-trapping performance by increasing the HfO_x layer thickness. To improve the charge trapping behavior of the flash memory devices, band engineering using atomic layer deposited HfO_2/Al_2O_3 nanolaminates was demonstrated by Lan et al. (W.-C. Li et al., 2010b). For increasing the dielectric constant, sublayer thickness scaling of the HfO_2/Al_2O_3 nanolaminates was used in Lan et al. (2013) and resulted in a dielectric constant value ϵ_r =17.7 for HfO_2:Al_2O_3 with 1:1 nm, as compared to ϵ_r =13.3 with 50:50 nm. Banerjee et al. (W.-C. Li et al., 2010a) used an ultrathin 1-nm HfO_2 layer as an additional tunnel oxide to SiO_2. The stack tunnel oxide layer was proven to be beneficial for improving the operation speed and data retention by band engineering methodology. As charge trapping layer, HfO_2/Al_2O_3 nanolaminates with a 4:1 ratio were deposited. After going through the N_2 annealing process at 900°C for 1 min, the $HfAlO_x$ nanocrystals were formed. It was reported that as compared to a continuous charge-trapping layer, the nanocrystals can improve device performance by serving as a discrete charge-trapping node. The nanocrystals flash memory can effectively improve the operation speed by enhancing the localized electric field. Additionally, the discrete charge-trapping ability can improve data retention by minimizing the possibility of leakage during retention. As compared to HfO_2's charge-trapping memory, HfAlO nanocrystal flash can improve the effective memory window margin (WM) by enhancing electron/hole trapping probability and data retention. Apart from the high-κ nanocrystals, other semiconductor and metal nanocrystals have been investigated to improve the performance of HfO_2-based flash devices (T. C. Chang et al., 2011; Goda, 2021; Monzio Compagnoni & Spinelli, 2019). As compared to various metal nanocrystals, the semiconductor nanocrystals or high-κ (HfO_2-/Al_2O_3-based) nanocrystals are compatible with CMOS technology. Even after decades of intensive efforts for their development, flash memory devices are confronted with major issues in terms of thickness scaling of oxide layers, as it can severely degrade the data retention properties. Additionally, the performance of flash memory in terms of speed and endurance is much lower than the volatile DRAM or SRAM devices. Emerging NVMs are essential alternatives to mitigate the performance gap and can be regarded as storage-class memory (SCM).

To fulfill the need of the hour, a new prototype NVM like PCM is an emerging solution. Unlike flash memory, HfO_x cannot control the basic switching principle of PCM devices. In general, the memory in PCM is constructed with a changing phase from crystalline to amorphous in chalcogenide glass. The HfO_x thin films are used to improve the performance of such switching. Lu et al. (J. S. Lee, 2011; Y. Lu, Song, Song, Ren, Peng, et al., 2011) asserted that the doping of HfO_x in a PCM film can stabilize the phase and improve the device's lifetime. The activation energy of crystallization in HfO_x-incorporated PCM device increases from 2.36 eV to 4.69 eV with improved data retention from 108°C to 187°C upon increasing HfO_x concentration. Suri et al. (Y. Lu, Song, Song, Ren, Liu, et al., 2011) demonstrated that an ultrathin 2-nm HfO_x layer can act as an interfacial layer between the heating plug and GST layer and improve the intermediate resistance states of the devices, which is suitable for neuromorphic applications. As HfO_x cannot directly participate in the PCM switching process, the use of HfO_x in PCM is very limited. Nevertheless, the HfO_2 thin films have been thoroughly investigated for designing emerging NVMs such as RRAM and ferroelectric memory devices. On this note, before going to discuss the role of HfO_2 in RRAM and ferroelectric memories, it is essential to understand the manufacturing conditions, which will be discussed in the following section.

1.4 DEPOSITION AND GROWTH OF HFO$_2$ FILM

Usually both physical vapor deposition (PVD) and chemical vapor deposition (CVD) are employed to deposit the desired quality of HfO_2 for RRAM applications. Apart from that, the sol-gel coating method for HfO_2 thin film is also reported. However, as compared to PVD and CVD HfO_2 thin films, the sol-gel method can't guarantee a much uniform film, so this method is not suggested for volume production. In PVD systems, as compared to electron beam evaporation deposited HfO_2, sputtered HfO_2 is reported by several groups (Hua et al., 2019). The HfO_2 layer can be deposited from a pure HfO_2 target with or without an oxygen (O_2) environment. In the case of the deposition of HfO_2, RF magnetron sputtering was used by Goux et al. (2012) under an argon flow of 30 sccm at the working pressure of 4 mTorr at room temperature with 150 W power. Others have used different conditions to deposit HfO_2 film. Point to be noted: by varying the O_2 flow in the deposition chamber the composition of

HfO$_2$ can be modulated. Therefore, depending on the manufacturing process the film quality can be altered. However, the sputtering process can easily widen the process window with the flexibility of the deposition of desired nonstoichiometry structure. But the emerging high aspect ratio structures are pushing the manufacturing process to adopt the atomic layer deposition (ALD), a type of CVD process, with better conformal deposition and fine control of ultrathin film. Several studies identify the HfO$_2$ film growth methodology using various ALD systems (S. Lee et al., 2015; H. Tian et al., 2014). Generally, tetrakis-ethyl methylaminohafnium precursor is used to grow the HfO$_2$ thin film using a thermal ALD system. During the film growth the deposition temperature was fixed at 250°C and the precursor temperature at 90°C. Here we must discuss the impact of the manufacturing system on the film qualities. In general, the HfO$_2$ films prepared using the ALD system are nearly stoichiometry with a higher density as compared to the sputtering system. A comparative study shows that different manufacturing systems can produce HfO$_2$ with different quality (Hua et al., 2019). The Hf:O ratio can change from 1:1.8, 1:1.91, and 1:2 with densities of 6.3, 9.4, and 9.8 g cm^{-3}, respectively, for the film deposited by sputtering, thermal ALD, and plasma-enhanced ALD, respectively. Other CVD systems, like MOCVD, MBE, etc., are rarely reported to manufacture HfO$_2$ for RRAM application. Hence sputtering and ALD are the most common systems in RRAM, and depending on the requirement the fabrication methodology must be selected. For example, in the case of sputtered HfO$_x$, an Ag/HfO$_x$ RRAM can show MS performance, but with ALD HfO$_x$ the same Ag/HfO$_x$ RRAM can show TS performance (S. Lee et al., 2015). Therefore the design of the experiment is one of the key factors to control the performance of the RRAM device.

1.5   USE OF HAFNIUM OXIDE FOR RESISTIVE RANDOM ACCESS MEMORY DEVICES

In general, depending on the current-voltage (*I–V*) performance, an RRAM can be operated under two different switching modes, i.e., unipolar *I–V* and bipolar *I–V*. The field-dependent migration of V$_O$ can play a crucial role in the whole switching process in any kind of RRAM device. Considering O_O as the oxygen ions, the formation of V$_O$ can be understood using Equation 1.2 (Govoreanu et al., 2013):

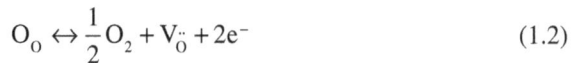

$$O_O \leftrightarrow \frac{1}{2}O_2 + V_{\ddot{O}} + 2e^- \qquad (1.2)$$

The positively charged V$_O$s tend to migrate toward the negatively charged cathode side, and the negatively charged O_Os tend to migrate toward the positively charged anode side. In RRAM, the switching process can be conducted by the formation of conducting filament (CF), i.e., filamentary switching type, or the forward and backward movement of ions at the metal/oxide interface, i.e., interface switching type. However, for any type, the transition from the high-resistance state (HRS) to the low-resistance state (LRS) is known as the SET process, whereas the reverse is known as RESET. The forming process of CF is triggered by the electric field-driven dielectric breakdown process. A dielectric breakdown can be divided into five regimes: (1) stress-induced leakage current (SILC); (2) time-dependent dielectric breakdown (TDDB); (3) digital breakdown (Di-BD); (4) analog breakdown (An-BD); and (5) thermal runaway dependent breakdown or hard breakdown (HBD). Regimes (1) and (2) are area-dependent behaviors, while (3)–(5) are localized. It is worth noting that the current limitation can play a key role in (5), and a very high current can degrade the dielectric. A limited current within several tens to hundreds of microamperes can be sufficient to obtain resistive switching behavior. However, lower current levels are suitable for low-power applications.

In the unipolar type, both the SET and RESET processes are observed under external voltage pulses with only one polarity. The unipolar-type is only possible after the formation of CF, which can control the following resistive switching events. The electric field–dependent movement of O_O toward the anodic interface generates the V_O-filament. An excessive current that can increase the temperature of the CF surrounding area is known as the Joule heating process, which can be several hundred Kelvins near the filament and can be described analytically using Equation 1.3 (Govoreanu et al., 2011):

$$T = T_O + \frac{R_{Th}}{R} V^2 \qquad (1.3)$$

where T_O is the room temperature and R_{Th} is the effective thermal resistance. Previous studies have demonstrated that the CF in Cu/HfO_2 can have a temperature of around 700 K (Banerjee & Hwang, 2020b). Two different forces – Soret force and Fick force – may arise because of Joule heating (Q. Wu, Wang, et al., 2018). During forming and SET process, the former is in action. Under the Joule heating condition, the filament is the most heated region in the oxide layer. Therefore, the O_O near the filament area is expected to move toward the cooler region, and thus, the new V_Os can easily form within the filament area. The Fick force is the opposite of the Soret force and can be in action during the RESET operation. In RESET, the V_O tends to move from the region of higher concentration to the lower concentration and eventually breaks the CF.

In bipolar type, the SET and RESET processes occur at opposite bias polarities. In general, the bipolar type can show filamentary or interface switching. The mechanism indicates that some charged species react under the electrical field. The forming process depends on the film thickness. Thin films (<2–3 nm) may not require forming process and mostly show interfacial switching. Therefore, the bipolar type is known to take place near the metal–oxide interface. Even in the filamentary bipolar type, the actual SET/RESET process is expected to take place near the metal–oxide interface only. The actual location of the switching is likely to vary with the type of cycle and device. Nevertheless, depending on the structure design and electrode materials, the HfO_2-based RRAM devices can function as unipolar or bipolar type (Banerjee, Lu, et al., 2018). Depending on the filament nature, the HfO_2-based RRAM devices are classified into two broad VCM and ECM groups, as discussed below.

1.5.1 VALENCE CHANGE MEMORY

The transition metal oxides with inert/oxidizable or inert/interelectrode combinations are useful for VCM devices (Banerjee, Cai, Zhao, Liu, et al., 2017; Banerjee, Lu, et al., 2018; Banerjee, Wu, et al., 2018; Banerjee, Xu, Lv, Liu, et al., 2017a, 2017b; Banerjee, Zhang, et al., 2018; Giri et al., 2021; N. Lu et al., 2015; Pi et al., 2019; Roy et al., 2020; Sassine et al., 2016; Q. Wu, Banerjee, et al., 2018). Several oxide materials, such as TiO_x, NiO_x, HfO_x, TaO_x, AlO_x, WO_x, and nitrides, such as AlN and NiN, have been studied rigorously as the resistive switching layer. Pt, Au, and Ir can be used as an inert electrode material. The Ti electrode can act as an oxygen-scavenging layer or oxidizable electrode. Other metals such as TiN and W also can act as the oxygen-scavenger at the metal–oxide interface. In the oxide-based filamentary VCM, the V_O–CF and, in nitride-based filamentary VCM, nitrogen vacancies (V_N)-CF are the reason for resistive switching. Herein, we focus only on the HfO_x-based devices, and thus, further discussion is based on the field-induced movement of V_O. In general, unlike other oxides such as TiO_x, the VCM devices based on HfO_x are most likely to show filamentary switching (Banerjee, Xu, Lv, Liu, et al., 2017b; Pi et al., 2019). Under a positive voltage on top electrode, the O_O drift toward the top metal–HfO_x interface and accumulates or forms an interfacial layer depending on the nature of the metal electrode, thereby generating an oxygen reservoir at the top interface. The field-dependent movement of O_O produces V_O–CF during the forming process. The current limiter can define the morphology of the filament, i.e., the lower the current, the thinner the CF. Therefore, in this section, the discussion is limited to the understanding of the impact of V_O, other doping or alloying processes in HfO_x, and the impact of the interface in HfO_x–VCM.

1.5.2 IMPACT OF OXYGEN VACANCY

Generally, at a temperature lower than 2000 K, HfO_2 is in the monoclinic phase, which includes two types of V_Os, namely V_{O3} and V_{O4}. The activation energy (E_a) for each type can be derived using the following equation:

$$E_a = E_p^{-1} + E_D^0 - E_p^0 - E_D^{-1} + \delta \qquad (1.4)$$

where E_p^0 and E_p^{-1} represent the total energy of the perfect supercell at a charge 0 and −1, respectively; E_D^0 and E_D^{-1} represent the energy of the defective supercell at a charge 0 and −1, respectively; and δ is the correction of the position of the bottom of the conductance band. The energy calculations are performed using first-principles studies by Lu et al. (Sassine et al., 2016). The density of states (DoS) calculations reveal that as compared to the DoS of the perfect cell (Figure 1.3(a)), the DoS of the defective cell shifts toward the deeper energy and generates new DOS near the Fermi level (Figure 1.3(b,c)). This suggests the generation of new V_Os, which can improve the carrier transport in HfO_x–VCM devices. The calculated E_a of V_{O3} and V_{O4} are 1.233 eV and 1.341 eV, respectively, resulting in a shallower energy level in V_{O3} as compared to V_{O4}. Considering the hopping theory to be in action, the total resistance of the hopping process can be described mathematically as:

$$\exp\left(\frac{qE_{a(path)}}{k_B T}\right) = \exp\left(2\alpha R_{ij(1)} + \frac{qE_{a(1)}}{k_B T}\right) + ... + \exp\left(2\alpha R_{ij(i)} + \frac{qE_{a(i)}}{k_B T}\right) \quad (1.5)$$

where $E_{a(path)}$ is the activation energy of the carrier transport path, $E_{a(i)}$ is the activation energy of the carrier hopping of the i-th number of times, $R_{ij(i)}$ is the carrier hopping distance, and α is the inverse of localization length.

After the formation of CF in VCM, in LRS, the carrier jumps among the series of defect sites with variable energies, which can lead to LRS fluctuation. The lower the resistance of the CF, the higher probability of carrier transport and higher stability. For a sufficiently low LRS, the Ohmic conduction can observe. Unlike LRS, the HRS contains a discontinuous path, and the electric field distribution obeys Poisson's equation in the hopping process. The experimental and simulated I–V performance of a 5-nm HfO_2-based VCM at 300 K is depicted in Figure 1.3(d). The defect energy landscape in the CF with different operations is shown in Figure 1.3(e–g). Depending on the different defect energy levels, the carrier transport path can change the I–V characteristics, as shown in Figure 1.3(d). During the resistive switching process, the different carrier transport probability through different carrier paths is one of the many reasons for the switching parameters fluctuation. It is worth noting that apart from V_{O3} and V_{O4}, the generation of other defects (V_{OD}s) such as interstitial and ion defects cannot be neglected. Depending on the above consideration, the switching parameter fluctuation (ΔI) in HfO_2–VCM can be defined as:

$$\Delta I = \left(1 - \frac{I_{min}}{I_{max}}\right) \times 100\% \quad (1.6)$$

where I_{min} and I_{max} are the minimum and maximum values of the current at a constant voltage, respectively. Figure 1.3(i) shows the ΔI as a function of E_a in V_{OD}. The results indicate that the ΔI is maintaining a constant value in LRS and HRS when the E_a in V_{OD} is within the range from 1.233 eV to 1.341 eV. In general, the shallower energy level has a higher carrier transport probability than its deeper counterpart. The higher the E_a in V_{OD}, the deeper the V_{OD} level. When the E_a in V_{OD} is greater than 1.341 eV, the carriers only jump through the V_Os, resulting in constant ΔI. Contrastingly, when the E_a in V_{OD} is less than 1.233 eV, the probability of carrier transport through V_{OD} increases and the ΔI depends on both V_Os and V_{OD}s. Figure 1.3(j) shows ΔI as a function of the concentration of V_O. A different scenario is achieved with V_{O3} and V_{O4}. It can be seen that with adequate tuning and the concentration of V_{O3} and/or V_{O4}, ΔI can be reduced. This is directly related to the shallower and deeper energy levels of V_{O3} and or V_{O4}, respectively. Figure 1.3(k) shows ΔI as a function of the HfO_2

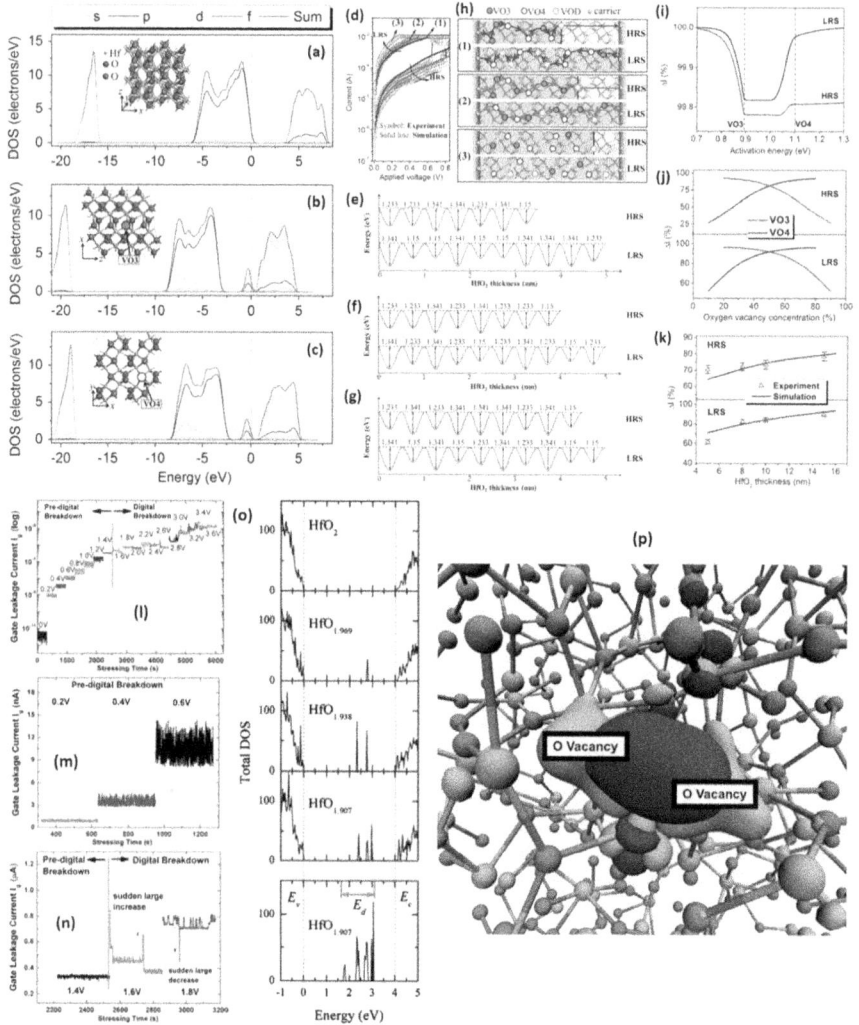

FIGURE 1.3 (a–c) The DoS of *m*-HfO$_2$ phase by first-principles calculations for $2 \times 2 \times 2$ supercell with perfect and defect structure. (d) The simulated and experimental *I–V*. (e–g) The energy landscape of different defect levels. (h) Defect-based carrier transport path in HfO$_2$. The impact of (i) activation energy of carrier transport; (j) V$_O$ concentration; and (k) HfO$_2$ thickness on the switching parameter fluctuation. (l) The evolution of post-TDDB current. (m) The fluctuations of current without any RTN signal. (n) The variation of current with stress time. (o) The DoS for HfO$_{2-x}$. (p) A divacancy of oxygen in HfO$_2$. Reproduced with permission (Banerjee, Kashir, et al., 2022). Copyright 2022, Wiley-VCH.

thickness, which indicates that the thicker HfO_2 can enhance parameter fluctuation. The thicker HfO_2 film increases the amount of defect and may even increase the distance of carrier transport during switching operation, resulting in switching parameter fluctuation. Therefore, an adequate amount of V_O in a VCM device is vital to have a controlled resistive switching process. In this case, the density and stoichiometry engineering of the HfO_2 film usually play a critical role in achieving the desired behavior (Raghavan, Fantini, et al., 2013).

The stress-induced V_O traps actively participate in the formation of the percolation path in HfO_2 and act as a determining factor of the time-dependent dielectric-breakdown (TDDB) lifetime in HfO_2 film (N. Lu et al., 2015). Figure 1.3(l) illustrates the evolution of the current in the HfO_2-based n-type MOS device in the post-TDDB regime. The voltage-dependent current evolution in Figure 1.3(m,n) shows that the digital fluctuations are triggered at higher voltages. In an HfO_2-based VCM device, the breakdown path mainly comprises V_Os. Hence, the first-principles calculations are widely used to understand the role of V_O and their effect on the electronic properties of HfO_{2-x}. Depending on the V_O concentration, the stoichiometry of HfO_2 can vary. Figure 1.3(o) shows the DoS of HfO_{2-x} near the gap region with different stoichiometry. It is evident that the formation of two or more V_O leads to newer peak formation and multiple defect energy levels in the gap region. Additionally, a defect subband (E_d) of 1.5 eV evolves for a composition of $HfO_{1.9}$. In our previous work, we have reported that the composition of $HfO_{1.9}$ is suitable for stabilizing the resistive switching behavior. Previously Gao et al. reported the divacancy model of HfO_2, as shown in Figure 1.3(p) (Banerjee, Zhang, et al., 2018). By comparing the energy required to produce the first and the second V_O, the interaction between the neutral vacancies can be quantified. If the formation energy of the first and the second V_O are E_{F1} and E_{F2}, respectively, then for $E_{F1} > E_{F2}$ the V_Os are said to be "attractive," and for $E_{F1} < E_{F2}$ the V_Os are said to be "repulsive." On average, E_{F2} is 0.1 eV lower than E_{F1}, i.e., most of the formation of new V_Os are attractive type. Likewise, the V_O formation energy of the third V_O is 0.2 eV lower than the E_{F2}. Hence, in the aggregation mechanism of V_O, its formation energy is lower for the new V_Os in the presence of the existing V_Os. The formation of new V_Os is possible in several ways: (1) new V_O is formed because of the electron trapping in existing V_O; (2) the distortion created by an electron trapping can act as a precursor, and (3) aggregation of several V_Os can enhance the local field and facilitate the creation of new V_O.

1.5.3 RESISTIVE SWITCHING PROPERTIES AND THE IMPACT OF DOPING/ALLOYING

Depending on the electrode material, the HfO_x–VCM devices can be employed in several ways. In general, the same electrode material in top and bottom can cause unipolar I–V and asymmetric electrode materials can cause bipolar I–V. For example, a $Pt/HfO_2/Pt$ structure can show unipolar switching, and a $Ti/HfO_2/Pt$ structure can show bipolar switching (Banerjee, Lu, et al., 2018). It is reported that the HfO_2 layer with a thickness of 4–5 nm can maintain the amorphous nature at 700°C (Banerjee, Wu, et al., 2018). Figure 1.4(a–c) depict the energy-dispersive spectroscopy (EDS)

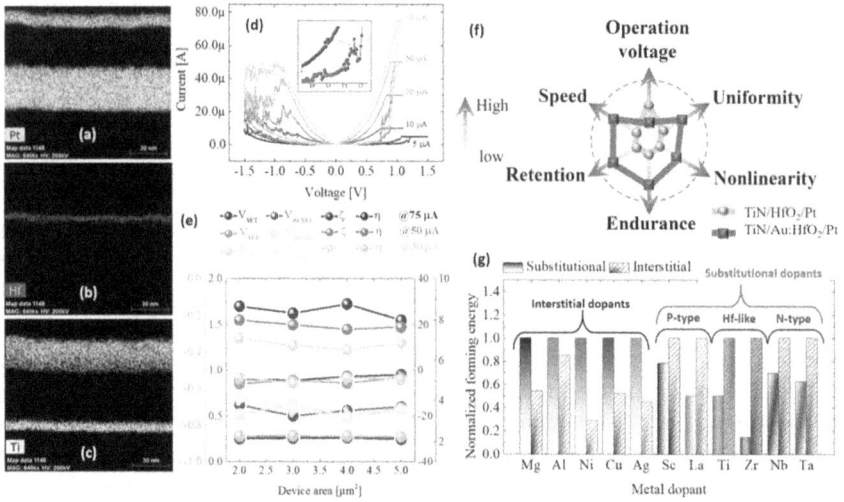

FIGURE 1.4 (a–c) The EDS image of the Ti/HfO$_x$/Pt-based VCM device. (d) The I–V characteristics and (e) area-dependent performance. (f) Improvement in resistive switching performance in HfO$_x$-based RRAM using doping. (g) Impact of different doping elements to define interstitial and substitutional systems. Reproduced with permission (Banerjee, Kashir, et al., 2022). Copyright 2022, Wiley-VCH.

image of the HfO$_2$–VCM with Pt bottom and Ti top electrodes (Banerjee, Xu, Lv, Liu, et al., 2017b). The HfO$_2$ layer has a thickness of 5 nm, and the initial resistance of the stack can be broken with a forming voltage (V_{Form}) of +2.5 V. The forming process of the CF is area dependent, in which the V_{Form} increases upon decreasing device area. In VCM, during the forming process, the V_O–CF formation can change the resistance of the device. As the number of defects can decrease with area scaling, the V_{Form} increases. In bipolar switching, the forming process changes the initial resistance to an LRS and the reverse bias can break the CF, which changes the resistance once again from LRS to HRS. Subsequent to that, the actual resistive switching process can be performed in the device. The typical resistive switching SET/RESET processes are very much dependent on the SET current level, as shown in Figure 1.4(d). The impact of the area on different switching parameters is summarized in Figure 1.4(e). The observations can be concluded as: (1) unlike the CF forming process, the SET (V_{SET}) and RESET voltages (V_{RESET}) are not area-dependent phenomena that describe the filamentary switching process; (2) the WM (HRS/LRS) weakly depends on the area but strongly depends on the SET current (I_{SET}) levels; and (3) the LRS nonlinearity

$$\eta = \frac{I @ V_{read}}{I @ \frac{1}{2} V_{read}}$$ weakly depends on device area. It shall be noted that the WM and η

are crucial parameters for increasing the array size of the RRAM device. Further improvements in WM and η are reported by using the doped HfO$_2$-based devices. Wu et al. reported that Au-nanoparticle-doped TiN-/HfO$_2$-/Pd-based VCM can achieve fast nonlinear switching operation in less than 100 ns, with excellent endurance of greater than 10^8 and high retention at 85°C, as summarized in Figure 1.4(f) (Z. Wu et al., 2019).

Zhao et al. performed density-functional theory (DFT)–based calculations and reported the impact of metal dopants in HfO_2-based RRAM devices (Q. Wu, Banerjee, et al., 2018). Depending on the formation energy, the metal dopants are classified into two groups – interstitial and substitutional – as shown in Figure 1.4(g). Among various dopants, Mg, Al, Ni, Cu, and Ag have lower formation energy at the interstitial site, whereas Sc, La, Ti, Zr, Nb, and Ta have lower formation energy at the substitutional site. HfO_2 with interstitial doping is called the interstitial system, and the HfO_2 with substitutional doping is called the substitutional system. As compared to the undoped HfO_2, in the interstitial system, the Fermi energy moves up into the bandgap. In this situation, the impurity energy is located just below the conduction band. This decreases the level spacing for electron transition and increases the conductivity of the whole system. In all interstitial dopants, the impurity energy is mainly composed of s states, while p and d states have a minor contribution. Additionally, in the interstitial system HfO_2, s, p, and d states have a negligible contribution to the impurity energy. Therefore, the major macroscopic conductivity changes in an interstitial system originate from the interstitial dopants. This demonstrates that interstitial dopants can directly engage in the formation of CF in HfO_x. Compared to that, in the substitutional system, the substitution metal atoms can replace the Hf atoms and bond chemically with oxygen. Depending on the valence electron numbers, the substitutional dopants can be categorized as P-type (valence electron number: 3), Hf-like (valence electron number: 4), and N-type (valence electron number: 5). Similar to the undoped HfO_2 supercell, the Fermi energy stays on the top of the valence band in the substitutional system. The P-type substitutional dopants, for example, Sc or La, and Hf-like substitutional dopants, such as Ti or Zr, do not introduce impurity energy to the bandgap, which crosses with the Fermi level. Therefore, P-type and Hf-like dopants cannot participate directly in the formation of CF. In contrast, the N-type substitutional dopants, such as Nb and Ta, can pull up the Fermi energy, and the impurity energy lies below the conduction band. The major difference between the interstitial dopants and the N-type substitutional dopants is the composition of s states. Unlike the interstitial dopants, the impurity energy of the N-type substitutional system is mainly composed of Hf s states. In the case of Nb, the impurity energy is associated with only Hf s states, and in the case of Ta, the impurity energy is mainly composed of Ta d states and Hf s states. The DFT studies show that the N-type substitutional dopants cannot produce CF themselves but can assist in promoting the formation of CF in the HfO_2 system. The resistive switching process in the doped-HfO_2 system depends not only on the material of dopants but also on the concentration of dopants. Roy et al. reported that a 16.5% Al doping in the HfO_x system can show enhanced resistive switching performance as compared to higher or lower concentrations (Giri et al., 2021). However, the ideal doping concentration is likely to vary with HfO_x thicknesses and stoichiometry.

Unlike metal dopants, various oxide layers can combine with HfO_x to improve resistive switching performance. Additionally, Hf-based mixed oxides such as HfAlO switching oxide have been found to be promising for the application of RRAM with enhanced data retention and endurance (Maikap & Banerjee, 2020; Milo et al., 2019; H. W. Park et al., 2021; Roy et al., 2020; Traore et al., 2015; Tsai et al., 2016). Several studies have confirmed the possibility of interfacial switching in RRAM (Banerjee, Cai, Zhao, Liu, et al., 2017; Govoreanu et al., 2016; Luo et al., 2018; J. Park et al., 2019; Traore et al., 2015; Yong et al., 2021). Goux et al. reported an ultrathin 1-nm Al_2O_3 interface in TiN/HfO$_2$/Hf RRAM, which can minimize I_{SET} to sub-500 nA (H. W. Park et al., 2021). The higher HRS in these devices originates from the large bandgap of the Al_2O_3 layer. Additionally, Al_2O_3/HfO$_2$/Hf stacking can improve the V_O profile modulation as compared to HfO$_2$/Al$_2$O$_3$/Hf stacking. Luo et al. reported the resistive switching at the HfO$_2$/WO$_x$ interface in W/WO$_x$/HfO$_2$/Pt structure (Govoreanu et al., 2016). In this system, a 3-nm HfO$_2$ serves as an oxygen reservoir. As compared to filamentary switching, interfacial switching devices are highly scalable (Banerjee, Cai, Zhao, Liu, et al., 2017). Nevertheless, the interstitial doping of the HfO$_2$ can lead to the design of the metal-CF-based ECM devices, which is elaborated in the following section.

1.5.4 ELECTROCHEMICAL METALLIZATION MEMORY

To design an ECM device, one electrode must be an active metal (AM) and the other an inert metal, which is the counter electrode (CE) (Lin et al., 2021). For the former, Cu and Ag are central to the designing of ECM devices. Under the electric field, because of the electrochemical dissolution process at the AM, the mobile metal cations such as Cu$^+$ and Ag$^+$ can directly participate in constructing the metal–CF and can thus act as interstitial dopants. Not only do these cations dissolve easily electrochemically but their oxide formation energy is also lower than other metals, such as Ir, Pt, and Ni. Considering M as metal atoms and M^{z+} as metal ions, in such an ECM system, the CF formation is a four-step process including (1) oxidation of anode, i.e., AM ($M \rightarrow M^{z+} + Ze^-$); (2) electromigration of M^{z+} ions from the anode to CE or cathode; (3) nucleation on CE, i.e., the reduction of M^{z+} ions on top of CE, which forms metal atoms ($M^{z+} + Ze^- \rightarrow M$); and finally (4) the filament growth.

1.5.5 UNDERSTANDING FILAMENT FORMATION

The most common challenge concomitant to ECM devices is the control of metal ion diffusion in the switching matrix. Additionally, the random injection of metal cations can lead to the formation of unstable filaments, which is not beneficial for the long-lasting nonvolatile performance of ECM devices. In contrast, a well-localized strong filament is an essential requirement for fulfilling the application criteria. To protect the switching oxide layer from over diffusive metal cations, several proposals have been posited, for example, the design of the interstitial system using doped oxide layer and metal electrode engineering with different nanostructures, insert an ion diffusion barrier layer. Among these emerging concepts, a thin metal layer (e.g., TiN, Ti) or graphene ion barrier layers for HfO$_2$-based ECM devices have been widely studied (Banerjee & Hwang, 2019; Grisafe et al., 2019). Unlike VCM devices, the visualization of metal CF is the most evident in ECM devices. Several tools have been employed in this mission such as high-resolution transmission electron micros-copy (HRTEM), in situ TEM, conductive atomic force microscopy (C-AFM), and scalpel-AFM.

By using C-AFM, Zhao et al. investigated the localized cation injection in the nanoscale regime guided by a nanohole in an impermeable ion barrier on the top of the HfO_2 switching layer (Grisafe et al., 2019). Because of its excellent imper-meability for molecules, atoms, or ions, the two-dimensional (2D) graphene layer was used as the ion barrier. The C-AFM is a widely used tool employed for understanding the filament formation and dissolution morphology in the nanoscale region. Moreover, to clearly identify the filament, the Olympic ring structure was patterned on the graphene layer. Figure 1.5(a) presents the schematic of the Olympic ring–patterned HfO_2–ECM device. Two clearly distinguishable regions can be seen in the scanning electron microscopy (SEM) image of Figure 1.5(b), in which the structure of region 1 is HfO_2/graphene/Ag and the structure of region 2 is HfO_2/Ag because of the Olympic ring–patterned graphene. Figure 1.5(c) depicts the typical *I–V* characteristics acquired by C-AFM. In region 1, when the C-AFM tip is in contact with HfO_2/graphene/Ag structure, no such current growth can be observed. In con-trast, when the tip is connected to region 2, the tip/HfO_2/Ag structure shows bipolar resistive switching operation. At a read voltage (V_{read}) of 1 V, the initial current is lower than 50 nA, indicating a smooth surface and the initial HRS before the CF forming process (Figure 1.5(d,i)). During the CF-forming process at 4 V, region 1 remains at low current to some nanoampere range; however, region 2 experiences high current at several microampere ranges, which justifies the formation of CF only in region 2 through the grain boundary or breakpoints of graphene film (Figure 1.5(e,j)). After formation, region 2 still maintains a high current of ~1.2 μA, which indicates a non-volatile CF (Figure 1.5(f,k)). During the RESET process with −3 V and after the RESET process, region 2 transformed from LRS to HRS (Figure 1.5(g,h,l,m)). The experimental evidence established the fact that the CF formation and dissolution pro-cess only occurs at the patterned region because the graphene layer acts as a barrier to metal cation injection. Additionally, the CF in the HfO_2–ECM is also localized, which can improve the electrical performance.

To understand the growth kinetics of the CF, the HRTEM analysis of the Cu–CF in Cu/HfO_2/Pt device was studied by Lv et al. (Zhao et al., 2017). Figure 1.5(n) depicts the Cu/HfO_2/Pt device in one transistor–one resistor (1T1R) structure and its cross-sectional magnification view in Figure 1.5(o). The high-resolution images from (1) to (4) are shown in Figure 1.5(p–s). All the HRTEM analyses were performed after programming the 1T1R device at a gate voltage of 2.5 V and a current of 1 mA. A conical region was found in HfO_2 near the left corner and was investigated with EDS analysis, as shown in Figure 1.5(t,u). The high Cu signals indicate the forma-tion of Cu-rich CF in the Cu/HfO_2/Pt device. More detailed analysis with the fast Fourier transformation (FFT) images showed crystalline Cu and amorphous HfO_2 layers. Different research groups have reported crystalline filament in other oxide systems (Lv, Xu, Liu, et al., 2015; C. Wang et al., 2018). However, in regions (2) and (3), no clear lattice fringes can be observed. From the refined FFT in Figure 1.5(s), the crystalline phase is only at the center of CF and the fringe spacing of 0.2 nm is well matched with face-centered cubic Cu with orientation (Azzaz et al., 2016).

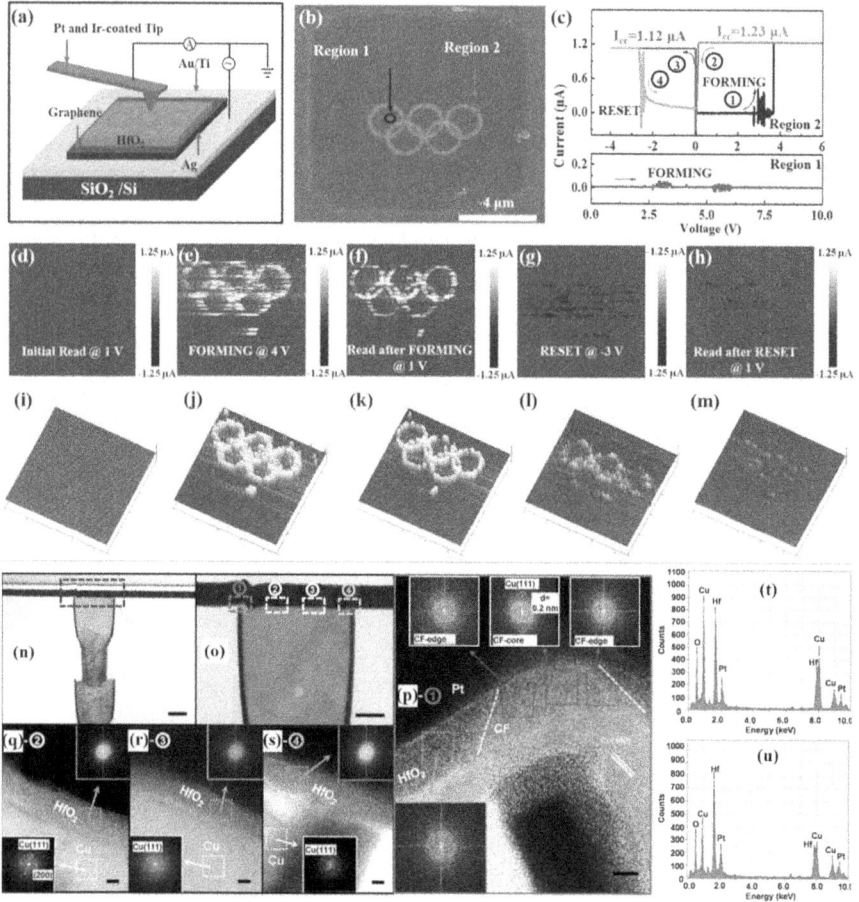

FIGURE 1.5 (a–m) Analysis of the resistive switching performance in Ag/HfO$_2$-based device with Olympic ring–patterned graphene layer. (n–u) Analysis of the resistive switching performance in Cu/HfO$_2$-based device with 1T1R structure. Reproduced with permission (Banerjee, Kashir, et al., 2022). Copyright 2022, Wiley-VCH.

The results suggest that Cu atoms can occupy some positions in the HfO$_2$ lattice and play a key role in the formation of CF. However, for other structures based on Cu/Ag, it has been reported that the filament can be constructed from pure metal atoms. Hence, a proper understanding of the CF in ECM devices can generate the possibility of scaling to the atomic dimension.

1.5.6 QUANTUM CONDUCTANCE AND DEVICE SCALING

In the ECM devices, the scaling potential of the metallic CF is projected to be in atomic contact. If the cross-section of a metallic CF is scaled down to a comparable dimension of the Fermi wavelength of the electrons, the transport properties change from diffusive to ballistic type. The single-atom contact often experiences conductance quantum, where $G_O = \dfrac{2e^2}{h}$ is the quantized unit of electrical conductance, e is the elementary charge, and h is the Planck constant. Electronic switches constructed with atomic-sized functional blocks are considered to be the ultimate goal of device scaling; however, it is indeed a mammoth task. Zhirnov et al. predicted that to have a WM of at least 10, a minimum 1-nm oxide film is essential (Roy et al., 2020). Therefore, a single-atomic contact is not easy to realize in a real device. However, for simplicity, let us assume the formation of a single-atom filament in an oxide matrix in which the classical resistance is given as follows:

$$R_{atom-bridge} = \rho_O \left(\frac{\lambda_O}{\theta^2} N + \frac{2}{\theta} \right); \; R_{atom} = \rho_O \left(\frac{\lambda_O}{\theta^2} + \frac{2}{\theta} \right), \text{ when } N = 1 \qquad (1.7)$$

where R_{atom} is the classical resistance of the single-atom contact; ρ_O is the bulk resistivity; λ_O and θ are material-specific parameters; and N is the number of constituting metal atoms. For single-atom contact, the R_{atom} is 12.87 kΩ. A barrier with transmission probability $\tau < 1$ present in the system will increase the resistance and decrease the conductance. If the single-atom contact is regarded as the LRS, then the removal of single-atom is the origin of the barrier and HRS, i.e., $LRS = \tau HRS$. In the presence of such a barrier, different electron transport mechanisms are possible such as thermionic emission, direct tunneling, and Fowler–Nordheim tunneling.

$$E_b \approx \varphi_O - \frac{e^2 \ln 2}{16\pi\varepsilon_O K} \frac{a}{x(a-x)} - \frac{eV_{gap} x}{a} \qquad (1.8)$$

where E_b is the potential barrier, φ_O is the metal work function, K is the dielectric constant of the oxide, a is the gap length, and V_{gap} is the gap voltage. The theoretical analysis shows that φ_O and the K can actually determine the minimum oxide thickness. Depending on the design of the material stack, the oxide thickness can be as small as a three-atom gap. In some situations, the atomic path formation is not so straightforward and the CF path can be a discontinuous bridge of atoms. In such a system, if g is the number of gaps, then:

$$I_g = \frac{1}{g^2} \left(\frac{2e^2}{h} V\tau \right) \qquad (1.9)$$

The prediction depicts the fact that a minimum two-atom gap with a single-atom in the interelectrode regime is sufficient to produce a detectable WM in ECM devices.

The conductance quantum phenomena in HfO$_x$-based RRAM devices are reported frequently (Zhirnov et al., 2011). Considering the above, a switching device based on an electrically controllable break junction (ECBJ) was proposed in Waser et al. (2009) with Cu/Ti/HfO$_2$/TiN structure in which a minimum 2 G_O contact can switch the

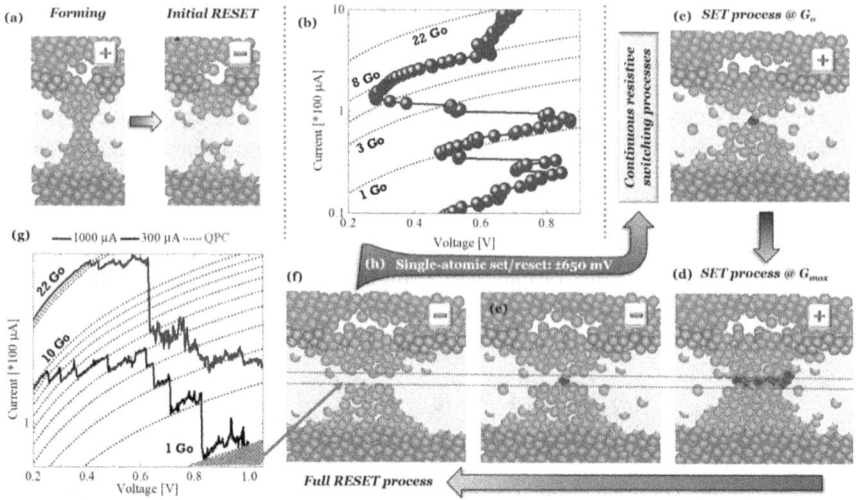

FIGURE 1.6   (a) The variation in maximum conductance of the filament with temperature. A temperature of 125°C is suitable to maintain the maximum conductance < 2 G_o. (b) The conductance change during the SET process. (c–f) Schematic switching mechanism. (g) The RESET I–V and fitting with the quantum point contact model. Reproduced with permission (Banerjee, Kashir, et al., 2022). Copyright 2022, Wiley-VCH.

device back and forth. In this system, the Ti layer is 2 nm thick and acts as a diffusion barrier to the Cu ion migration process, and the resistive switching event is conducted in the 2-nm HfO$_2$ layer. A high-temperature CF-forming process was introduced as the controlling parameter for metal ion migration in such a system. Therefore, the CF is in the nanometer range in which the ballistic transport originated owing to two main reasons; the first is the direct force, i.e., the direct impact of the electric field on Cu ions, and the second is wind force due to the movements of electrons. In such nanometer-scaled CF, the electromigration process increases the temperature locally. The increase in the thermal energy can cause an additional increment of the single-atomic junction to a fewer-atomic junction. Figure 1.6(a) suggests that a temperature around 125°C balances the ion migration process and lower energy barrier, which is suitable for controlling the thinnest junction of the CF to 1–2 G_0. Any temperature higher or lower than 125°C can degrade the single-atom switching condition chemically and thermodynamically. For obvious reasons, the CF morphology is dependent on the I_{SET}. Following the quantum point contact (QPC) model, the atomic level incremental conductance at 1 mA current level is shown in Figure 1.6(b). It is worth noting that the current growth process can be explained with integer or half-integer values of G_0, as follows:

$$I = G_0 N_{sub-Fermi} V + G_0 \frac{N_{contact}}{2} V \qquad (1.10)$$

where $N_{sub\text{-}Fermi}$ is the number of subbands below the Fermi level and $N_{contact}$ is the number of subbands at the contact Fermi level. Hence, when $N_{contact} = 1$, half-integer G_O could be observed. In $Cu/Ti/HfO_2/TiN$ structure, although the integer values are well in agreement, in practical scenarios, this kind of I–V characteristic cannot be fully explained using one curve. During the SET switching, the filament condition for 1 G_O and several G_O are schematically shown in Figures 1.6(c) and 1.6(d), respectively. During the RESET switching, the change of conductance from 1 G_O to a broken junction is schematically shown in Figures 1.6(e) and 1.6(f), respectively. The corresponding change in the RESET I–V is nicely fitted to the QPC model, as shown in Figure 1.6(g). In general, almost 650 mV is necessary to switch 1 G_O contact in $Cu/Ti/HfO_2/TiN$ structure. Such precisely controlled conductance in HfO_2–ECM provided a multilevel cell (MLC) storage capacity of 6 bits/cell (Waser et al., 2009). Although some of the VCM devices can be fitted with QPC, most of the ECM devices can be explained with QPC. Therefore, the electrode material and the device design is one of the big factors in RRAM devices.

1.5.7 IMPACT OF METAL ELECTRODES

Metal electrode materials can play the defining role of a switch in HfO_x-based RRAM devices. Additionally, the location of the metal electrode, process condition of HfO_x deposition along with HfO_x film density, and stoichiometry can have a massive impact on the performance of the resistive switching device.

1.5.8 DIFFERENT ELECTRODE MATERIALS AND THE IMPACT OF LOCATION

The nature of constituent materials and the location of the electrode (either top or bottom side) have been researched extensively. Figure 1.7(a) depicts the EDS line scan of the $TiN/HfO_2/Ti/Cu$ structure (Y. Li, Long, et al., 2015). The 6-nm-thick HfO_2 layer was deposited using the atomic layer deposition (ALD) process at 250°C on the top of the TiN bottom electrode. It is evident from the EDS line scan that at the interface of TiN–HfO_2, a 4-nm-thick TiO_2 layer is present; the same is also evidenced after further investigation using an HRTEM image. Owing to the ALD growth of HfO_x film, the oxidation only takes place only at the bottom TiN–HfO_2 interface. Though it was originally intended to design 6-nm HfO_2-based RRAM, the final outcome turned out to be a 4-nm TiO_2/6-nm HfO_2-based bilayer RRAM with $TiN/TiO_2/HfO_2/Ti/Cu$ structure. A similar problem has also been reported for other RRAM devices based on the W electrode, which generally produces a WO_x layer because of surface oxidation during device processing (J. Park et al., 2019). This kind of situation has to be circumvented during the fabrication of HfO_x-based RRAM or any other RRAM devices. Although Ti or W can create some interface at the bottom electrode, they are mostly safe at the top electrode side.

Figure 1.7(b)-(1) illustrates the impact of different metal electrodes in the resistive switching process in $M/HfO_x/Pt$ structure, where M's are W, Cu, Ag, and AgTe, and the devices can be categorized as VCM type (W-based) and ECM type (Cu-, Ag-, AgTe-based) (Raghavan, Fantini, et al., 2013). Depending on the electrode materials, the HfO$_x$ device can behave as memory (W, Cu) or selector (Ag, AgTe). Generally, the bond dissociation energies are 666/720, 201/287.4, 162.9/221, and 195 kJ/mol for W–W/W–O, Cu–Cu/Cu–O, Ag–Ag/Ag–O, and Ag–Te, respectively, which indicates bond weakening from W- to Ag-based electrodes and results in changing switching behavior from MS to TS. Therefore, in accordance with the material's nature and its structural arrangement, the resistive switching process can be altered in HfO$_2$-based RRAM. Figure 1.7(b)-(2) depicts the typical TS process in the Ag device. The density and stoichiometry of HfO$_x$ film are very critical to understanding the performance of Ag/HfO$_x$-based TS devices (Raghavan, Fantini, et al., 2013). Details about the HfO$_x$-based TS device and its engineering process are going to be discussed later. It shall be noted that the scaling of the metal electrode materials can actually reduce the effective area of the RRAM device. Pi et al. reported 2 x 2 nm^2 Pt-based nanofin crossbar RRAM devices (Banerjee, Cai, Zhao, Liu, et al., 2017). Recently, emerging 2D materials have also been used for a wide array of purposes in HfO$_x$-based RRAM devices, for example, as a metal electrode and ion diffusion barrier layer (Banerjee & Hwang, 2020b).

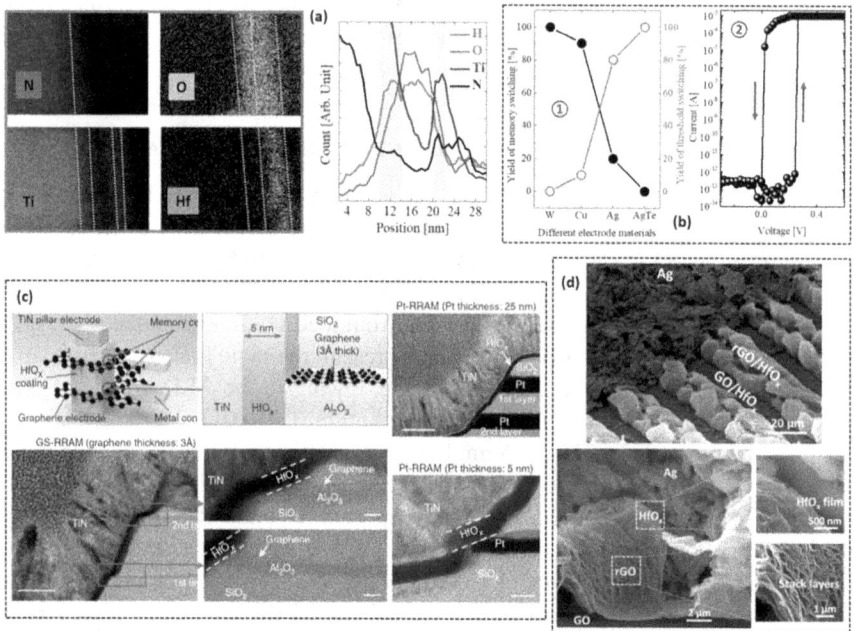

FIGURE 1.7 (a) The impact of the TiN electrode in the bottom electrode can introduce TiO$_x$ interfacial layer. (b) Impact of different metal electrodes in the resistive switching process in HfO$_x$. W and Cu can show memory-switching behavior. The Ag-based electrode can show threshold-switching behavior. (c) The use of graphene as electrode material. (d) The use of reduced graphene oxide as electrode material. Reproduced with permission (Banerjee, Kashir, et al., 2022). Copyright 2022, Wiley-VCH.

1.6 EMERGING TWO-DIMENSIONAL MATERIALS AND THEIR IMPACT ON RESISTIVE SWITCHING

2D materials have become the center of attraction in contemporary research. Among several others, graphene is one of the long-standing 2D materials, which has been demonstrated to be useful in designing RRAM devices in a number of ways (Banerjee, Kim, Lee, Lee, Lee, et al., 2021; Banerjee & Hwang, 2020b; Ding et al., 2020; H. Tian et al., 2013, 2014). It is an excellent material, especially when the electrode requires scaling for integrating RRAM devices in three-dimensional vertical RRAM (3D-VRRAM). The scaling of conventional metal electrodes can approach the scale of a few nanometers. However, the use of graphene can scale it further down to the angstrom scale. Lee et al. reported 5-nm HfO_x-based RRAM with graphene electrode (see Figure 1.7(c)) (Ding et al., 2020). Atomically thin 3-Å monolayer graphene can serve the purpose of electrode scaling in a 3D-VRRAM stack. The simulation projected using graphene electrode with 200 stack layers is possible as compared to just 60 layers of the stack of bulk-metal-electrode in 3D-VRRAM. Statistically, TiN/HfO_x/graphene device can scale down the minimum I_{SET} to several microamperes and switching is accommodated by the formation and dissolution of V_O–CF. As compared to other devices, the graphene electrode-based HfO_x-based RRAM can perform switching with ultralow energy consumption to 230 fJ. Furthermore, the prospect of reduced graphene oxide (rGO) has been examined for flexible RRAM applications. Tian et al. reported a laser-scribed transfer-free rGO in Ag/HfO_x flexible RRAM (H. Tian et al., 2014). Figure 1.7(d) shows a 10-nm HfO_x/rGO-based RRAM structure. The formation of Ag-based metal–CF conducts the resistive switching process with satisfactory electrical performance. Therefore, the use of scaled graphene-based electrodes can control the resistive switching process in HfO_x-based RRAM devices using both the V_O–CF and metal–CF.

1.7 DESIGN OF THE HYBRID FILAMENT IN HAFNIUM OXIDE

Apart from V_O–CF or metal–CF, the performance of the RRAM device can be altered with the formation of (metal + V_O)–CF – a hybrid filament (HF) (Banerjee, Karpov, et al., 2020a; Banerjee, Kim, Lee, Lee, & Hwang, 2021; S. Fujii et al., 2019; S. Lee et al., 2015; Molas et al., 2015; Raghavan, Fantini, et al., 2013; Sassine et al., 2019). Molas et al. reported that the presence of V_O can facilitate the migration of Cu cations in the Cu/HfO_x system (Banerjee, Kim, Lee, Lee, & Hwang, 2021). To understand the favorable nature of the filament, atomistic simulations were performed on Cu/HfO_2, Cu/Ta_2O_5, Cu/Al_2O_3, and Cu/GdO_x systems (Banerjee, Karpov, et al., 2020a). The results indicated a V_O-rich Cu_iV_O filament in Cu/HfO_2 and Cu/Ta_2O_5 systems and a Cu_iV_O filament in Cu/Al_2O_3 and Cu/GdO_x systems. In any situation, the studies confirm that the presence of V_O can facilitate cation movements in the switching oxide layer. However, the key challenge must be the control of the concentration of V_O. Therefore, it can be one of the determining factors of the thickness of CF in any RRAM system. In our previous study, we reported the methodology to control the V_O concentration in Ag/HfO_x/Pt structure (S. Lee et al., 2015; Molas et al., 2015; Raghavan, Fantini, et al., 2013). A performing step can guide the formation of vacancy-induced-percolation (VIP) path in a 4-nm HfO_x film. The stoichiometry can be understood using X-ray photoelectron spectroscopy (XPS) analysis. The spin–orbit components in Hf $4f$ core-level spectra are separated by 1.7 eV (Figure 1.8(a)). Owing to the presence of a lesser number of Hf–O bonds in HfO_x, the lower binding energy peak in O $1s$ spectra (Figure 1.8(b)) is shifted 1 eV toward the higher side as compared to conventional low binding energy peak. The C–O bond is represented by the higher binding energy peak at 533.58 eV. Therefore, the stoichiometry of the film depends on the atomic percentages of Hf and lattice O, i.e., 30.51% and 58.55%, respectively, equivalent to the Hf:O ratio of 1:1.91. Hence, the nature of the HfO_x is slightly nonstoichiometric with the presence of a small amount of V_O, which can create VIP under preforming conditions with negative voltage at the Ag electrode. After the formation of VIP, a positive voltage was applied to the Ag electrode to move Ag cations in the V_O location. In such conditions, the CF in $Ag/HfO_x/Pt$ system is not a simple Ag–CF but an HF, which is a combination of Ag interstitial (Ag_i) and V_O,

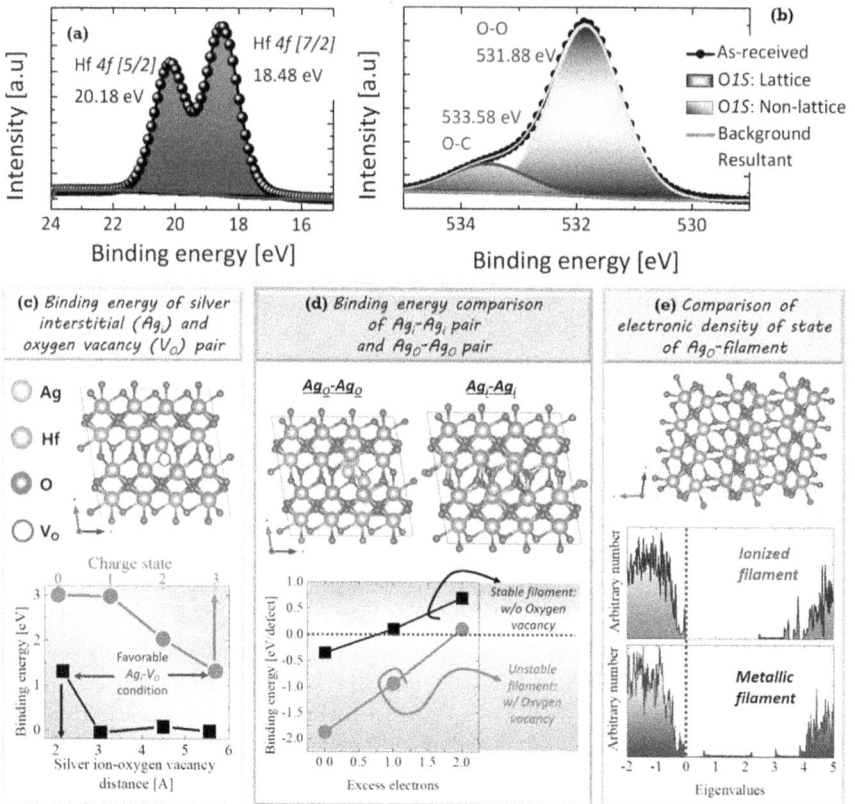

FIGURE 1.8 (a,b) The XPS analysis of the HfO_x-based RRAM device. (c) The binding energy and charge state variation depending on Ag_i and V_O distance. (d) The metallic Ag is more stable than the hybrid filament. (e) The DoS variation with and without excess electrons. Reproduced with permission (Banerjee, Kashir, et al., 2022). Copyright 2022, Wiley-VCH.

hence defined as Ag_O. The fundamentals of the HF formation have been studied by calculating the binding energy ($E_{binding}$) of Ag_i–V_O pair in HfO_2 using DFT methodology, which can be represented by the energy difference between Ag_i and V_O, to the nearest and farthest positions.

$$E_{binding} = E_{total}^{far} - E_{total}^{near} \qquad (1.11)$$

Figure 1.8(c) shows that the Ag_i is most preferably located next to V_O with a maximum binding energy of 1.31 eV and charge +3, which rapidly decreases to 0 eV with the increasing distance between Ag_i and V_O. Thus, the nature of the filament is HF type. Figure 1.8(d) shows that the $E_{binding}$ of two Ag_O–Ag_O pairs is lower than the $E_{binding}$ of two Ag_i–Ag_i pairs, symbolizing the fact that Ag_O-based HF is suitable for designing volatile switching, and Ag_i-based CF is suitable for designing nonvolatile switching. Figure 1.8(e) shows the HF model in $Ag/HfO_x/Pt$ device. In ionized conditions, the $E_{binding} = -0.63$ eV/defect and HF exhibits an insulating DoS. In neutralized conditions with excess electrons, HF exhibits an $E_{binding} = +1.27$ eV/defect and the metallic DoS. However, the nature of the HF depends on the amount of V_O in Ag_O. In the HfO_x matrix, a higher amount of V_O concentration can produce thicker HF and a lower amount of V_O concentration can produce thinner HF, and both can be utilized in different ways.

1.7.1 HYBRID FILAMENT-BASED MEMORY

In general, the Ag/HfO_x systems exhibit poor nonvolatility. Therefore, it is a challenging task to design Ag/HfO_x structure with enhanced nonvolatility. Previously, Fujii et al. proposed that the proper density and stoichiometry of the oxide layer can improve the MS performance in Ag/SiO_x structure (Sassine et al., 2019). To further improve the performance of the Ag/HfO_x structure, VIP at moderately high current (I_{VIP}) has been proposed in our earlier work (S. Lee et al., 2015). Figure 1.9(a) depicts the variation of HRS after MS as a function of LRS after preforming with different I_{VIP} values. The I_{VIP} of 100 pA has a negligible impact on the V_O concentration. Increasing I_{VIP} from 1 nA to 10 µA increases the probability of MS from <20% to >90%, respectively. The HRS conductance after MS increases with increasing I_{VIP}, which further minimizes the window margin of the device. Because of the filamentary switch, the switching voltages are similar in all conditions. In an optimized device, the I_{VIP} of 1 µA is considered suitable for the superior functioning of Ag/HfO_x devices along with sufficiently larger WM. The understanding of increasing V_O concentration and its impact on the formation of HF was also studied using current evolution with time (I–t) during the HF-forming process. Under the biased condition, the O–Hf–O bonds can break with the help of hot electrons from the anode side, which produces the V_O percolation path formation and guides the electron-trapping process. Compared to low I_{VIP}, the higher I_{VIP} can have more hot electrons. Gao et al. reported that in the HfO_2 matrix, the formation of new V_O can readily occur near the preexisting V_O (Banerjee, Zhang, et al., 2018). Therefore, the VIP with a higher current can have a higher V_O concentration, which further produces a stronger filament. Following VIP formation, Ag is injected through the VIP and forms the HF filament with a positive bias at the Ag electrode. After the HF forming process, because of the higher V_O concentration, the RESET switching is gradual for the higher current of 1 μA, as shown in Figure 1.9(a)-(5). In most of the studies in the literature, Ag-based RRAM devices have been reported to have poor data retention at low current. In our previous work, we discussed the phenomenal improvement in data retention in Ag/HfO$_x$ devices upon tuning the V_O concentration. As compared to other reported devices, HF-based devices can achieve better data retention with 1.16 eV activation energy and endurance of 10^8 cycles. The extrapolation indicates that such devices can retain data for >10 years at 70°C, which is similar to the data retention of density-engineered Ag/SiO$_x$ devices manufactured by Toshiba (Sassine et al., 2019). Hence, to design such devices, the density and stoichiometry of the switching layer are some of the most crucial factors.

1.7.2 HYBRID FILAMENT–BASED SELECTOR

Generally, the Ag/HfO$_x$ devices are a viable candidate for designing TS for selector applications. Unlike MS, in TS, the voltage for changing the OFF state to ON is known as the "threshold voltage," and the voltage for changing the ON state to OFF is known as the "hold voltage." Other critical parameters for TS include OFF/ON currents, subthreshold swing, selectivity, endurance, and switching speed. The Ag/HfO$_x$ devices with superior TS performance have been reported several times for steep-slope applications (Molas et al., 2015; Raghavan, Fantini, et al., 2013). Shukla et al. reported the performance of Ag/HfO$_x$ devices on the p$^+$-silicon with limited selectivity (W. Wang et al., 2019). The most noteworthy drawback of the Ag electrode is the overdiffusion phenomenon, which limits the selectivity and degrades the stability of the OFF state. Therefore, the stuck-ON failure of the OFF state is a highly prevalent problem associated with such devices. Several studies have reported that the barrier layers limit Ag injection in HfO$_x$ and other oxide layers (Banerjee & Hwang, 2019). Hua et al. reported patterned Ag nanodots to limit the localized injection of Ag in the HfO$_2$ matrix (Shukla et al., 2017). Therefore, the research mostly focuses on addressing the issue of the overinjection of Ag in the oxide layer. Banerjee et al. reported improvement in TS performance by controlling appropriate V_O concentration and for HF filament (Raghavan, Fantini, et al., 2013). The VIP process with 1-nA I_{VIP} is the most preferable condition to form Ag-based HF in the HfO$_x$ matrix. Moreover, the overdiffusion problem with the Ag electrode was optimized with the AgTe alloy electrode. Figure 1.9(b)-(1) shows that because of the more available Ag ions in Ag/HfO$_x$, the threshold voltage is lower as compared to AgTe devices. Both devices experience a similar OFF current of ~0.4 pA. However, the ON current in Ag is higher than AgTe, as shown in Figure 1.9(b)-(2). Although the Ag devices can achieve higher selectivity, such devices exhibit limited endurance because of the remaining Ag in HfO$_x$ after each switching cycle. The problem can be overcome with the application of a slightly negative voltage. Figure 1.9(b)-(3) shows that with a small negative bias, the subthreshold slope is <2 mV/dec, which is five times lower than its nonnegative voltage counterpart. Finally, a remarkably long endurance of >10^9 cycles is achieved with AgTe/HfO$_x$-based RRAM. Furthermore,

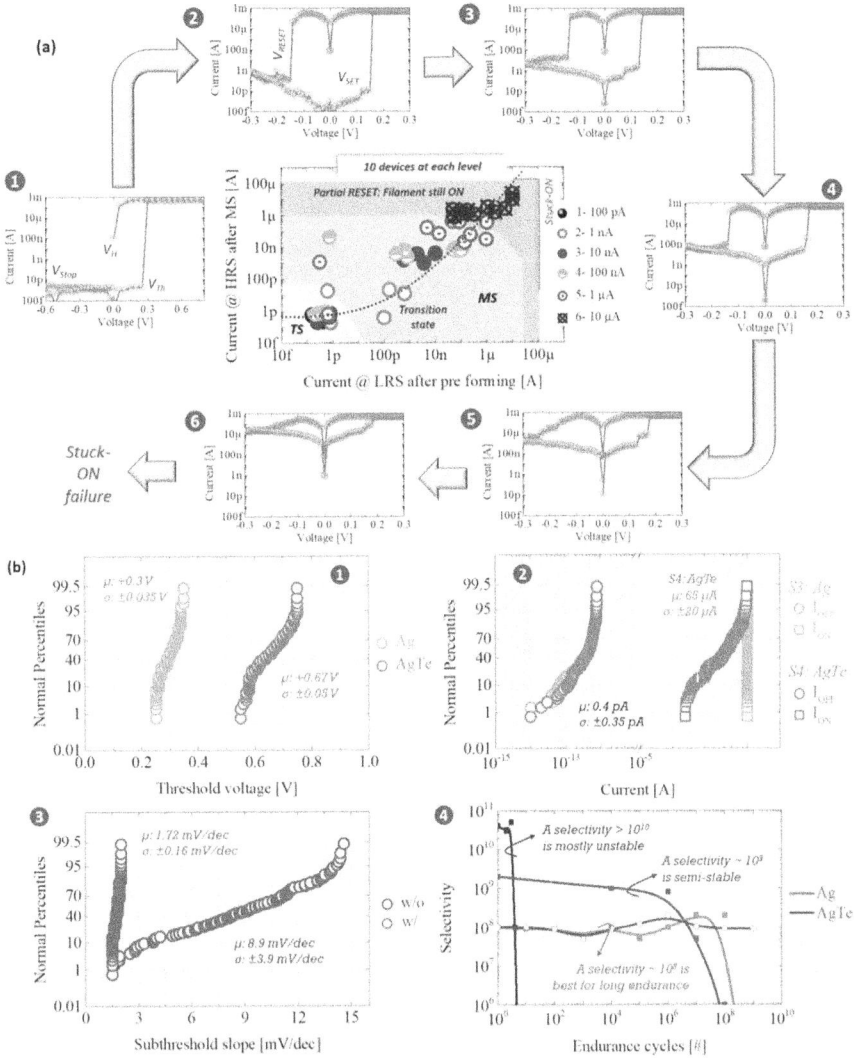

FIGURE 1.9 (a) The tuning of V_O and the change of switching from (1) threshold switching to (6) memory switching in Ag/HfO$_x$ structure. (b) The threshold switching in Ag/HfO$_x$ structure. Impact of Ag and AgTe electrode on (1) threshold voltage, (2) ON current, and (4) device endurance. (3) Impact of negative voltage on steep-slope behavior. Reproduced with permission (Banerjee, Kashir, et al., 2022). Copyright 2022, Wiley-VCH.

Lee et al. reported that the turn-OFF speed is largely dependent on the density and stoichiometry of the HfO$_x$ film (Hua et al., 2019). Additionally, a density-graded HfO$_x$-based RRAM device can limit the switching region to a 0.5-nm-thick stoichiometry HfO$_2$ layer. A HfO$_{1.9}$/HfO$_2$ bilayer design can achieve a turn-OFF < 100 ns with excellent endurance. Point to be noted: the material parameters and device modeling are interdisciplinary topics. Several groups have explored metrology parameters along the device- and system-level modeling approach to understand the design and optimization of switching devices and arrays. In the field of NVMs simulation tools can be used to understand the physical design parameters, experimental data, process optimizations, performance prediction, and so on. However, a single model can't fulfill all the requirements. Therefore at least three hierarchical modeling approach is necessary. 1) From the very beginning one must focus to understand the materials properties and defect physics using atomistic models like ab initio methods. 2) The material parameters then can be used in device modeling like kinetic Monte Carlo (KMC) and finite elements method (FEM) to understand the electrical performance of the devices. 3) In the final state semiempirical or compact models can be used to understand the array and circuital behavior. For a detailed understanding of modeling can be found in the previous works. So far, the focus of this discussion has been the impact of HfO_x materials and their dielectric properties on the design of emerging RRAM devices. The next section is going to focus on the ferroelectric properties of HfO_2 and the use of the same in designing emerging ferroelectric memory devices.

1.8 EMERGING APPLICATIONS

RRAM is useful in various combinations such as 1T1R and one selector-1R (1S1R). Figure 1.10(a) shows the schematic of 32 × 32 1T1R-type array (Saadi et al., 2016). The gates of the transistors are connected to the word line on the bottom side, and the top electrode of the RRAM cell is connected to the bit line. The corresponding HRTEM image of the integrated Cu/HfO_2-based RRAM with 1T (fabricated with 0.13-μm CMOS technology) is depicted in Figure 1.10(b). The switching device is integrated between the bit line and metal 4. The HfO_2 layer has a thickness of 4 nm. These devices have been extensively used in a wide array of applications. However, the horizontal array can use a larger space on the chip. Therefore, the concept of 3D V-RRAM is one of the promising candidates for high-density applications. Luo et al. reported the vertical integration scheme of RRAM devices (Lv, Xu, Liu, et al., 2015). Figure 1.10(c) shows the schematic process flow of the HfO_2-based 3D V-RRAM array. The 60-nm thick TiN bottom electrodes are separated by the 100-nm thick SiO_2 insulators. In such a device, the smoothness of the vertical sidewalls is one of the major concerns. Figure 1.10(d) shows the smooth sidewall of the array devices owing to the controlled etching process. After that, the $HfO_2/CuGeS$ bilayer and top electrode were deposited, as shown in Figure 1.10(e). Figure 1.10(f) shows the optical image of the fabricated 3D V-RRAM with four layers and 32 × 8 arrays. The high WM and high η can secure a 10-Mb array of the HfO_2-based 3D V-RRAM in the worst-case scenario.

The next-generation devices are looking for neuromorphic computing, which enables an efficient computing by parallel computation and adaptive learning. The conventional "von Neumann" computing systems in which processing and memory units are separated require shuttling of information between the components, which slows processing and increases energy consumption. In contrast, neuromorphic computing systems combine processing and memory in a single unit, which requires

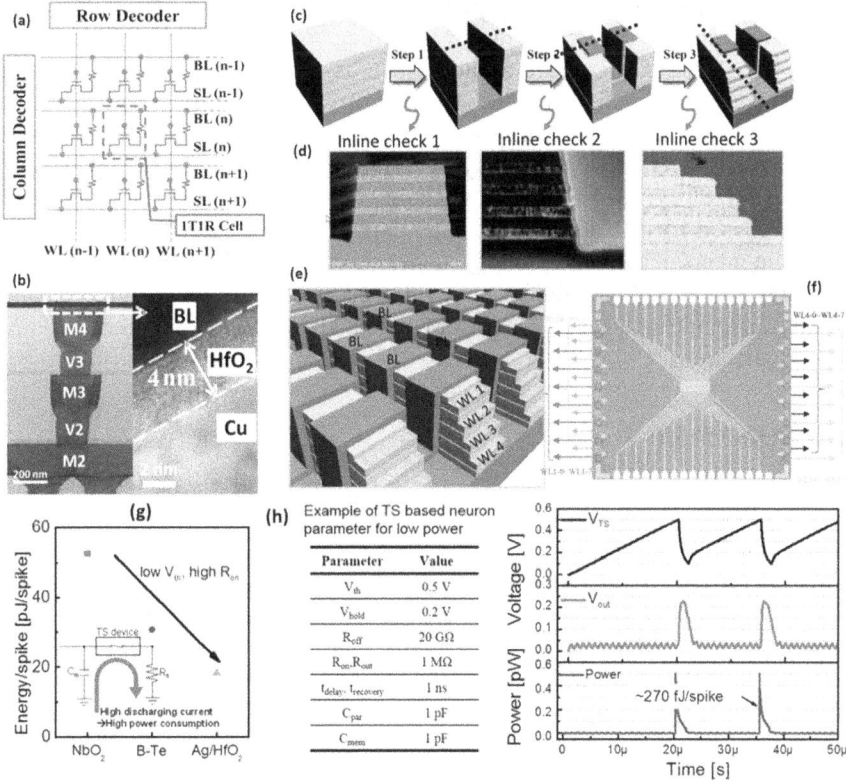

FIGURE 1.10 (a,b) The HfO$_x$-based 1T1R array structure. (c–f) The design of HfO$_x$-based 3D-VRRAM devices for high-density memory applications (Lv, Xu, Liu, et al., 2015). (g) The low energy consumption of Ag/HfO$_x$-based neurons. (h) Simulation of the low power consumption neuron. Reproduced with permission (Banerjee, Kashir, et al., 2022). Copyright 2022, Wiley-VCH.

synaptic devices that can achieve analog updates of synaptic weights similar to the neural network of the brain. The artificial neural network is expected to be one of the solutions in the post–von Neumann computing era in which the spiking neural network is extremely advantageous for reducing power consumption. The HfO$_x$-based resistive switching devices have been well studied for such neuromorphic applications. Especially the self-rupturing nature of the Ag-filament has been adopted for integrating and fire neuron applications (Banerjee, Liu, Lv, et al., 2017; Luo et al., 2016). Figure 1.10(g) shows a comparative energy consumption per spike in different TS devices based on NbO$_2$, B-Te, and Ag/HfO$_2$ (Q. Wang et al., 2019). Low energy consumption is achieved in Ag/HfO$_2$ devices because of the high LRS and low threshold voltage-controlled suppressed discharging current. Figure 1.10(h) shows the estimation of low power consumption. The Ag/HfO$_2$ devices are useful as low-power neurons with a low energy consumption of 270 fJ per spike, which is lower than the conventional CMOS-based neurons.

Apart from the high-density memory and neuromorphic applications, the random distribution of defects in the switching oxide is the entropy source to design RRAM-based hardware security such as true random number generator (TRNG) and hash function (D. Lee et al., 2019). So far, several researchers have identified the usefulness of different RRAM structures for hardware security applications (Niu et al., 2016). Pebay-Peyroula et al. reported TRNG by using the entropy source in HfO_2-based RRAM (Jiang et al., 2017). The entropy source in the HfO_2-based RRAM device was found during the SET switching process. The random numbers generated by this system were further verified by standard NIST test suites SP800-90B and SP800-22. A 100% success rate of the test ensures high-quality randomness in the HfO_2-RRAM-based TRNG for the Internet of things (IoT) applications.

1.9   SUMMARY

The evolution of materials is an important part of the development of technology. HfO_2 is a long-standing, all-purpose oxide, especially when it comes to the memory domain. Dielectric properties of HfO_2 have been extensively explored in baseline flash memory devices, which can effectively reduce EOT and introduce nanolaminate structures with MPB composition. However, because of the retention issue along with the scaling limits of flash memory devices, the impact of HfO_2 has been investigated in emerging RRAM and ferroelectric memories. As a host material, the scope of HfO_2 has been explored in RRAM devices such as VCM, ECM, filamentary, interfacial, hybrid, and other devices. Furthermore, the integration of HfO_2 with 2D materials and 3D-VRRAM has been demonstrated. Excellent resistive switching in HfO_2-based RRAM devices has been achieved with subnanosecond speed, 6 bits/cell MLC, 10^{10} endurance cycles, and 10 years of high-temperature retention. The emergent ferroelectric behavior discovered in HfO_2-based ultrathin films provides a highly promising opportunity to miniaturize nonvolatile ferroelectric memory.

NOTE

1   This chapter is a part of the earlier publication Small, 2107575, 2022 and Electronics, 9(6), 1029, 2020.

REFERENCES

Azzaz, M., Vianello, E., Sklenard, B., Blaise, P., Roule, A., Sabbione, C., Bernasconi, S., Charpin, C., Cagli, C., Jalaguier, E., Jeannot, S., Denorme, S., Candelier, P., Yu, M., Nistor, L., Fenouillet-Beranger, C., & Perniola, L. (2016). Endurance/Retention Trade off in HfOx and TaOx Based RRAM. *2016 IEEE 8th International Memory Workshop, IMW 2016.* https://doi.org/10.1109/IMW.2016.7495268

Banerjee, W. (2020). Challenges and Applications of Emerging Nonvolatile Memory Devices. *Electronics*, 9(6), 1029. https://doi.org/10.3390/electronics9061029

Banerjee, W., Cai, W. F., Zhao, X., Liu, Q., Lv, H., Long, S., & Liu, M. (2017). Intrinsic anionic rearrangement by extrinsic control: Transition of RS and CRS in thermally elevated TiN/ HfO2/Pt RRAM. *Nanoscale*, 9(47). https://doi.org/10.1039/c7nr06628g

Banerjee, W., & Hwang, H. (2019). Quantized Conduction Device with 6-Bit Storage Based on Electrically Controllable Break Junctions. *Advanced Electronic Materials*, *5*(12). https://doi.org/10.1002/aelm.201900744

Banerjee, W., & Hwang, H. (2020a). Evolution of 0.7 conductance anomaly in electric field driven ferromagnetic CuO junction based resistive random access memory devices. *Applied Physics Letters*, *116*(5). https://doi.org/10.1063/1.5136290

Banerjee, W., & Hwang, H. (2020b). Understanding of Selector-Less 1S1R Type Cu-Based CBRAM Devices by Controlling Sub-Quantum Filament. *Advanced Electronic Materials*, *6*(9). https://doi.org/10.1002/aelm.202000488

Banerjee, W., Karpov, I. v., Agrawal, A., Kim, S., Lee, S., Lee, S., Lee, D., & Hwang, H. (2020a). Highly-stable (< 3% fluctuation) Ag-based Threshold Switch with Extreme-low off Current of 0.1 pA, Extreme-high Selectivity of 109and High Endurance of 109Cycles. *Technical Digest – International Electron Devices Meeting, IEDM*, *2020-December*. https://doi.org/10.1109/IEDM13553.2020.9371960

Banerjee, W., Karpov, I. v., Agrawal, A., Kim, S., Lee, S., Lee, S., Lee, D., & Hwang, H. (2020b). Highly-stable (<3% fluctuation) Ag-based Threshold Switch with Extreme-low OFF Current of 0.1 pA, Extreme-high Selectivity of 10 9 and High Endurance of 10 9 Cycles. *2020 IEEE International Electron Devices Meeting (IEDM)*, 28.4.1–28.4.4. https://doi.org/10.1109/IEDM13553.2020.9371960

Banerjee, W., Kashir, A., & Kamba, S. (2022). Hafnium Oxide (HfO 2) – A Multifunctional Oxide: A Review on the Prospect and Challenges of Hafnium Oxide in Resistive Switching and Ferroelectric Memories. *Small*, *18*(23), 2107575. https://doi.org/10.1002/smll.202107575

Banerjee, W., Kim, S. H., Lee, S., Lee, D., & Hwang, H. (2021). An Efficient Approach Based on Tuned Nanoionics to Maximize Memory Characteristics in Ag-Based Devices. *Advanced Electronic Materials*, *7*(4). https://doi.org/10.1002/aelm.202100022

Banerjee, W., Kim, S. H., Lee, S., Lee, S., Lee, D., & Hwang, H. (2021). Deep Insight into Steep-Slope Threshold Switching with Record Selectivity (>4 × 1010) Controlled by Metal-Ion Movement through Vacancy-Induced-Percolation Path: Quantum-Level Control of Hybrid-Filament. *Advanced Functional Materials*, *31*(37). https://doi.org/10.1002/adfm.202104054

Banerjee, W., Liu, Q., & Hwang, H. (2020). Engineering of defects in resistive random access memory devices. *Journal of Applied Physics*, *127*(5). https://doi.org/10.1063/1.5136264

Banerjee, W., Liu, Q., Long, S., Lv, H., & Liu, M. (2017). Crystal that remembers: several ways to utilize nanocrystals in resistive switching memory. *Journal of Physics D: Applied Physics*, *50*(30), 303002. https://doi.org/10.1088/1361-6463/aa7572

Banerjee, W., Liu, Q., Lv, H., Long, S., & Liu, M. (2017). Electronic imitation of behavioral and psychological synaptic activities using TiO:X/Al2O3-based memristor devices. *Nanoscale*, *9*(38). https://doi.org/10.1039/c7nr04741j

Banerjee, W., Lu, N., Yang, Y., Li, L., Lv, H., Liu, Q., Long, S., & Liu, M. (2018). Investigation of Retention Behavior of TiOx/Al2O3 Resistive Memory and Its Failure Mechanism Based on Meyer-Neldel Rule. *IEEE Transactions on Electron Devices*, *65*(3). https://doi.org/10.1109/TED.2017.2788460

Banerjee, W., Maikap, S., Chen, Y.-Y., & Yang, J. R. (2012). Unipolar Resistive Switching Memory Characteristics Using IrO x /Al 2 O 3 /SiO 2 /p-Si MIS Structure. *ECS Transactions*, *45*(3). https://doi.org/10.1149/1.3700898

Banerjee, W., Maikap, S., Lai, C.-S., Chen, Y.-Y., Tien, T.-C., Lee, H.-Y., Chen, W.-S., Chen, F. T., Kao, M.-J., Tsai, M.-J., & Yang, J.-R. (2012). Formation polarity dependent improved resistive switching memory characteristics using nanoscale (1.3 nm)

core-shell IrOx nano-dots. *Nanoscale Research Letters*, 7(1), 194. https://doi.org/10.1186/1556-276X-7-194

Banerjee, W., Maikap, S., Rahaman, S. Z., Prakash, A., Tien, T.-C., Li, W.-C., & Yang, J.-R. (2011). Improved Resistive Switching Memory Characteristics Using Core-Shell IrOx Nano-Dots in Al2O3/WOx Bilayer Structure. *Journal of The Electrochemical Society*, 159(2), H177. https://doi.org/10.1149/2.067202jes

Banerjee, W., Maikap, S., Tien, T. C., Li, W. C., & Yang, J. R. (2011). Impact of metal nano layer thickness on tunneling oxide and memory performance of core-shell iridium-oxide nanocrystals. *Journal of Applied Physics*, 110(7). https://doi.org/10.1063/1.3642961

Banerjee, W., Nikam, R. D., & Hwang, H. (2022). Prospect and challenges of analog switching for neuromorphic hardware. *Applied Physics Letters*, 120(6). https://doi.org/10.1063/5.0073528

Banerjee, W., Rahaman, S. Z., Prakash, A., & Maikap, S. (2011). High-κ Al2O3/WOx bilayer dielectrics for low-power resistive switching memory applications. *Japanese Journal of Applied Physics*, 50(10 PART 2). https://doi.org/10.1143/JJAP.50.10PH01

Banerjee, W., Rahaman, Sk. Z., & Maikap, S. (2012). Excellent Uniformity and Multilevel Operation in Formation-Free Low Power Resistive Switching Memory Using IrO_x/AlO_x/W Cross-Point. *Japanese Journal of Applied Physics*, 51, 04DD10. https://doi.org/10.1143/JJAP.51.04DD10

Banerjee, W., Wu, F., Hu, Y., Wu, Q., Wu, Z., Liu, Q., & Liu, M. (2018). Origin of negative resistance in anion migration controlled resistive memory. *Applied Physics Letters*, 112(13). https://doi.org/10.1063/1.5021019

Banerjee, W., Xu, X., Liu, H., Lv, H., Liu, Q., Sun, H., Long, S., & Liu, M. (2015). Occurrence of Resistive Switching and Threshold Switching in Atomic Layer Deposited Ultrathin (2 nm) Aluminium Oxide Crossbar Resistive Random Access Memory. *IEEE Electron Device Letters*, 36(4). https://doi.org/10.1109/LED.2015.2407361

Banerjee, W., Xu, X., Lv, H., Liu, Q., Long, S., & Liu, M. (2017a). Complementary Switching in 3D Resistive Memory Array. *Advanced Electronic Materials*, 3(12). https://doi.org/10.1002/aelm.201700287

Banerjee, W., Xu, X., Lv, H., Liu, Q., Long, S., & Liu, M. (2017b). Variability improvement of TiOx/Al2O3 bilayer nonvolatile resistive switching devices by interfacial band engineering with an ultrathin Al2O3 dielectric material. *ACS Omega*, 2(10). https://doi.org/10.1021/acsomega.7b01211

Banerjee, W., Zhang, X., Luo, Q., Lv, H., Liu, Q., Long, S., & Liu, M. (2018). Design of CMOS Compatible, High-Speed, Highly-Stable Complementary Switching with Multilevel Operation in 3D Vertically Stacked Novel HfO2/Al2O3/TiOx (HAT) RRAM. *Advanced Electronic Materials*, 4(2). https://doi.org/10.1002/aelm.201700561

Böttger, U., von Witzleben, M., Havel, V., Fleck, K., Rana, V., Waser, R., & Menzel, S. (2020). Picosecond multilevel resistive switching in tantalum oxide thin films. *Scientific Reports*, 10(1). https://doi.org/10.1038/s41598-020-73254-2

Bozano, L. D., Kean, B. W., Deline, V. R., Salem, J. R., & Scott, J. C. (2004). Mechanism for bistability in organic memory elements. *Applied Physics Letters*, 84(4). https://doi.org/10.1063/1.1643547

Burr, G. W., Shelby, R. M., Sebastian, A., Kim, S., Kim, S., Sidler, S., Virwani, K., Ishii, M., Narayanan, P., Fumarola, A., Sanches, L. L., Boybat, I., le Gallo, M., Moon, K., Woo, J., Hwang, H., & Leblebici, Y. (2017). Neuromorphic computing using non-volatile memory. In *Advances in Physics: X*, 2(1). https://doi.org/10.1080/23746149.2016.1259585

Chang, T. C., Jian, F. Y., Chen, S. C., & Tsai, Y. T. (2011). Developments in nanocrystal memory. In *Materials Today 14*(12). https://doi.org/10.1016/S1369-7021(11)70302-9

Chang, T., Jo, S. H., & Lu, W. (2011). Short-term memory to long-term memory transition in a nanoscale memristor. *ACS Nano*, *5*(9). https://doi.org/10.1021/nn202983n

Chen, A. (2016). A review of emerging non-volatile memory (NVM) technologies and applications. *Solid-State Electronics*, *125*. https://doi.org/10.1016/j.sse.2016.07.006

Chen, A. (2017). Memory selector devices and crossbar array design: a modeling-based assessment. *Journal of Computational Electronics*, *16*(4). https://doi.org/10.1007/s10 825-017-1059-7

Chen, C. Y., Goux, L., Fantini, A., Redolfi, A., Clima, S., Degraeve, R., Chen, Y. Y., Groeseneken, G., & Jurczak, M. (2015). Understanding the impact of programming pulses and electrode materials on the endurance properties of scaled Ta2O5 RRAM cells. *Technical Digest – International Electron Devices Meeting, IEDM*, *2015-February*. https://doi.org/10.1109/IEDM.2014.7047049

Chen, Y. Y., Goux, L., Clima, S., Govoreanu, B., Degraeve, R., Kar, G. S., Fantini, A., Groeseneken, G., Wouters, D. J., & Jurczak, M. (2013). Endurance/retention trade-off on HfO2\metal cap 1T1R bipolar RRAM. *IEEE Transactions on Electron Devices*, *60*(3). https://doi.org/10.1109/TED.2013.2241064

Cho, B., Song, S., Ji, Y., Choi, H. G., Ko, H. C., Lee, J. S., Jung, G. Y., & Lee, T. (2013). Demonstration of addressable organic resistive memory utilizing a PC-interface memory cell tester. *IEEE Electron Device Letters*, *34*(1). https://doi.org/10.1109/LED.2012.2226231

Choi, B. J., Torrezan, A. C., Strachan, J. P., Kotula, P. G., Lohn, A. J., Marinella, M. J., Li, Z., Williams, R. S., & Yang, J. J. (2016). High-Speed and Low-Energy Nitride Memristors. *Advanced Functional Materials*, *26*(29). https://doi.org/10.1002/adfm.201600680

Choi, S. J., Park, G. S., Kim, K. H., Cho, S., Yang, W. Y., Li, X. S., Moon, J. H., Lee, K. J., & Kim, K. (2011). In situ observation of voltage-induced multilevel resistive switching in solid electrolyte memory. *Advanced Materials*, *23*(29). https://doi.org/10.1002/adma.201100507

Choi, W., Gi, S. G., Lee, D., Lim, S., Lee, C., Lee, B. G., & Hwang, H. (2020). WOx-Based Synapse Device with Excellent Conductance Uniformity for Hardware Neural Networks. *IEEE Transactions on Nanotechnology*, *19*. https://doi.org/10.1109/TNANO.2020.3010070

Ding, X., Wang, X., Feng, Y., Shen, W., & Liu, L. (2020). Low operation current of Si/HfO2 double layers based RRAM device with insertion of Si film. *Japanese Journal of Applied Physics*, *59*(SG). https://doi.org/10.35848/1347-4065/ab6b7b

Du, C., Ma, W., Chang, T., Sheridan, P., & Lu, W. D. (2015). Biorealistic Implementation of Synaptic Functions with Oxide Memristors through Internal Ionic Dynamics. *Advanced Functional Materials*, *25*(27). https://doi.org/10.1002/adfm.201501427

Fantini, A., Goux, L., Clima, S., Degraeve, R., Redolfi, A., Adelmann, C., Polimeni, G., Chen, Y. Y., Komura, M., Belmonte, A., Wouters, D. J., & Jurczak, M. (2014). Engineering of Hf1-xAlxOy amorphous dielectrics for high-performance RRAM applications. *2014 IEEE 6th International Memory Workshop, IMW 2014*. https://doi.org/10.1109/IMW.2014.6849354

Fantini, A., Goux, L., Degraeve, R., Wouters, D. J., Raghavan, N., Kar, G., Belmonte, A., Chen, Y.-Y., Govoreanu, B., & Jurczak, M. (2013). Intrinsic switching variability in HfO$_2$ RRAM. *2013 5th IEEE International Memory Workshop*, 30–33. https://doi.org/10.1109/IMW.2013.6582090

Fujii, S., Ichihara, R., Konno, T., Yamaguchi, M., Seki, H., Tanaka, H., Zhao, D., Yoshimura, Y., Saitoh, M., & Koyama, M. (2019). Ag Ionic Memory Cell Technology for Terabit-Scale High-Density Application. *Digest of Technical Papers – Symposium on VLSI Technology*, *2019-June*. https://doi.org/10.23919/VLSIT.2019.8776568

Fujii, T., Arita, M., Takahashi, Y., & Fujiwara, I. (2011). In situ transmission electron micros-copy analysis of conductive filament during solid electrolyte resistance switching. *Applied Physics Letters*, *98*(21). https://doi.org/10.1063/1.3593494

Fumarola, A., Sidler, S., Moon, K., Jang, J., Shelby, R. M., Narayanan, P., Leblebici, Y., Hwang, H., & Burr, G. W. (2018). Bidirectional Non-Filamentary RRAM as an Analog Neuromorphic Synapse, Part II: Impact of Al/Mo/Pr 0.7 Ca 0.3 MnO 3 Device Characteristics on Neural Network Training Accuracy. *IEEE Journal of the Electron Devices Society*, *6*, 169–178. https://doi.org/10.1109/JEDS.2017.2782184

Gao, S., Song, C., Chen, C., Zeng, F., & Pan, F. (2013). Formation process of conducting fila-ment in planar organic resistive memory. *Applied Physics Letters*, *102*(14). https://doi.org/10.1063/1.4802092

Giri, A., Kumar, M., Kim, J., Pal, M., Banerjee, W., Nikam, R. D., Kwak, J., Kong, M., Kim, S. H., Thiyagarajan, K., Kim, G., Hwang, H., Lee, H. H., Lee, D., & Jeong, U. (2021). Surface Diffusion and Epitaxial Self-Planarization for Wafer-Scale Single-Grain Metal Chalcogenide Thin Films. *Advanced Materials*, *33*(35). https://doi.org/10.1002/adma.202102252

Goda, A. (2021). Recent progress on 3D NAND flash technologies. *Electronics (Switzerland)*, *10*(24). https://doi.org/10.3390/electronics10243156

Gogurla, N., Mondal, S. P., Sinha, A. K., Katiyar, A. K., Banerjee, W., Kundu, S. C., & Ray, S. K. (2013). Transparent and flexible resistive switching memory devices with a very high ON/OFF ratio using gold nanoparticles embedded in a silk protein matrix. *Nanotechnology*, *24*(34). https://doi.org/10.1088/0957-4484/24/34/345202

Goux, L., Fantini, A., Kar, G., Chen, Y.-Y., Jossart, N., Degraeve, R., Clima, S., Govoreanu, B., Lorenzo, G., Pourtois, G., Wouters, D. J., Kittl, J. A., Altimime, L., & Jurczak, M. (2012). Ultralow sub-500nA operating current high-performance TiN\Al<inf>2</inf>O<inf>3</inf>\HfO<inf>2</inf>\Hf\TiN bipolar RRAM achieved through understanding-based stack-engineering. *2012 Symposium on VLSI Technology (VLSIT)*, 159–160. https://doi.org/10.1109/VLSIT.2012.6242510

Govoreanu, B., Ajaykumar, A., Lipowicz, H., Chen, Y.-Y., Liu, J.-C., Degraeve, R., Zhang, L., Clima, S., Goux, L., Radu, I. P., Fantini, A., Raghavan, N., Kar, G.-S., Kim, W., Redolfi, A., Wouters, D. J., Altimime, L., & Jurczak, M. (2013). Performance and reli-ability of Ultra-Thin HfO<inf>2</inf>-based RRAM (UTO-RRAM). *2013 5th IEEE International Memory Workshop*, 48–51. https://doi.org/10.1109/IMW.2013.6582095

Govoreanu, B., di Piazza, L., Ma, J., Conard, T., Vanleenhove, A., Belmonte, A., Radisic, D., Popovici, M., Velea, A., Redolfi, A., Richard, O., Clima, S., Adelmann, C., Bender, H., & Jurczak, M. (2016). Advanced a-VMCO resistive switching memory through inner interface engineering with wide (>102) on/off window, tunable µa-range switching current and excellent variability. *Digest of Technical Papers – Symposium on VLSI Technology*, *2016-September*. https://doi.org/10.1109/VLSIT.2016.7573387

Govoreanu, B., Kar, G. S., Chen, Y.-Y., Paraschiv, V., Kubicek, S., Fantini, A., Radu, I. P., Goux, L., Clima, S., Degraeve, R., Jossart, N., Richard, O., Vandeweyer, T., Seo, K., Hendrickx, P., Pourtois, G., Bender, H., Altimime, L., Wouters, D. J., … Jurczak, M. (2011). 10×10nm2 Hf/HfO<inf>x</inf> crossbar resistive RAM with excel-lent performance, reliability and low-energy operation. *2011 International Electron Devices Meeting*, 31.6.1–31.6.4. https://doi.org/10.1109/IEDM.2011.6131652

Grisafe, B., Jerry, M., Smith, J. A., & Datta, S. (2019). Performance Enhancement of Ag/HfO2 Metal Ion Threshold Switch Cross-Point Selectors. *IEEE Electron Device Letters*, *40*(10). https://doi.org/10.1109/LED.2019.2936104

Grossi, A., Perez, E., Zambelli, C., Olivo, P., Miranda, E., Roelofs, R., Woodruff, J., Raisanen, P., Li, W., Givens, M., Costina, I., Schubert, M. A., & Wenger, C. (2018). Impact of the precursor chemistry and process conditions on the cell-to-cell variability in 1T-1R based HfO2 RRAM devices. *Scientific Reports*, 8(1). https://doi.org/10.1038/s41598-018-29548-7

Gubicza, A., Manrique, D. Z., Pósa, L., Lambert, C. J., Mihály, G., Csontos, M., & Halbritter, A. (2016). Asymmetry-induced resistive switching in Ag-Ag2S-Ag memristors enabling a simplified atomic-scale memory design. *Scientific Reports*, 6. https://doi.org/10.1038/srep30775

Guo, X., Schindler, C., Menzel, S., & Waser, R. (2007). Understanding the switching-off mechanism in Ag+ migration based resistively switching model systems. *Applied Physics Letters*, 91(13). https://doi.org/10.1063/1.2793686

Guy, J., Molas, G., Vianello, E., Longnos, F., Blanc, S., Carabasse, C., Bernard, M., Nodin, J. F., Toffoli, A., Cluzel, J., Blaise, P., Dorion, P., Cueto, O., Grampeix, H., Souchier, E., Cabout, T., Brianceau, P., Balan, V., Roule, A., ... de Salvo, B. (2013). Investigation of the physical mechanisms governing data-retention in down to 10nm nano-trench Al2O3/CuTeGe conductive bridge RAM (CBRAM). *Technical Digest – International Electron Devices Meeting, IEDM*. https://doi.org/10.1109/IEDM.2013.6724722

Hamzah, A., Ahmad, H., Tan, M. L. P., Alias, N. E., Johari, Z., & Ismail, R. (2019). Scaling Challenges of Floating Gate Non-Volatile Memory and Graphene as the Future Flash Memory Device: A Review. *Journal of Nanoelectronics and Optoelectronics*, 14(9). https://doi.org/10.1166/jno.2019.2204

Hang, C. Z., Wang, C., Gao, B., Chen, H., Xu, M. H., Hao, L., & Lu, H. L. (2019). Sub-nanosecond pulse programming and device design strategy for analog resistive switching in HfOx-based resistive random access memory. *Applied Physics Letters*, 114(11). https://doi.org/10.1063/1.5078782

Hong, X. L., Loy, D. J. J., Dananjaya, P. A., Tan, F., Ng, C. M., & Lew, W. S. (2018). Oxide-based RRAM materials for neuromorphic computing. In *Journal of Materials Science*, 53(12). https://doi.org/10.1007/s10853-018-2134-6

Hu, C., McDaniel, M. D., Posadas, A., Demkov, A. A., Ekerdt, J. G., & Yu, E. T. (2014). Highly controllable and stable quantized conductance and resistive switching mechanism in single-crystal TiO2 resistive memory on silicon. *Nano Letters*, 14(8). https://doi.org/10.1021/nl501249q

Hua, Q., Wu, H., Gao, B., Zhao, M., Li, Y., Li, X., Hou, X., Chang, M., Zhou, P., & Qian, H. (2019). Threshold Switching Selectors: A Threshold Switching Selector Based on Highly Ordered Ag Nanodots for X-Point Memory Applications (Adv. Sci. 10/2019). *Advanced Science*, 6(10). https://doi.org/10.1002/advs.201970058

Hui, F., Fang, W., Leong, W. S., Kpulun, T., Wang, H., Yang, H. Y., Villena, M. A., Harris, G., Kong, J., & Lanza, M. (2017). Electrical Homogeneity of Large-Area Chemical Vapor Deposited Multilayer Hexagonal Boron Nitride Sheets. *ACS Applied Materials and Interfaces*, 9(46). https://doi.org/10.1021/acsami.7b09417

Indiveri, G., Linares-Barranco, B., Legenstein, R., Deligeorgis, G., & Prodromakis, T. (2013). Integration of nanoscale memristor synapses in neuromorphic computing architectures. *Nanotechnology*, 24(38). https://doi.org/10.1088/0957-4484/24/38/384010

Islam, R., Li, H., Chen, P. Y., Wan, W., Chen, H. Y., Gao, B., Wu, H., Yu, S., Saraswat, K., & Philip Wong, H. S. (2019). Device and materials requirements for neuromorphic computing. *Journal of Physics D: Applied Physics*, 52(11). https://doi.org/10.1088/1361-6463/aaf784

Jameson, J. R., Blanchard, P., Dinh, J., Gonzales, N., Gopalakrishnan, V., Guichet, B., Hollmer, S., Hsu, S., Intrater, G., Kamalanathan, D., Kim, D., Koushan, F., Kwan, M., Lewis, D., Pedersen, B., Ramsbey, M., Runnion, E., Shields, J., Tsai, K., ... Gopinath, V. (2016).

(Invited) Conductive Bridging RAM (CBRAM): Then, Now, and Tomorrow. *ECS Transactions*, *75*(5). https://doi.org/10.1149/07505.0041ecst

Jameson, J. R., Dinh, J., Gonzales, N., Hollmer, S., Hsu, S., Kim, D., Koushan, F., Lewis, D., Runnion, E., Shields, J., Tysdal, A., Wang, D., & Gopinath, V. (2018). Towards Automotive Grade Embedded RRAM. *2018 48th European Solid-State Device Research Conference (ESSDERC)*, 58–61. https://doi.org/10.1109/ESSDERC.2018.8486890

Jameson, J. R., & Kamalanathan, D. (2016). Subquantum conductive-bridge memory. *Applied Physics Letters*, *108*(5). https://doi.org/10.1063/1.4941303

Jeon, W. (2020). Recent advances in the understanding of high- k dielectric materials deposited by atomic layer deposition for dynamic random-access memory capacitor applications. *Journal of Materials Research*, *35*(7), 775–794. https://doi.org/10.1557/jmr.2019.335

Jiang, H., Belkin, D., Savel'Ev, S. E., Lin, S., Wang, Z., Li, Y., Joshi, S., Midya, R., Li, C., Rao, M., Barnell, M., Wu, Q., Yang, J. J., & Xia, Q. (2017). A novel true random number generator based on a stochastic diffusive memristor. *Nature Communications*, *8*(1). https://doi.org/10.1038/s41467-017-00869-x

Kim, J., Hong, A. J., Kim, S. M., Shin, K. S., Song, E. B., Hwang, Y., Xiu, F., Galatsis, K., Chui, C. O., Candler, R. N., Choi, S., Moon, J. T., & Wang, K. L. (2011). A stacked memory device on logic 3D technology for ultra-high-density data storage. *Nanotechnology*, *22*(25). https://doi.org/10.1088/0957-4484/22/25/254006

Kol, S., & Oral, A. (2019). Hf-Based High-κ Dielectrics: A Review. *Acta Physica Polonica A*, *136*(6), 873–881. https://doi.org/10.12693/APhysPolA.136.873

Kozicki, M. N., & Mitkova, M. (2006). Mass transport in chalcogenide electrolyte films – materials and applications. *Journal of Non-Crystalline Solids*, *352*(6-7 SPEC. ISS.). https://doi.org/10.1016/j.jnoncrysol.2005.11.065

Lan, X., Ou, X., Lei, Y., Gong, C., Yin, Q., Xu, B., Xia, Y., Yin, J., & Liu, Z. (2013). The interface inter-diffusion induced enhancement of the charge-trapping capability in HfO2/Al2O3 multilayered memory devices. *Applied Physics Letters*, *103*(19). https://doi.org/10.1063/1.4829066

Lanza, M., Wong, H. S. P., Pop, E., Ielmini, D., Strukov, D., Regan, B. C., Larcher, L., Villena, M. A., Yang, J. J., Goux, L., Belmonte, A., Yang, Y., Puglisi, F. M., Kang, J., Magyari-Köpe, B., Yalon, E., Kenyon, A., Buckwell, M., Mehonic, A., … Shi, Y. (2019). Recommended Methods to Study Resistive Switching Devices. In *Advanced Electronic Materials*, *5*(1). https://doi.org/10.1002/aelm.201800143

Lee, D., Kwak, M., Moon, K., Choi, W., Park, J., Yoo, J., Song, J., Lim, S., Sung, C., Banerjee, W., & Hwang, H. (2019). Various Threshold Switching Devices for Integrate and Fire Neuron Applications. *Advanced Electronic Materials*, *5*(9). https://doi.org/10.1002/aelm.201800866

Lee, H. Y., Chen, P. S., Wu, T. Y., Chen, Y. S., Wang, C. C., Tzeng, P. J., Lin, C. H., Chen, F., Lien, C. H., & Tsai, M. J. (2008). Low power and high speed bipolar switching with a thin reactive Ti buffer layer in robust HfO2 based RRAM. *Technical Digest – International Electron Devices Meeting, IEDM*. https://doi.org/10.1109/IEDM.2008.4796677

Lee, H. Y., Chen, Y. S., Chen, P. S., Gu, P. Y., Hsu, Y. Y., Wang, S. M., Liu, W. H., Tsai, C. H., Sheu, S. S., Chiang, P. C., Lin, W. P., Lin, C. H., Chen, W. S., Chen, F. T., Lien, C. H., & Tsai, M. J. (2010). Evidence and solution of over-RESET problem for HfOX based resistive memory with sub-ns switching speed and high endurance. *Technical Digest – International Electron Devices Meeting, IEDM*. https://doi.org/10.1109/IEDM.2010.5703395

Lee, J. S. (2011). Review paper: Nano-floating gate memory devices. In *Electronic Materials Letters*, *7*(3). https://doi.org/10.1007/s13391-011-0901-5

Lee, J. S., Lee, S., & Noh, T. W. (2015). Resistive switching phenomena: A review of statistical physics approaches. In *Applied Physics Reviews*, 2(3). https://doi.org/10.1063/1.4929512

Lee, M. J., Lee, C. B., Lee, D., Lee, S. R., Chang, M., Hur, J. H., Kim, Y. B., Kim, C. J., Seo, D. H., Seo, S., Chung, U. I., Yoo, I. K., & Kim, K. (2011). A fast, high-endurance and scalable non-volatile memory device made from asymmetric Ta2O5-xx/TaO2-xbilayer structures. *Nature Materials*, 10(8). https://doi.org/10.1038/nmat3070

Lee, M. K. F., Cui, Y., Somu, T., Luo, T., Zhou, J., Tang, W. T., Wong, W. F., & Goh, R. S. M. (2019). A system-level simulator for RRAM-based neuromorphic computing chips. *ACM Transactions on Architecture and Code Optimization*, 15(4). https://doi.org/10.1145/3291054

Lee, S., Banerjee, W., Lee, S., Sung, C., & Hwang, H. (2021). Improved Threshold Switching and Endurance Characteristics Using Controlled Atomic-Scale Switching in a 0.5 nm Thick Stoichiometric HfO2 Layer. *Advanced Electronic Materials*, 7(2). https://doi.org/10.1002/aelm.202000869

Lee, S., Sohn, J., Jiang, Z., Chen, H. Y., & Philip Wong, H. S. (2015). Metal oxide-resistive memory using graphene-edge electrodes. *Nature Communications*, 6. https://doi.org/10.1038/ncomms9407

Li, W. C., Banerjee, W., Maikap, S., & Yang, J. R. (2011). Particle size and morphology of iridium oxide nanocrystals in non-volatile memory device. *Materials Transactions*, 52(3). https://doi.org/10.2320/matertrans.MB201020

Li, W.-C., Banerjee, W., Maikap, S., & Yang, J.-R. (2010a). Characteristics of ALD High-k HfAlO x Nanocrystals in Memory Capacitors Annealed at High Temperatures. *ECS Transactions*, 33(3). https://doi.org/10.1149/1.3481623

Li, W.-C., Banerjee, W., Maikap, S., & Yang, J.-R. (2010b). Density and Grain Size of the IrO x Metal Nanocrystals in n-Si/SiO$_2$/Al$_2$O$_3$/IrO$_x$/Al$_2$O$_3$ Memory Capacitors. *ECS Transactions*, 33(3), 333–337. https://doi.org/10.1149/1.3481621

Li, Y., Long, S., Liu, Y., Hu, C., Teng, J., Liu, Q., Lv, H., Suñé, J., & Liu, M. (2015). Conductance Quantization in Resistive Random Access Memory. In *Nanoscale Research Letters*, 10(1). https://doi.org/10.1186/s11671-015-1118-6

Li, Y., Xu, L., Zhong, Y. P., Zhou, Y. X., Zhong, S. J., Hu, Y. Z., Chua, L. O., & Miao, X. S. (2015). Memristors: Associative Learning with Temporal Contiguity in a Memristive Circuit for Large-Scale Neuromorphic Networks (Adv. Electron. Mater. 8/2015). In *Advanced Electronic Materials*, 1(8). https://doi.org/10.1002/aelm.201570026

Li, Y., Zhong, Y., Xu, L., Zhang, J., Xu, X., Sun, H., & Miao, X. (2013). Ultrafast synaptic events in a chalcogenide memristor. *Scientific Reports*, 3. https://doi.org/10.1038/srep01619

Lin, J., Wang, S., & Liu, H. (2021). Multi-level switching of al-doped HfO2 RRAM with a single voltage amplitude set pulse. *Electronics (Switzerland)*, 10(6). https://doi.org/10.3390/electronics10060731

Liu, Q., Sun, J., Lv, H., Long, S., Yin, K., Wan, N., Li, Y., Sun, L., & Liu, M. (2012). Real-time observation on dynamic growth/dissolution of conductive filaments in oxide-electrolyte-based ReRAM. *Advanced Materials*, 24(14). https://doi.org/10.1002/adma.201104104

Liu, S., Lu, N., Zhao, X., Xu, H., Banerjee, W., Lv, H., Long, S., Li, Q., Liu, Q., & Liu, M. (2016). Eliminating Negative-SET Behavior by Suppressing Nanofilament Overgrowth in Cation-Based Memory. *Advanced Materials*, 28(48). https://doi.org/10.1002/adma.201603293

Lu, N., Li, L., Sun, P., Wang, M., Liu, Q., Lv, H., Long, S., Banerjee, W., & Liu, M. (2015). Carrier-transport-path-induced switching parameter fluctuation in oxide-based resistive switching memory. *Materials Research Express*, *2*(4). https://doi.org/10.1088/2053-1591/2/4/046304

Lu, Y., Song, S., Song, Z., Ren, K., Liu, B., & Feng, S. (2011). Sb2Te3-HfO2 composite films for low-power phase change memory application. *Applied Physics A: Materials Science and Processing*, *105*(1). https://doi.org/10.1007/s00339-011-6478-x

Lu, Y., Song, S., Song, Z., Ren, W., Peng, C., Cheng, Y., & Liu, B. (2011). Investigation of HfO2 doping on GeTe for phase change memory. *Solid State Sciences*, *13*(11). https://doi.org/10.1016/j.solidstatesciences.2011.08.021

Luo, Q., Xu, X., Liu, H., Lv, H., Gong, T., Long, S., Liu, Q., Sun, H., Banerjee, W., Li, L., Gao, J., Lu, N., Chung, S. S., Li, J., & Liu, M. (2015). Demonstration of 3D vertical RRAM with ultra low-leakage, high-selectivity and self-compliance memory cells. *Technical Digest – International Electron Devices Meeting, IEDM, 2016-February*. https://doi.org/10.1109/IEDM.2015.7409667

Luo, Q., Xu, X., Liu, H., Lv, H., Gong, T., Long, S., Liu, Q., Sun, H., Banerjee, W., Li, L., Gao, J., Lu, N., & Liu, M. (2016). Super non-linear RRAM with ultra-low power for 3D vertical nano-crossbar arrays. *Nanoscale*, *8*(34). https://doi.org/10.1039/c6nr02029a

Luo, Q., Xu, X., Liu, H., Lv, H., Gong, T., Long, S., Liu, Q., Sun, H., Banerjee, W., Li, L., Lu, N., & Liu, M. (2015). Cu BEOL compatible selector with high selectivity (> 107), extremely low off-current (~pA) and high endurance (> 1010). *Technical Digest – International Electron Devices Meeting, IEDM, 2016-February*. https://doi.org/10.1109/IEDM.2015.7409669

Luo, Q., Zhang, X., Hu, Y., Gong, T., Xu, X., Yuan, P., Ma, H., Dong, D., Lv, H., Long, S., Liu, Q., & Liu, M. (2018). Self-Rectifying and Forming-Free Resistive-Switching Device for Embedded Memory Application. *IEEE Electron Device Letters*, *39*(5). https://doi.org/10.1109/LED.2018.2821162

Lv, H., Xu, X., Liu, H., Liu, R., Liu, Q., Banerjee, W., Sun, H., Long, S., Li, L., & Liu, M. (2015). Evolution of conductive filament and its impact on reliability issues in oxide-electrolyte based resistive random access memory. *Scientific Reports*, *5*. https://doi.org/10.1038/srep07764

Lv, H., Xu, X., Sun, P., Liu, H., Luo, Q., Liu, Q., Banerjee, W., Sun, H., Long, S., Li, L., & Liu, M. (2015). Atomic view of filament growth in electrochemical memristive elements. *Scientific Reports*, *5*. https://doi.org/10.1038/srep13311

Maikap, S., & Banerjee, W. (2020). In Quest of Nonfilamentary Switching: A Synergistic Approach of Dual Nanostructure Engineering to Improve the Variability and Reliability of Resistive Random-Access-Memory Devices. *Advanced Electronic Materials*, *6*(6). https://doi.org/10.1002/aelm.202000209

Maikap, S., Banerjee, W., Tien, T. C., Wang, T. Y., & Yang, J. R. (2011). Temperature-dependent physical and memory characteristics of atomic-layer-deposited RuOx Metal Nanocrystal Capacitors. *Journal of Nanomaterials*, *2011*. https://doi.org/10.1155/2011/810879

Mallol, M. M., Gonzalez, M. B., & Campabadal, F. (2017). Impact of the HfO2/Al2O3 stacking order on unipolar RRAM devices. *Microelectronic Engineering*, *178*. https://doi.org/10.1016/j.mee.2017.05.024

Menzel, S., von Witzleben, M., Havel, V., & Böttger, U. (2019). The ultimate switching speed limit of redox-based resistive switching devices. *Faraday Discussions*, *213*. https://doi.org/10.1039/c8fd00117k

Merolla, P. A., Arthur, J. v., Alvarez-Icaza, R., Cassidy, A. S., Sawada, J., Akopyan, F., Jackson, B. L., Imam, N., Guo, C., Nakamura, Y., Brezzo, B., Vo, I., Esser, S. K., Appuswamy, R., Taba, B., Amir, A., Flickner, M. D., Risk, W. P., Manohar, R., & Modha, D. S. (2014).

A million spiking-neuron integrated circuit with a scalable communication network and interface. *Science, 345*(6197). https://doi.org/10.1126/science.1254642

Milo, V., Zambelli, C., Olivo, P., Pérez, E., K. Mahadevaiah, M., G. Ossorio, O., Wenger, C., & Ielmini, D. (2019). Multilevel HfO2-based RRAM devices for low-power neuromorphic networks. *APL Materials, 7*(8). https://doi.org/10.1063/1.5108650

Molas, G., Sassine, G., Nail, C., Alfaro Robayo, D., Nodin, J.-F., Cagli, C., Coignus, J., Blaise, P., & Nowak, E. (2018). (Invited) Resistive Memories (RRAM) Variability: Challenges and Solutions. *ECS Transactions, 86*(3). https://doi.org/10.1149/08603.0035ecst

Molas, G., Vianello, E., Dahmani, F., Barci, M., Blaise, P., Guy, J., Toffoli, A., Bernard, M., Roule, A., Pierre, F., Licitra, C., de Salvo, B., & Perniola, L. (2015). Controlling oxygen vacancies in doped oxide based CBRAM for improved memory performances. *Technical Digest – International Electron Devices Meeting, IEDM, 2015-February*. https://doi.org/10.1109/IEDM.2014.7046993

Monzio Compagnoni, C., & Spinelli, A. S. (2019). Reliability of NAND Flash Arrays: A Review of What the 2-D-to-3-D Transition Meant. *IEEE Transactions on Electron Devices, 66*(11). https://doi.org/10.1109/TED.2019.2917785

Moon, K., Fumarola, A., Sidler, S., Jang, J., Narayanan, P., Shelby, R. M., Burr, G. W., & Hwang, H. (2018). Bidirectional non-filamentary RRAM as an analog neuromorphic synapse, Part I: Al/Mo/Pr0.7Ca0.3MnO3 material improvements and device measurements. *IEEE Journal of the Electron Devices Society, 6*(1). https://doi.org/10.1109/JEDS.2017.2780275

Nikam, R. D., Kwak, M., Lee, J., Rajput, K. G., Banerjee, W., & Hwang, H. (2019). Near ideal synaptic functionalities in Li ion synaptic transistor using Li3POxSex electrolyte with high ionic conductivity. *Scientific Reports, 9*(1). https://doi.org/10.1038/s41598-019-55310-8

Nikam, R. D., Lee, J., Choi, W., Banerjee, W., Kwak, M., Yadav, M., & Hwang, H. (2021). Ionic Sieving Through One-Atom-Thick 2D Material Enables Analog Nonvolatile Memory for Neuromorphic Computing. *Small, 17*(44), 2103543. https://doi.org/10.1002/smll.202103543

Niu, G., Calka, P., Auf Der Maur, M., Santoni, F., Guha, S., Fraschke, M., Hamoumou, P., Gautier, B., Perez, E., Walczyk, C., Wenger, C., di Carlo, A., Alff, L., & Schroeder, T. (2016). Geometric conductive filament confinement by nanotips for resistive switching of HfO2-RRAM devices with high performance. *Scientific Reports, 6*. https://doi.org/10.1038/srep25757

Padovani, A., & Larcher, L. (2018). Time-dependent dielectric breakdown statistics in SiO_2 and HfO_2 dielectrics: Insights from a multi-scale modeling approach. *2018 IEEE International Reliability Physics Symposium (IRPS)*, 3A.2-1–3A.2-7. https://doi.org/10.1109/IRPS.2018.8353552

Parat, K., & Dennison, C. (2015). A floating gate based 3D NAND technology with CMOS under array. *Technical Digest – International Electron Devices Meeting, IEDM, 2016-February*. https://doi.org/10.1109/IEDM.2015.7409618

Park, H. W., Lee, J., & Hwang, C. S. (2021). Review of ferroelectric field-effect transistors for three-dimensional storage applications. *Nano Select, 2*(6). https://doi.org/10.1002/nano.202000281

Park, I.-S., Ryu, K., Jeong, J., & Ahn, J. (2013). Dielectric Stacking Effect of Al_2O_3 and HfO_2 in Metal–Insulator–Metal Capacitor. *IEEE Electron Device Letters, 34*(1), 120–122. https://doi.org/10.1109/LED.2012.2228162

Park, J., Lee, C., Kwak, M., Chekol, S. A., Lim, S., Kim, M., Woo, J., Hwang, H., & Lee, D. (2019). Microstructural engineering in interface-type synapse device for enhancing linear and symmetric conductance changes. *Nanotechnology*, *30*(30). https://doi.org/10.1088/1361-6528/ab180f

Park, S., Noh, J., Choo, M. L., Sheri, A. M., Chang, M., Kim, Y. B., Kim, C. J., Jeon, M., Lee, B. G., Lee, B. H., & Hwang, H. (2013). Nanoscale RRAM-based synaptic electronics: Toward a neuromorphic computing device. *Nanotechnology*, *24*(38). https://doi.org/10.1088/0957-4484/24/38/384009

Pearson, C., Bowen, L., Lee, M. W., Fisher, A. L., Linton, K. E., Bryce, M. R., & Petty, M. C. (2013). Focused ion beam and field-emission microscopy of metallic filaments in memory devices based on thin films of an ambipolar organic compound consisting of oxadiazole, carbazole, and fluorene units. *Applied Physics Letters*, *102*(21). https://doi.org/10.1063/1.4808026

Pebay-Peyroula, F., Dalgaty, T., & Vianello, E. (2020). Entropy source characterization in HfO2 RRAM for TRNG applications. *2020 15th Design & Technology of Integrated Systems in Nanoscale Era (DTIS)*, 1–2. https://doi.org/10.1109/DTIS48698.2020.9081294

Peng, C. N., Wang, C. W., Chan, T. C., Chang, W. Y., Wang, Y. C., Tsai, H. W., Wu, W. W., Chen, L. J., & Chueh, Y. L. (2012). Resistive switching of Au/ZnO/Au resistive memory: An in situ observation of conductive bridge formation. *Nanoscale Research Letters*, *7*. https://doi.org/10.1186/1556-276X-7-559

Peng, P., Xie, D., Yang, Y., Zang, Y., Gao, X., Zhou, C., Feng, T., Tian, H., Ren, T., & Zhang, X. (2012). Resistive switching behavior in diamond-like carbon films grown by pulsed laser deposition for resistance switching random access memory application. *Journal of Applied Physics*, *111*(8). https://doi.org/10.1063/1.3703063

Petzold, S., Zintler, A., Eilhardt, R., Piros, E., Kaiser, N., Sharath, S. U., Vogel, T., Major, M., McKenna, K. P., Molina-Luna, L., & Alff, L. (2019). Forming-Free Grain Boundary Engineered Hafnium Oxide Resistive Random Access Memory Devices. *Advanced Electronic Materials*, *5*(10). https://doi.org/10.1002/aelm.201900484

Pi, S., Li, C., Jiang, H., Xia, W., Xin, H., Yang, J. J., & Xia, Q. (2019). Memristor crossbar arrays with 6-nm half-pitch and 2-nm critical dimension. *Nature Nanotechnology*, *14*(1). https://doi.org/10.1038/s41565-018-0302-0

Prakash, A., Maikap, S., Banerjee, W., Jana, D., & Lai, C. S. (2013). Impact of electrically formed interfacial layer and improved memory characteristics of IrOx/high-κx/W structures containing AlOx, GdOx, HfOx, and TaOx switching materials. *Nanoscale Research Letters*, *8*(1). https://doi.org/10.1186/1556-276X-8-379

Prakash, A., Park, J., Song, J., Woo, J., Cha, E. J., & Hwang, H. (2015). Demonstration of low power 3-bit multilevel cell characteristics in a TaOx-Based RRAM by stack engineering. *IEEE Electron Device Letters*, *36*(1). https://doi.org/10.1109/LED.2014.2375200

Raghavan, N., Degraeve, R., Fantini, A., Goux, L., Strangio, S., Govoreanu, B., Wouters, D. J., Groeseneken, G., & Jurczak, M. (2013). Microscopic origin of random telegraph noise fluctuations in aggressively scaled RRAM and its impact on read disturb variability. *IEEE International Reliability Physics Symposium Proceedings*. https://doi.org/10.1109/IRPS.2013.6532042

Raghavan, N., Fantini, A., Degraeve, R., Roussel, P. J., Goux, L., Govoreanu, B., Wouters, D. J., Groeseneken, G., & Jurczak, M. (2013). Statistical insight into controlled forming and forming free stacks for HfOx RRAM. *Microelectronic Engineering*, *109*. https://doi.org/10.1016/j.mee.2013.03.065

Rajendran, G., Banerjee, W., Chattopadhyay, A., & Aly, M. M. S. (2021). Application of Resistive Random Access Memory in Hardware Security: A Review. *Advanced Electronic Materials*, *7*(12), 2100536. https://doi.org/10.1002/aelm.202100536

Ray, S. K., Maikap, S., Banerjee, W., & Das, S. (2013). Nanocrystals for silicon-based light-emitting and memory devices. *Journal of Physics D: Applied Physics*, *46*(15). https://doi.org/10.1088/0022-3727/46/15/153001

Roy, S., Niu, G., Wang, Q., Wang, Y., Zhang, Y., Wu, H., Zhai, S., Shi, P., Song, S., Song, Z., Ye, Z. G., Wenger, C., Schroeder, T., Xie, Y. H., Meng, X., Luo, W., & Ren, W. (2020). Toward a Reliable Synaptic Simulation Using Al-Doped HfO2 RRAM. *ACS Applied Materials and Interfaces*, *12*(9). https://doi.org/10.1021/acsami.9b21530

Saadi, M., Gonon, P., Vallée, C., Mannequin, C., Grampeix, H., Jalaguier, E., Jomni, F., & Bsiesy, A. (2016). On the mechanisms of cation injection in conducting bridge memories: The case of HfO2 in contact with noble metal anodes (Au, Cu, Ag). *Journal of Applied Physics*, *119*(11). https://doi.org/10.1063/1.4943776

Sassine, G., la Barbera, S., Najjari, N., Minvielle, M., Dubourdieu, C., & Alibart, F. (2016). Interfacial versus filamentary resistive switching in TiO 2 and HfO 2 devices. *Journal of Vacuum Science & Technology B, Nanotechnology and Microelectronics: Materials, Processing, Measurement, and Phenomena*, *34*(1). https://doi.org/10.1116/1.4940129

Sassine, G., Nail, C., Blaise, P., Sklenard, B., Bernard, M., Gassilloud, R., Marty, A., Veillerot, M., Vallée, C., Nowak, E., & Molas, G. (2019). Hybrid-RRAM toward Next Generation of Nonvolatile Memory: Coupling of Oxygen Vacancies and Metal Ions. *Advanced Electronic Materials*, *5*(2). https://doi.org/10.1002/aelm.201800658

Seok, J. Y., Song, S. J., Yoon, J. H., Yoon, K. J., Park, T. H., Kwon, D. E., Lim, H., Kim, G. H., Jeong, D. S., & Hwang, C. S. (2014). A review of three-dimensional resistive switching cross-bar array memories from the integration and materials property points of view. *Advanced Functional Materials*, *24*(34). https://doi.org/10.1002/adfm.201303520

Sheridan, P. M., Cai, F., Du, C., Ma, W., Zhang, Z., & Lu, W. D. (2017). Sparse coding with memristor networks. *Nature Nanotechnology*, *12*(8). https://doi.org/10.1038/nnano.2017.83

Shi, Y., Nguyen, L., Oh, S., Liu, X., Koushan, F., Jameson, J. R., & Kuzum, D. (2018). Neuroinspired unsupervised learning and pruning with subquantum CBRAM arrays. *Nature Communications*, *9*(1). https://doi.org/10.1038/s41467-018-07682-0

Shin, K. Y., Kim, Y., Antolinez, F. v., Ha, J. S., Lee, S. S., & Park, J. H. (2016). Controllable Formation of Nanofilaments in Resistive Memories via Tip-Enhanced Electric Fields. *Advanced Electronic Materials*, *2*(10). https://doi.org/10.1002/aelm.201600233

Shirota, R. (2019). 3D-NAND Flash memory and technology. In *Advances in Non-Volatile Memory and Storage Technology* (pp. 283–319). Elsevier. https://doi.org/10.1016/B978-0-08-102584-0.00009-7

Shubhakar, K., Bosman, M., Neucheva, O. A., Loke, Y. C., Raghavan, N., Thamankar, R., Ranjan, A., O'Shea, S. J., & Pey, K. L. (2015). An SEM/STM based nanoprobing and TEM study of breakdown locations in HfO2/SiOx dielectric stacks for failure analysis. *Microelectronics Reliability*, *55*(9–10). https://doi.org/10.1016/j.microrel.2015.07.027

Shukla, N., Grisafe, B., Ghosh, R. K., Jao, N., Aziz, A., Frougier, J., Jerry, M., Sonde, S., Rouvimov, S., Orlova, T., Gupta, S., & Datta, S. (2017). Ag/HfO2 based threshold switch with extreme non-linearity for unipolar cross-point memory and steep-slope phase-FETs. *Technical Digest – International Electron Devices Meeting, IEDM*. https://doi.org/10.1109/IEDM.2016.7838542

Sokolov, A. S., Son, S. K., Lim, D., Han, H. H., Jeon, Y.-R., Lee, J. H., & Choi, C. (2017). Comparative study of Al 2 O 3, HfO 2, and HfAlO *x* for improved self-compliance bipolar resistive switching. *Journal of the American Ceramic Society*, *100*(12), 5638–5648. https://doi.org/10.1111/jace.15100

Stathopoulos, S., Khiat, A., Trapatseli, M., Cortese, S., Serb, A., Valov, I., & Prodromakis, T. (2017). Multibit memory operation of metal-oxide Bi-layer memristors. *Scientific Reports*, *7*(1). https://doi.org/10.1038/s41598-017-17785-1

Sun, H., Liu, Q., Long, S., Lv, H., Banerjee, W., & Liu, M. (2014). Multilevel unipolar resistive switching with negative differential resistance effect in Ag/SiO2/Pt device. *Journal of Applied Physics*, *116*(15). https://doi.org/10.1063/1.4898807

Sung, C., Hwang, H., & Yoo, I. K. (2018). Perspective: A review on memristive hardware for neuromorphic computation. *Journal of Applied Physics*, *124*(15). https://doi.org/10.1063/1.5037835

Suri, M., Bichler, O., Hubert, Q., Perniola, L., Sousa, V., Jahan, C., Vuillaume, D., Gamrat, C., & Desalvo, B. (2013). Addition of HfO2 interface layer for improved synaptic performance of phase change memory (PCM) devices. *Solid-State Electronics*, *79*. https://doi.org/10.1016/j.sse.2012.09.006

Tian, H., Chen, H. Y., Gao, B., Yu, S., Liang, J., Yang, Y., Xie, D., Kang, J., Ren, T. L., Zhang, Y., & Wong, H. S. P. (2013). Monitoring oxygen movement by Raman spectroscopy of resistive random access memory with a graphene-inserted electrode. *Nano Letters*, *13*(2). https://doi.org/10.1021/nl304246d

Tian, H., Chen, H. Y., Ren, T. L., Li, C., Xue, Q. T., Mohammad, M. A., Wu, C., Yang, Y., & Wong, H. S. P. (2014). Cost-effective, transfer-free, flexible resistive random access memory using laser-scribed reduced graphene oxide patterning technology. *Nano Letters*, *14*(6). https://doi.org/10.1021/nl5005916

Tian, X., Yang, S., Zeng, M., Wang, L., Wei, J., Xu, Z., Wang, W., & Bai, X. (2014). Bipolar electrochemical mechanism for mass transfer in nanoionic resistive memories. *Advanced Materials*, *26*(22). https://doi.org/10.1002/adma.201400127

Traoré, B., Blaise, P., Vianello, E., Grampeix, H., Bonnevialle, A., Jalaguier, E., Molas, G., Jeannot, S., Perniola, L., DeSalvo, B., & Nishi, Y. (2015). Microscopic understanding of the low resistance state retention in HfO2 and HfAlO based RRAM. *Technical Digest – International Electron Devices Meeting, IEDM*, *2015-February*. https://doi.org/10.1109/IEDM.2014.7047097

Traore, B., Blaise, P., Vianello, E., Grampeix, H., Jeannot, S., Perniola, L., de Salvo, B., & Nishi, Y. (2015). On the Origin of Low-Resistance State Retention Failure in HfO2-Based RRAM and Impact of Doping/Alloying. *IEEE Transactions on Electron Devices*, *62*(12). https://doi.org/10.1109/TED.2015.2490545

Tsai, T. L., Jiang, F. S., Ho, C. H., Lin, C. H., & Tseng, T. Y. (2016). Enhanced Properties in Conductive-Bridge Resistive Switching Memory with Oxide-Nitride Bilayer Structure. *IEEE Electron Device Letters*, *37*(10). https://doi.org/10.1109/LED.2016.2602886

Wang, C., Feng, D., Tong, W., Liu, J., Li, Z., Chang, J., Zhang, Y., Wu, B., Xu, J., Zhao, W., Li, Y., & Ren, R. (2019). Cross-point Resistive Memory: Nonideal properties and solutions. *ACM Transactions on Design Automation of Electronic Systems*, *24*(4). https://doi.org/10.1145/3325067

Wang, C., Wu, H., Gao, B., Zhang, T., Yang, Y., & Qian, H. (2018). Conduction mechanisms, dynamics and stability in ReRAMs. *Microelectronic Engineering*, 187–188. https://doi.org/10.1016/j.mee.2017.11.003

Wang, Q., Niu, G., Roy, S., Wang, Y., Zhang, Y., Wu, H., Zhai, S., Bai, W., Shi, P., Song, S., Song, Z., Xie, Y. H., Ye, Z. G., Wenger, C., Meng, X., & Ren, W. (2019). Interface-engineered reliable HfO2-based RRAM for synaptic simulation. *Journal of Materials Chemistry C*, *7*(40). https://doi.org/10.1039/c9tc04880d

Wang, W., Li, Y., Wang, M., Wang, L., Liu, Q., Banerjee, W., Li, L., & Liu, M. (2016). A hardware neural network for handwritten digits recognition using binary RRAM as synaptic weight element. *2016 IEEE Silicon Nanoelectronics Workshop, SNW 2016*. https://doi.org/10.1109/SNW.2016.7577980

Wang, W., Wang, M., Ambrosi, E., Bricalli, A., Laudato, M., Sun, Z., Chen, X., & Ielmini, D. (2019). Surface diffusion-limited lifetime of silver and copper nanofilaments in resistive switching devices. *Nature Communications*, *10*(1). https://doi.org/10.1038/s41467-018-07979-0

Wang, Z., Griffin, P. B., McVittie, J., Wong, S., McIntyre, P. C., & Nishi, Y. (2007). Resistive switching mechanism in ZNxCd1-xS nonvolatile memory devices. *IEEE Electron Device Letters*, *28*(1). https://doi.org/10.1109/LED.2006.887640

Wang, Z., Joshi, S., Savel'ev, S. E., Jiang, H., Midya, R., Lin, P., Hu, M., Ge, N., Strachan, J. P., Li, Z., Wu, Q., Barnell, M., Li, G. L., Xin, H. L., Williams, R. S., Xia, Q., & Yang, J. J. (2017). Memristors with diffusive dynamics as synaptic emulators for neuromorphic computing. *Nature Materials*, *16*(1). https://doi.org/10.1038/nmat4756

Wang, Z., Joshi, S., Savel'Ev, S., Song, W., Midya, R., Li, Y., Rao, M., Yan, P., Asapu, S., Zhuo, Y., Jiang, H., Lin, P., Li, C., Yoon, J. H., Upadhyay, N. K., Zhang, J., Hu, M., Strachan, J. P., Barnell, M., ... Yang, J. J. (2018). Fully memristive neural networks for pattern classification with unsupervised learning. *Nature Electronics*, *1*(2). https://doi.org/10.1038/s41928-018-0023-2

Wang, Z., Kang, J., Yu, Z., Fang, Y., Ling, Y., Cai, Y., Huang, R., & Wang, Y. (2017). Modulation of nonlinear resistive switching behavior of a TaOx-based resistive device through interface engineering. *Nanotechnology*, *28*(5). https://doi.org/10.1088/1361-6528/28/5/055204

Wang, Z., Li, C., Song, W., Rao, M., Belkin, D., Li, Y., Yan, P., Jiang, H., Lin, P., Hu, M., Strachan, J. P., Ge, N., Barnell, M., Wu, Q., Barto, A. G., Qiu, Q., Williams, R. S., Xia, Q., & Yang, J. J. (2019). Reinforcement learning with analogue memristor arrays. *Nature Electronics*, *2*(3). https://doi.org/10.1038/s41928-019-0221-6

Wang, Z. Q., Xu, H. Y., Li, X. H., Yu, H., Liu, Y. C., & Zhu, X. J. (2012). Synaptic learning and memory functions achieved using oxygen ion migration/diffusion in an amorphous InGaZnO memristor. *Advanced Functional Materials*, *22*(13). https://doi.org/10.1002/adfm.201103148

Wang, Z., Rao, M., Midya, R., Joshi, S., Jiang, H., Lin, P., Song, W., Asapu, S., Zhuo, Y., Li, C., Wu, H., Xia, Q., & Yang, J. J. (2018). Threshold Switching of Ag or Cu in Dielectrics: Materials, Mechanism, and Applications. *Advanced Functional Materials*, *28*(6). https://doi.org/10.1002/adfm.201704862

Waser, R., & Aono, M. (2007). Nanoionics-based resistive switching memories. *Nature Materials*, *6*(11). https://doi.org/10.1038/nmat2023

Waser, R., Dittmann, R., Staikov, C., & Szot, K. (2009). Redox-based resistive switching memories nanoionic mechanisms, prospects, and challenges. *Advanced Materials*, *21*(25–26). https://doi.org/10.1002/adma.200900375

Wong, T M, Preissl, R., Datta, P., Flickner, M., Singh, R., Esser S. K., McQuinn, E., Appuswamy, R., Risk, W. P., Simon, H. D., Modha, D. S. (2012, November 13). Ten to Power 14. *IBM Technical Paper RJ10502*.

Wong, H. S. P., & Salahuddin, S. (2015). Memory leads the way to better computing. *Nature Nanotechnology*, *10*(3). https://doi.org/10.1038/nnano.2015.29

Wu, Q., Banerjee, W., Cao, J., Ji, Z., Li, L., & Liu, M. (2018). Improvement of durability and switching speed by incorporating nanocrystals in the HfOx based resistive random access memory devices. *Applied Physics Letters*, *113*(2). https://doi.org/10.1063/1.5030780

Wu, Q., Wang, H., Luo, Q., Banerjee, W., Cao, J., Zhang, X., Wu, F., Liu, Q., Li, L., & Liu, M. (2018). Full imitation of synaptic metaplasticity based on memristor devices. *Nanoscale*, *10*(13). https://doi.org/10.1039/c8nr00222c

Wu, Z., Zhao, X., Yang, Y., Wang, W., Zhang, X., Wang, R., Cao, R., Liu, Q., & Banerjee, W. (2019). Transformation of threshold volatile switching to quantum point contact originated nonvolatile switching in graphene interface controlled memory devices. *Nanoscale Advances*, *1*(9). https://doi.org/10.1039/c9na00409b

Xia, Q., & Yang, J. J. (2019). Memristive crossbar arrays for brain-inspired computing. *Nature Materials, 18*(4). https://doi.org/10.1038/s41563-019-0291-x

Xu, Z., Bando, Y., Wang, W., Bai, X., & Golberg, D. (2016). Erratum: Real-time in situ HRTEM-resolved resistance switching of Ag_2S nanoscale ionic conductor (ACS Nano (2010) 4:5 (2515-2522) https://doi.org/10.1021/nn100483a). *ACS Nano 10*(2). https://doi.org/10.1021/acsnano.6b00098

Yan, X. B., Hao, H., Chen, Y. F., Li, Y. C., & Banerjee, W. (2014). Highly transparent bipolar resistive switching memory with In-Ga-Zn-O semiconducting electrode in In-Ga-Zn-O/ Ga2O3/In-Ga-Zn-O structure. *Applied Physics Letters, 105*(9). https://doi.org/10.1063/1.4894521

Yan, Z. B., & Liu, J. M. (2013). Coexistence of high performance resistance and capacitance memory based on multilayered metal-oxide structures. *Scientific Reports, 3*. https://doi.org/10.1038/srep02482

Yang, J. J., Strukov, D. B., & Stewart, D. R. (2013). Memristive devices for computing. *Nature Nanotechnology, 8*(1). https://doi.org/10.1038/nnano.2012.240

Yang, Y. C., Pan, F., Zeng, F., & Liu, M. (2009). Switching mechanism transition induced by annealing treatment in nonvolatile Cu/ZnO/Cu/ZnO/Pt resistive memory: From carrier trapping/detrapping to electrochemical metallization. *Journal of Applied Physics, 106*(12). https://doi.org/10.1063/1.3273329

Yong, Z., Persson, K. M., Saketh Ram, M., D'Acunto, G., Liu, Y., Benter, S., Pan, J., Li, Z., Borg, M., Mikkelsen, A., Wernersson, L. E., & Timm, R. (2021). Tuning oxygen vacancies and resistive switching properties in ultra-thin HfO2 RRAM via TiN bottom electrode and interface engineering. *Applied Surface Science, 551*. https://doi.org/10.1016/j.apsusc.2021.149386

Yuan, F. Y., Deng, N., Shih, C. C., Tseng, Y. T., Chang, T. C., Chang, K. C., Wang, M. H., Chen, W. C., Zheng, H. X., Wu, H., Qian, H., & Sze, S. M. (2017). Conduction Mechanism and Improved Endurance in HfO2-Based RRAM with Nitridation Treatment. *Nanoscale Research Letters, 12*. https://doi.org/10.1186/s11671-017-2330-3

Zhao, X., Liu, S., Niu, J., Liao, L., Liu, Q., Xiao, X., Lv, H., Long, S., Banerjee, W., Li, W., Si, S., & Liu, M. (2017). Confining Cation Injection to Enhance CBRAM Performance by Nanopore Graphene Layer. *Small, 13*(35). https://doi.org/10.1002/smll.201603948

Zhirnov, V. v., Meade, R., Cavin, R. K., & Sandhu, G. (2011). Scaling limits of resistive memories. *Nanotechnology, 22*(25). https://doi.org/10.1088/0957-4484/22/25/254027

Zhu, L. Q., Wan, C. J., Guo, L. Q., Shi, Y., & Wan, Q. (2014). Artificial synapse network on inorganic proton conductor for neuromorphic systems. *Nature Communications, 5*. https://doi.org/10.1038/ncomms4158

Zhu, X., Su, W., Liu, Y., Hu, B., Pan, L., Lu, W., Zhang, J., & Li, R. W. (2012). Observation of conductance quantization in oxide-based resistive switching memory. *Advanced Materials, 24*(29). https://doi.org/10.1002/adma.201201506

Zidan, M. A., Strachan, J. P., & Lu, W. D. (2018). The future of electronics based on memristive systems. *Nature Electronics, 1*(1). https://doi.org/10.1038/s41928-017-0006-8

2 III-V Materials and Their Transistor Application

Ananda Sankar Chakraborty

2.1 THE SHORT BACKGROUND STORY

Relentless downsizing of conventional Si-MOSFET technology for the past 60 years has resulted in ever-increasing transistor performance and IC density (Kuhn, 2012), although problems like reduced gate control, increased mobility degradation etc., crept in more. To address the problem, research has progressed through newer materials (Ghani et al., 2003; Gault et al., 1986) and new device architectures with time. Parallel research efforts on improvised channel materials and transistor architecture research started in the 1970s in order to cater to both high-speed and low-power circuits with a smaller size as well as growing applications of photovoltaics and optoelectronics (Antonio L. Luque, 2007).With the carrier mobility varying inversely with the effective mass, a higher drive current can be generated if the silicon in the channel is replaced by low-effective-mass materials, which can be synthesized and integrated into the existing silicon technology easily. Among such wonder materials, III-V compounds are right now probably one of the most sought-after topics in electronic material research around the globe (Vurgaftman et al., 2001).

2.2 III-V MATERIALS

As discussed earlier the major gain of using low-effective-mass channel material lies in their inherent low-effective-mass (Melloch et al., 1995,Singh et al., 2014),because low field mobility(μ) and effective mass (m^*) of a material are related to the ON current (I_{ON}) of the material, by the relation (2.1):

$$I_{ON} \propto \mu \propto \frac{1}{m^*} \tag{2.1}$$

The relationship (2.1) shows that I_{ON} can be enhanced through boosting μ, by choosing a material having lower m^*. A comprehensive list of transistor channel materials is provided in Table 2.1, which corroborates the fact. The bottom three are compound semiconductors, known as III-V materials—a combination of group III (B, Al, Ga, In, Ti) and group V (N, P, As, Sb, Bi) elements of the periodic table.

DOI: 10.1201/9781003323518-3

TABLE 2.1
Table Showing Various Transistor Channel Materials

Material	m_e^*/m_0	m_h^*/m_0	μ_n	μ_p	vsat	E_g
Si	0.19	0.16	1500	450	1E+07	1.12
Ge	0.085	0.045	3900	1900	0.7E+07	0.66
$In_{0.7}Ga_{0.3}As$	0.041	0.05	12,000	250	0.48E+07	0.75
GaAs	0.066	0.09	8500	400	1.1E+07	1.42
InAs	0.023	0.41	35,000	500	0.6E+07	0.354

Source: Vurgaftman et al., 2001, Kim et al., 2010, Chakraborty, 2019.

A salient reason behind this popularity is the relatively easier integration of III-V materials with existing Si-technology (Cho, 1983; Li et al., 2016). The explanation can be put forward in this manner—group III and group V elements have either one electron shortage or one electron in excess compared to tetravalent silicon. So compared to other compound semiconductors like II-VI (two electrons shortage or in excess) the probability of a bond formation with tetravalent silicon atom is much more for III-V compounds. Moreover, a small change in the mole fraction leads to a different compound with different electrical and optical properties for III-V semiconductors—which also have both direct and indirect band gap varieties. Naturally they find immense applications in both optoelectronics and electronics domains.

2.3   DEVELOPING THE MATHEMATICAL MODEL

In this section we intend to develop a mathematical model for III-V MOSFET, which is a two-step process—developing the potential distribution and developing the transport. As double-gate architecture is the best hand-off between gate control and the fabrication cost (Chakraborty, 2019), henceforth our focus will be on this architecture unless otherwise mentioned.

2.4   MODELING THE SURFACE POTENTIAL

An effective compact model is expected to converge flawlessly under all circumstances in any circuit simulation environment. In the III-V material channel, the highly prominent quantum confinement due to the low carrier effective mass, results in unique capacitance characteristics. Hence it is important for us to first understand the potential distribution model first.

Figure 2.1 shows the device structure. It is obvious that a direct analytical solution of a coupled differential equation system (Schrodinger + Poisson) signifying both quantum confinement and MOS electrostatics inside the device—is not obtainable. We need to find out a simple yet elegant analytical modeling technique for this.

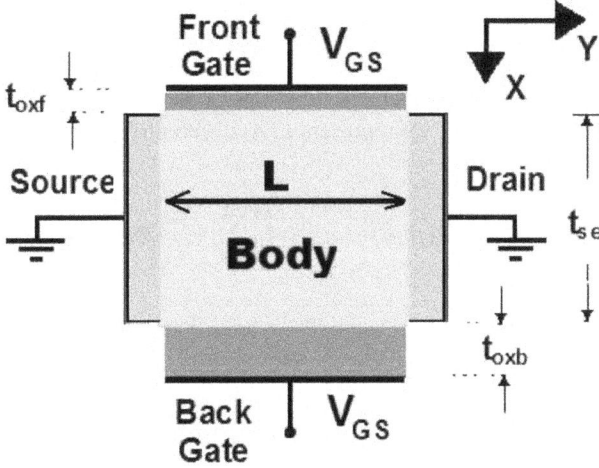

FIGURE 2.1 Schematic of a CDGMOSFET with asymmetric gate-oxide thickness.

The expression for the coupled Schrodinger and Poisson equations system for low-effective-mass channel devices is given as:

$$-\frac{\hbar^2}{2m^*}\frac{d^2\psi}{dx^2}+(qV(x)-E)\psi=0 \tag{2.2}$$

$$\varepsilon\frac{d}{dx}F(x)=-\rho(x) \tag{2.3}$$

where q = electronic charge, ψ = electronic wavefunction, ϵ = material permittivity, $\hbar$ = reduced Planck constant, qV(x) = self-energy profile in the oxide-semiconductor system, E = subband energy, F(x) = electric field-profile, $\rho(x)$ = the charge density, is given as (Chakraborty, 2019):

$$\rho(x)=\frac{m^*kT}{\pi\hbar^2}q\sum_{i=1}^{n_{max}}|\psi_i(x)|^2\ ln\left(1+e^{\frac{E_{F,i}-E_i}{kT}}\right) \tag{2.4}$$

where, k = Boltzmann constant, $E_{F,i}$ = Fermi energy level for i^{th} subband, E_i = i^{th} energy subband, T = absolute temperature and n_{max} = maximum number of energy subbands taken into account, which for this type of device according to "International Technology Roadmap for Semiconductors" (ITRS) varies from 1 to 3 (Chakraborty, 2019). Integrating (2.3) with $\epsilon=\epsilon_{se}$, $E_{F,i}=E_F$ and $m^*=m_{se}$ substitution, we obtain

$$F(x=x_0)-F(x)=\frac{qm_{se}kT}{\pi\hbar^2\varepsilon_{se}}\int_{x_0}^{x}\sum_{i=1}^{n_{max}}|\psi_i(x)|^2\ ln\left(1+e^{\frac{E_F-E_i}{kT}}\right)dx \tag{2.5}$$

Here x_0 = any arbitrary point inside the channel and $\epsilon_{se}(m_{se})$ = permittivity (electron effective mass) of the quasi-2D semiconductor material. The Poisson equation in (2.4) gives the expression for electric field. Equations (2.2) and (2.4) are coupled through $\psi_i(x)$ and E_i, both of which are functions of electrostatic potential. To tackle the mathematical complexity, we represent (2.4) differently as:

$$F\left(x = x_0\right) - F(x) = F^{(r)} \int_{x_0}^{x} \sum_{i=1}^{n_{max}} \left| \psi_i^{(0)}(x) \right|^2 dx \qquad (2.6)$$

Here $\psi_i^{(0)}(x)$ = ground-state $\psi_i(x)$ and $F^{(r)}$ is a resultant quantity, bearing the cumulative effect of all subbands. Thus the proposed solution finally leads to a surface potential profile similar to that of a bulk/multigate MOSFET structure (Chakraborty, 2019).

In any MOS device architecture, the total gate charge per unit surface area (Q_g) should equate to the charge per unit surface area inside the semiconductor (Q_{se})—given that the interface trapped charge is negligible. Again Q_g has two components Q_{gf} and Q_{gb}, due to front and back gates, respectively. Quite evidently, we may assume that individual components of gate charges constitute the charge at corresponding oxide-semiconductor interfaces, or $Q_{gf} = Q_{sf}$ and $Q_{gb} = Q_{sb}$. Hence, $Q_g = Q_{sf} + Q_{sb}$.

For the purpose of handling the charge centroid near the oxide-semiconductor interface (Chakraborty and Mahapatra, 2017), we define two equivalent oxide capacitances c_{oxf}^e and c_{oxb}^e, such that $Q_{sf(b)} = C_{oxf(b)}(V_g - \Delta\varphi - \varphi_{sf(b)}) = C_{oxf(b)}^e(V_g - \Delta\varphi - \varphi_m)$

Here $\Delta\varphi$ is the oxide-semiconductor workfunction difference. We may obtain the complete potential profile inside the semiconductor from (2.5) by integration. Thereafter we may model the heavy quantum confinement using the perturbation technique discussed in (Chakraborty and Mahapatra, 2017). We may arrive at the final charge density balance equation as

$$\left(C_{oxf}^e + C_{oxb}^e\right)\left(\frac{V_g - \Delta\varphi - \varphi_m}{U_t}\right) = \sum_{i=1}^{n_{max}} C_{qi} ln\left[1 + \exp\left(\frac{p_0 + p_1\varphi_m + p_2\varphi_m^2}{U_t}\right)\right] \qquad (2.7)$$

For $i \in \{1, 2, 3\}$, C_{qi} = quantum capacitance for i^{th} subband (Chakraborty, 2019). The centroid potential φ_m is obtained from the numerical solution of (2.7), which linearly varies with surface potential (Chakraborty and Mahapatra, 2017). $U_t = kT / q$ = the thermal voltage, $p_{0,1,2}$ encompasses the coefficients for first and second order perturbations in i^{th} subband, the band gap and the applied bias (Chakraborty, 2019).

2.5   MODELING THE DRAIN CURRENT

In the previous section we have presented a surface potential model for low-effective-mass channel CDGMOSFET. The low effective mass results in low density of states, and hence, there is less charge available for current transport. These devices furnish remarkably different capacitance and current characteristics compared to conventional silicon transistors due to their inherently strong quantum confinement (Chakraborty and Mahapatra, 2018).

Circuit designers require computationally efficient, portable, and predictable models to work on design problems. It is thus imperative for us to find out an accurate compact model for drain current for this kind of device.

To address the quantum confinement in the channel, the normal drift-diffusion (DD) formalism is modified into quantum drift-diffusion (QDD), which combines the solution of the coupled Schrodinger-Poisson equation in the transverse direction with the classical DD model in the transport direction. Individual quasi-Fermi levels for each subband are in thermal equilibrium within themselves (Baccarani et al., 2008; Datta, 2000; Lundstrom and Ren, 2002).The total current density due to all the subbands is obtained by modifying the "drift" component in the semiclassical DD equation, as follows (Baccarani et al., 2008, Chakraborty and Mahapatra, 2018):

$$J_n = \sum_{i=1}^{n_{max}} \left[\mu_{n,i} n_i \frac{dE_i(y)}{dy} + qD_{n,i} \frac{dn_i}{dy} \right] \tag{2.8}$$

Here $\mu_{n,i}$ = the mobility for i^{th} subband. Applying the charge linearization technique described in (Chakraborty, 2019) and integrating with respect to the transport direction (y) we arrive at the final drain current equation:

$$I_{DS} = \mu_{n,eff} \left(\frac{W}{L} \right) \sum_{i=1}^{n_{max}} K_{Pr,i} \left[Q^2_{se,i(S)} - Q^2_{se,i(D)} \right] \tag{2.9}$$

where $\mu_{n,eff}$ = field-dependent effective mobility corresponding to all the subbands, $Q_{se,i(S(D))}$ = charge per unit area at source(drain), $K_{Pr,i}$ = transconductance parameter for the ith subband (Chakraborty, 2019).

2.6 MODEL VALIDATION AND SPICE IMPLEMENTATION

Figure 2.2 furnishes the surface potential vs gate voltage plot for both symmetric and asymmetric transistors. It may be noted that due to heavy quantum confinement and Fermi-Dirac distribution of carriers, the surface potential does not attain early saturation with increasing gate voltage—unlike with Si-MOSFETs. The gate capacitance per unit width—which is proportional to the derivative of the surface potential φ_m with respect to gate voltage V_g—is shown in Figure 2.3. The staircase nature of the capacitance plot shows the efficacy of the proposed model in predicting the subband quantization of the charge density—which is expected in such low-effective-mass channel ultra-thin-body devices with two degrees of freedom. Figure 2.4 shows the normalized drain current of symmetric device vs gate to source voltage. Similarly, Figure 2.5 shows the normalized drain current of symmetric device vs drain to source voltage.

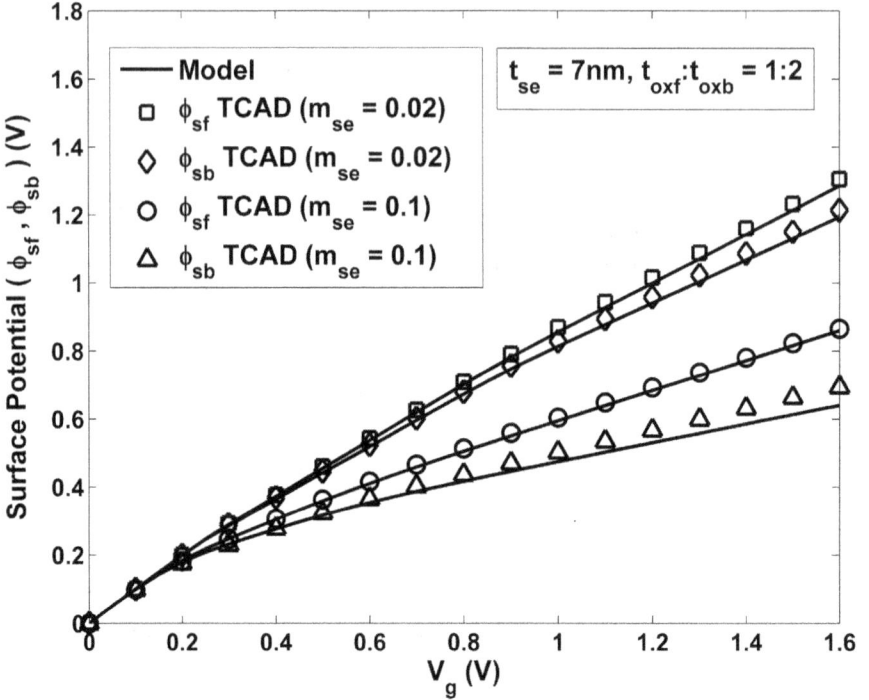

FIGURE 2.2 Front and back surface potential variation with effective gate voltage V_g for asymmetry in oxide thickness (Line = model, Symbol = TCAD, $\Delta\phi$= 0).

Figure 2.6 shows the transient characteristics of a CMOS inverter implemented using the CDG MOSFET models described so far. For testing the model we use L= 1μm, $\Delta\varphi = 0$ hereafter unless otherwise mentioned.

Figures 2.7 and 2.8 describe a two-input NAND gate and a NAND-based level triggered D flip-flop being simulated with this transistor model.

2.7 CONCLUSION

We are at the end of a journey through a short developmental cycle of a top-down compact model of low-effective-mass channel CDGMOSET. We have learned the importance of the III-V materials in today's microelectronics and proceeded upto standard circuit implementation. The gist of learning in this chapter is given as follows:

1. III-V or any other low-effective-mass channel material offers high mobility but high quantum confinement—affecting the overall charge profile. We have seen how this leads to entirely different capacitance characteristics.
2. It is worth noting from Figures 2.6 to 2.8 that transistors of large dimensions like W =1μm, L=1μm could run at near-GHz frequency—which is way faster than their silicon counterparts—thanks to these wonder materials.

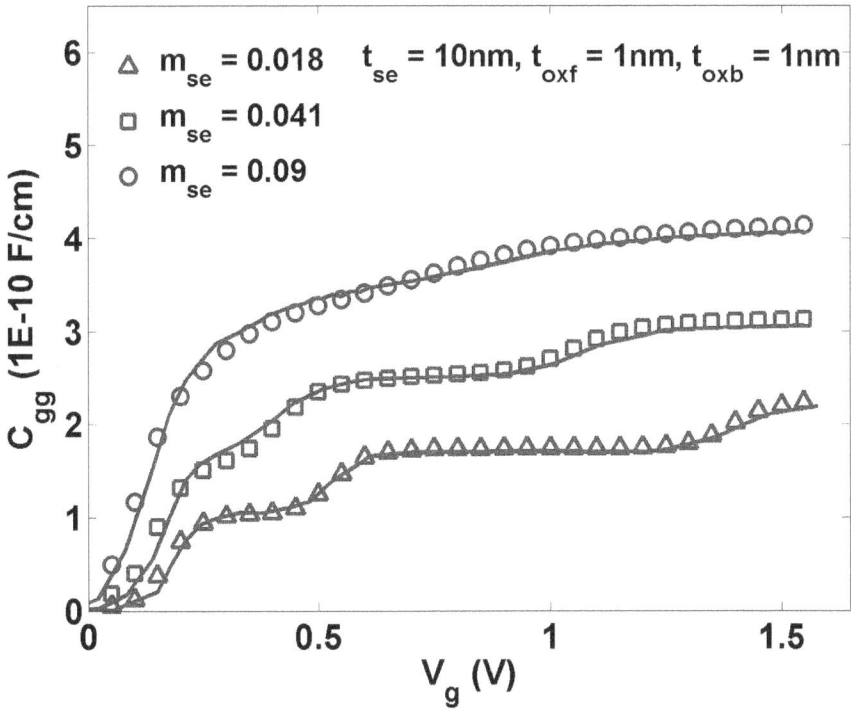

FIGURE 2.3 Variation of gate capacitance per unit width C_{gg} with gate voltage V_g for 10nm thick device (n_{max}= 2, Line = model, Symbol=TCAD, $\Delta\phi$= 0).

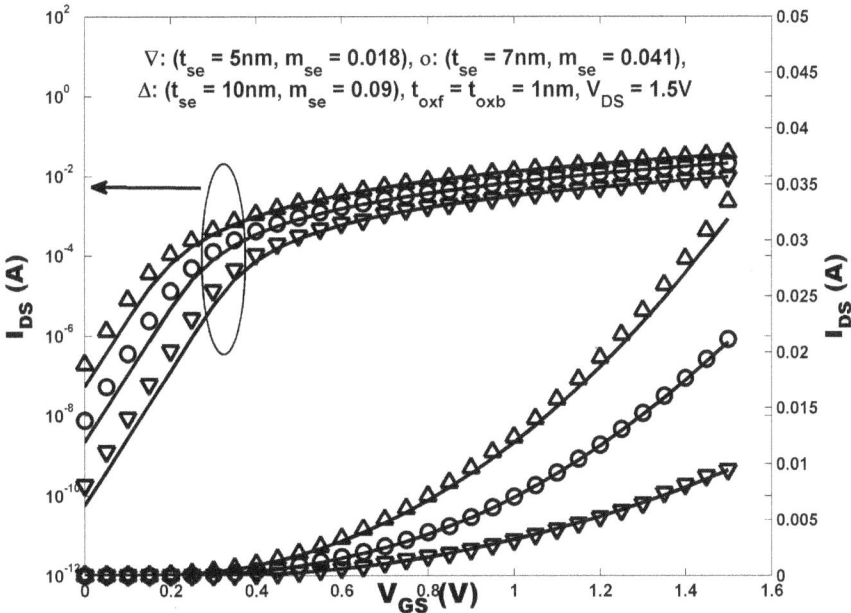

FIGURE 2.4 Normalized drain current of symmetric device vs gate to source voltage (W=1 µm, L= 1 µm, Line = model, Symbol=TCAD).

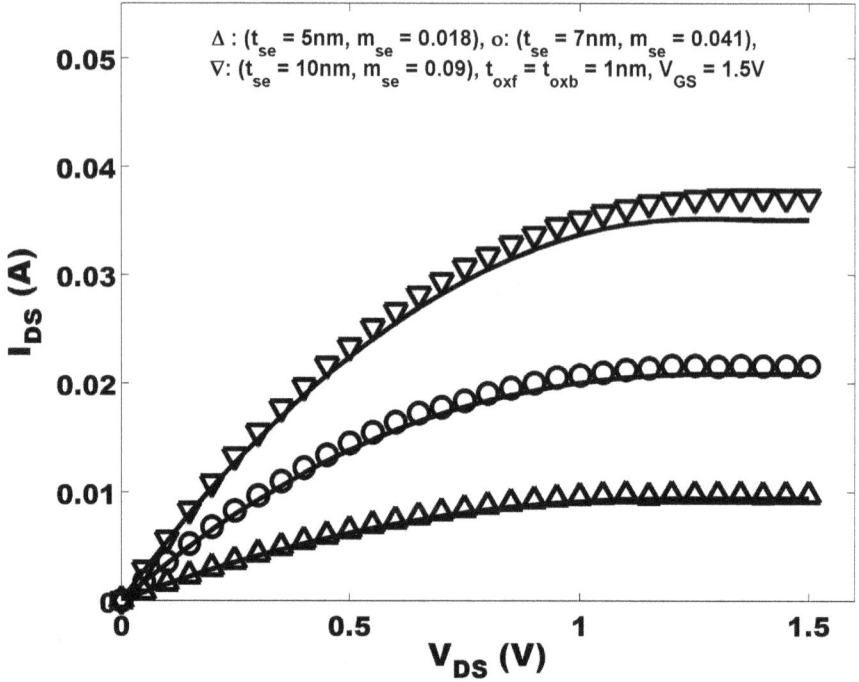

FIGURE 2.5 Normalized drain current of symmetric device vs drain to source voltage (W=1 μm, L= 1μm, Line = model, Symbol=TCAD).

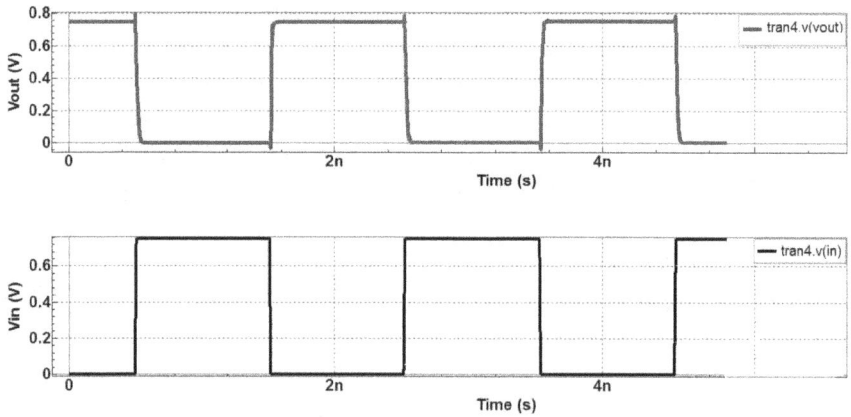

FIGURE 2.6 CMOS inverter input and output waveform (W=1μm, L=1μm, $\Delta\phi$= 0).

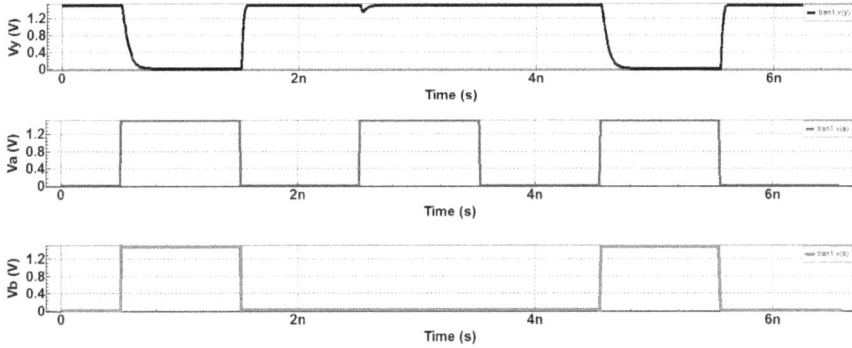

FIGURE 2.7   Dual-input NAND gate input (1st and 2nd row) and output (3rd row) waveform.

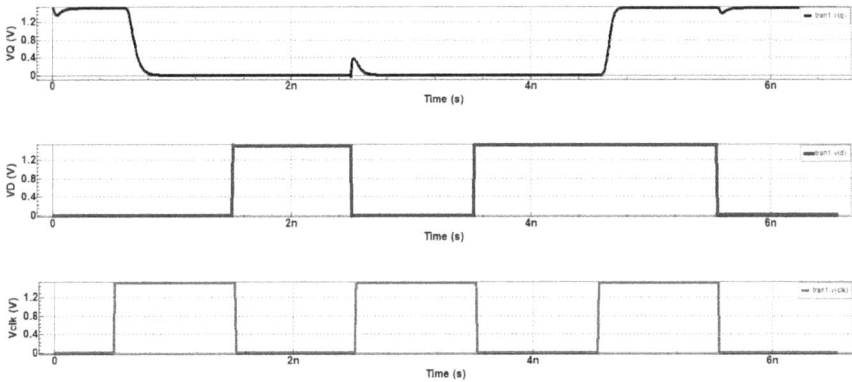

FIGURE 2.8   D flip-flop (level triggered) waveforms. Clock (1st row), input (2nd row), output (3rd row), inverted output (4th row).

3. We have witnessed analytic techniques to solve the coupled Schrodinger-Poisson equations taking into account the asymmetry in gate oxide thickness, wave-function penetration and charge centroid effect. The bias dependency of the subband energy is modeled using accurate perturbation on the ground state energy.

4. It is interesting to note that the drain current model is at-par with those of the established silicon MOSFET compact models—and also bears significant similarity.

5. We have done implementation of standard circuits using this charge and drain current model.

REFERENCES

Antonio L. & Luque, A. V.(2007). Concentrator Photovoltaics. *Conc. Photovoltaics.* https:// doi. org/10.1007/978-3-540-68798-6

Baccarani, G., Gnani, E., Gnudi, A., Reggiani, S., & Rudan, M. (2008). Theoretical foundations of the quantum drift-diffusion and density-gradient models. *Solid. State. Electron.*, 52(4), 526–532. https://doi.org/10.1016/j.sse.2007.10.051

Chakraborty, A. S. (2019). Quantum-Drift-Diffusion Formalism Based Compact Model For Low Effective Mass Channel MOSFET. https://etd.iisc.ac.in/handle/2005/4338

Chakraborty, A. S., & Mahapatra, S.(2018). Compact Model for Low Effective Mass Channel Common Double-Gate MOSFET. *IEEE Trans. Electron Devices*, 65(3), 888–894.https:// doi.org/10.1109/TED.2018.2794381

Chakraborty, A. S., & Mahapatra, S. (2017). Surface Potential Equation for Low Effective Mass Channel Common Double-Gate MOSFET. *IEEE Trans. Electron Devices*, 64(4), 1519–1527. https://doi.org/10.1109/TED.2017.2661798

Chakraborty, A. S., Jandhyala, S., & Mahapatra, S. (2018). Analytical surface potential solution for low effective mass channel common double gate MOSFET. TechConnect Briefs 2018. *Adv. Mater.*, 4, 224–227.

Cho, A. Y. (1983). Growth of III-V semiconductors by molecular beam epitaxy and their properties. Thin Solid Films, 100(4), 291–317.https://doi.org/10.1016/0040-6090(83)90154-2

Datta, S. (2000). Nanoscale device modeling: the Green's function method. *Superlattices Microstruct.*, 28(4), 253–278.https://doi.org/10.1006/spmi.2000.0920

Gault, W. A., Monberg, E. M., & Clemans, J. E. (1986). A novel application of the vertical gradient freeze method to the growth of high quality III-V crystals. *J. Cryst. Growth*, 74(3), 491–506. https://doi.org/10.1016/0022-0248(86)90194-6

Ghani, T., Armstrong, M., Auth, C., Bost, M., Charvat, P., Glass, G., Hoffmann, T., Johnson, K., Kenyon, C., Klaus, J., McIntyre, B., Mistry, K., Murthy, A., Sandford, J., Silberstein, M., Sivakumar, S., Smith, P., Zawadzki, K., Thompson, S., & Bohr, M. (2003). A 90nm high volume manufacturing logic technology featuring novel 45nm gate length strained silicon CMOS transistors. *Tech. Dig. Int. Electron Devices Meet.*, 978–980. https://doi.org/10.1109/iedm.2003.1269442

Kim, Y. S., Marsman, M., Kresse, G., Tran, F., & Blaha, P. (2010). Towards efficient band structure and effective mass calculations for III-V direct band-gap semiconductors. *Phys. Rev. BCondens. Matter Mater. Phys.*, 82(20), 205212. https://doi.org/10.1103/ PhysRevB.82.205212

Kuhn, K. J. (2012). Considerations for ultimate CMOS scaling. *IEEE Trans. Electron Devices*, 59(7), 1813–1828. https://doi.org/10.1109/TED.2012.2193129

Li, T., Mastro, M., & Dadgar, A. (2016). III-V COMPOUND SEMICONDUCTORS: Integration with silicon-based microelectronics. *CRC Press (Taylor and Francis)*, 2011. https://doi. org/10.1201/b10390

Lundstrom, M., & Ren, Z. (2002). Essential physics of carrier transport in nanoscale MOSFETs. IEEE Trans. Electron Devices, 49(1), 133–141. https://doi.org/10.1109/16.974760

Melloch, M. R., Woodall, J. M., Harmon, E. S., Otsuka, N., Pollak, F. H., Nolte, D. D., Feenstra, R. M., & Lutz, M. A. (1995). Low-temperature grown III-V materials. *Annu. Rev. Mater. Sci.*, 25(1), 547–600. https://doi.org/10.1146/annurev.ms.25.080195.002555

Singh, A. K., Zhuang, H. L., & Hennig, R. G. (2014). Ab initio synthesis of single-layer III- V materials. *Phys. Rev. BCondens. Matter Mater. Phys.*, 89(24), 245431.https: //doi.org/ 10.1103/PhysRevB.89.245431

Vurgaftman, I., Meyer, J. R., & Ram-Mohan, L. R. (2001). Band parameters for III-V compound semiconductors and their alloys. *J. Appl. Phys.*, 89(11 I), 5815–5875.https: //doi. org/10.1063/1.1368156

3 Transition Metal Dichalcogenides Properties, Synthesis, and Application in Nanoelectronics Devices

Bibek Chettri, Prasanna Karki, Pronita Chettri,
Sanat Kr. Das, and Bikash Sharma

3.1 INTRODUCTION

Graphene is quite popular due to its remarkable characteristics; however, due to zero electronic bandgap it has sparked research interest in two-dimensional (2D) materials with tunable electronic properties (Jariwala et al. 2014). Due to graphene's lack of a bandgap, its high electron mobility is constrained, making it ineffective as an active element in FETs (Novoselov et al. 2005). Limited success has been achieved in opening the bandgap of graphene utilizing nanoribbons and chemical doping; the bandgap opening only occurs at a very small scale (Han et al. 2007, Xiao et al. 2010, Lee, Duong, et al. 2015). Following the enormous success of graphene, there has been an equally remarkable rush to develop alternative 2D materials with interesting properties (Butler et al. 2013). Transition metal dichalcogenides (TMDs) are atomically thin semiconductors with the formula MX_2, where M is an atom of a transition metal and X is an atom of a chalcogen (Choi et al. 2017). The structure of TMD is presented in Figure 3.1. The diverse optical, electrical, magnetic, and electronic properties of 2D TMD, particularly how those properties change as the number of layers is lowered, have piqued curiosity (Gibbon and Dhanak 2019a, Tamang et al. 2022). TMDs are suitable for fundamental investigations of unusual physical phenomena as well as applications spanning from nanoelectronics to sensing. The exhibition of the first transistor (Radisavljevic et al. 2011) and the finding of significant photoluminescence (PL) in MoS_2 monolayers have sparked a considerable upsurge of interest in 2D TMD (Mak et al. 2010, Splendiani et al. 2010). Linus Pauling discovered the structure of TMD in 1923 (Dickinson and Pauling 1923).

DOI: 10.1201/9781003323518-4

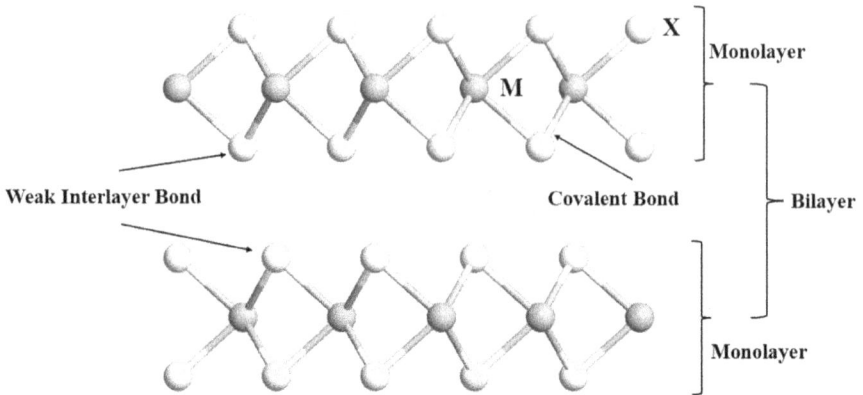

FIGURE 3.1   TMD with layered atomic structures.

As seen in Figure 3.2, quantum confinement effects take place when shifting from an indirect bandgap to a direct bandgap, resulting in the distinct electrical and optical characteristics of 2D TMD (Bao et al. 2013, Islam et al. 2014). Due to their tunable bandgap and strong PL, TMD is a possible choice for a variety of optoelectronic devices such as solar cells, LEDs, and phototransistors (Frey and Elani 1998, Mak et al. 2010, Dervin et al. 2016). Sensors based on 2D TMD have superior sensitivity due to the high surface-to-volume ratio and decreased power consumption (Varghese et al. 2015, Chettri, Thapa, et al. 2022). MoS_2-based FET devices could be used in gas and biosensor applications (Lee, Dak, et al. 2014). A field-effect transistor with HfO_2 as a gate insulator was recently realized using a single layer of MoS_2 (Radisavljevic et al. 2011). The transistor has a current on/off ratio of over 1×10^8 at room temperature and its mobility is comparable to thin silicon films. The ability to quickly isolate and layer with other TMD to produce a variety of van der Waals (vdW) heterostructures without the issues of lattice matching is another benefit of the weakly bound 2D TMD atomic layers (Geim and Grigorieva 2013, Choudhary et al. 2016). For instance, one-atom-thick sheets of various TMD can be piled on top of one another to form vertically stacked heterostructures. Now it is conceivable to achieve novel features and higher functionality that were before impossible. Numerous new electronic and optoelectronic devices, such as tunneling transistors and photodetectors, can be made of these vdW heterostructures (Lee et al. 2014).

3.1.1   CRYSTAL STRUCTURE OF TRANSITION METAL DICHALCOGENIDES

Transition metal complex bonding is commonly thought of as bonding between empty metal orbitals and a single pair of ligands. The four electrons needed to complete the bonding states in TMD are provided by metal atoms, and the formal charges for transition metal and chalcogen can be given as +4 and 2, respectively (Chhowalla et al. 2013). As shown in Figure 3.1, the layers are held together by weak vdW forces, and their interatomic interaction is covalent within each layer. TMDs have a plethora of crystal forms and properties. TMDs are more significant to basic investigations of

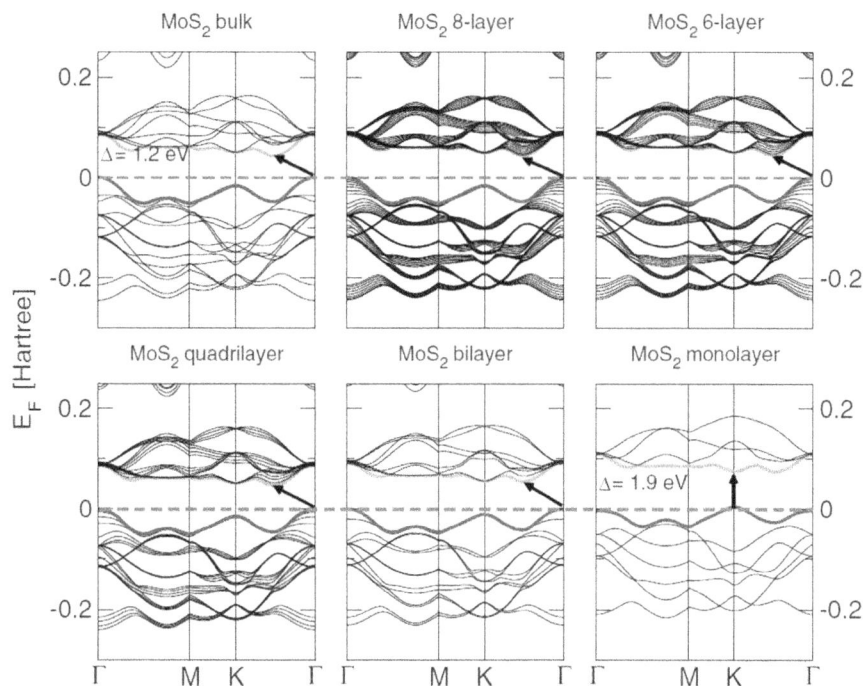

FIGURE 3.2 DFT/PBE-calculated band structures of bulk MoS$_2$, its monolayer, and its polylayer.

Source: Kuc et al. 2011.

numerous physical phenomena because of their phase-related properties. TMD can have a variety of structures depending on the arrangement of atoms, such as trigonal prismatic (hexagonal, H), octahedral (tetragonal, T), or distorted phase (T') with varying symmetry (see Figure 3.3). More than 40 distinct types of stable 2D TMD have been identified. The structural properties of all of these materials are similar (Wilson and Yoffe 1969, Rasmussen and Thygesen 2015). In H-phase materials, each metal atom has six branches that connect to two tetrahedrons in the +z and −z directions. As a result, the chalcogen-metal-chalcogen arrangement along the z-axis is considered a single layer, and the weak vdW interactions among each chalcogen-chalcogen layer enable mechanical exfoliation of bulk TMD to produce single layer flake (Choi et al. 2017). H-phase is the most prominent semiconducting ground state configuration in the TMD family (Huang et al. 2020). Due to its remarkable catalytic, optical, and electronic performance, it has drawn a lot of interest for use in catalysts (Cheng et al. 2017, Poorahong et al. 2017), supercapacitors (Habib et al. 2017), batteries (Li et al. 2017), and gas sensors (Chettri et al. 2021, Chettri, Thapa, et al. 2022). Alternative H-phase sequences can also be used to generate multilayer structures such as 2H (AB stacking) and 3R (ABC stacking) stacking patterns (Huang et al. 2020). Furthermore, because their resistance does not vary with voltage, H-phase counterparts exhibit ohmic activity (Cheng et al. 2016a).

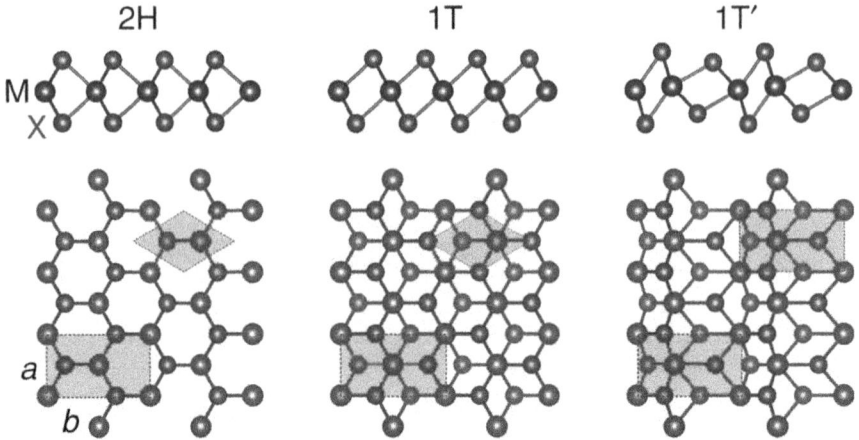

FIGURE 3.3   Typical structures of layered TMD.

Source: Li, Duerloo, et al. 2016.

T-phase has a trigonal chalcogen layer on top and a single layer with a 180° rotated structure at the bottom, resulting in a hexagonal arrangement of chalcogen atoms in the top view (Choi et al. 2017). In general, the T-phase has metallic properties, which improve charge transfer efficiency in a variety of energy-related applications and electrochemical reactions (Reshak and Auluck 2003, Lukowski et al. 2013, Wang, Lu, et al. 2014, Tan et al. 2018). Furthermore, the resistance of the T-phase can vary with voltage, indicating that T-phase can exhibit memristive activity (Cheng et al. 2016b). Metal atoms are further distorted or dimerized in one direction, which is known as the T′-phase, resulting in chalcogen atom displacement along the z-axis (Ali et al. 2014, Keum et al. 2015).

Since different electrical properties can occur in the same material, the phase structure transition in TMD has attracted study throughout the years. This opens up the possibility of controlling functionality in electrical and optoelectronic applications. To synthesize the metallic phase of TMD, various ways have been explored, including phase engineering and directly synthesizing them by changing the growth conditions. The T-phase of metastable materials like MoS_2 can spontaneously change to the H-phase when stimulated by external circumstances (Huang et al. 2020). As a result, regulated preparation of the metallic phase remains a significant issue, particularly in terms of obtaining samples of better quality and stability.

3.2   PROPERTIES OF TRANSITION METAL DICHALCOGENIDES

Graphene has immensely high electron mobility, 15,000 cm^2 V^{-1} s^{-1} at room temperature, but its application in FETs is limited due to its zero bandgap (Novoselov et al. 2005). Massive efforts have been made to tune the bandgap of graphene using nanoribbons and chemical doping, with only limited success, with the bandgap opening up to 200 meV (Han et al. 2007, Xiao et al. 2010, Lee, Duong, et al. 2015). This is still a critical issue that has fueled the development of 2D TMD with finite bandgap. The TMD in a nanostructure possesses quantum confinement that shows a variety of intriguing features not found in their bulk materials. When the thickness of semiconducting 2D TMD is reduced, novel electronic and optical bandgaps are observed, as well as a blue shift in absorbance and superior photoluminescence (PL) (Eda et al. 2011). Charge density wave (CDW) phases (Xu et al. 2013), superconducting phases (Pan et al. 2015, Qi et al. 2016, Zhou et al. 2016), Mott insulating phases, and topological insulating phases are all observed in many of the 2D TMD (Nie et al. 2015). TMD in the distorted 1T crystal structure (Qian et al. 2014, Choe et al. 2016), as well as nanoribbons (Xu et al. 2014, Yuan et al. 2014) have been predicted to have fascinating topological phases. Numerous potential room temperature topological phases have also been predicted based on fundamental principles (Ma, Gao, et al. 2016, Ma, Kou, et al. 2016). Despite the numerous theoretical predictions for 2D topological phases, experimentalists have yet to successfully demonstrate these phases. When thinned to a monolayer, MoS_2 becomes a direct band gap semiconductor, although it is an indirect band gap material in its bulk form (see Figure 3.2) (Ramakrishna Matte et al. 2010, Splendiani et al. 2010). As a result of quantum size effects, the band gaps of few-layer MoS_2 materials can be adjusted, and their values can vary by at least 500 meV from the bulk value (Kuc et al. 2011). Most MX_2 compounds are devoid of dangling bonds and dependent on the selection of the proper substrate and metal contacts as well as mobility suppression beyond grain boundaries, etc.; some of them exhibit significant mobility. For example, at room temperature, MoS_2 exhibits mobility of 700 cm^2 V^{-1} s^{-1} on a SiO_2/Si substrate with a scandium contact and mobility of 33–151 cm^2 V^{-1} s^{-1} on a BN/Si substrate (Das et al. 2013, Lee, Cui, et al. 2015).

The higher PL yield with decreasing layer number was the first experimental proof of the switch from an indirect to a direct gap (Mak et al. 2010, Splendiani et al. 2010). Spatially resolved optical absorption spectroscopy of MoS_2 flakes has also made it feasible to directly observe how the band gap changes with the varying layers (Castellanos-Gomez et al. 2016). A step function-like spectrum that is produced by the joint density of states and matrix elements close to the band borders often dominates the absorption spectra of a 2D material in the infrared-visible region of the electromagnetic spectrum. Due to excitonic effects, 2D TMD shows strong resonance features close to the absorption edge in practice (Gibbon and Dhanak 2019b). Doped TMD has also been observed to have trion quasiparticles (Lin et al. 2014, Lui et al. 2014, Singh et al. 2016, Wang, Zhang, et al. 2016), and biexcitons have been observed in monolayer TMD (Sie et al. 2015, 2016, You et al. 2015, Zhang et al. 2015, Lee et al. 2016). TMD excitonic properties can also increase monolayer absorbance, resulting in increased absorbance in the visible part of the electromagnetic spectrum (Bernardi et al. 2013). WS_2/MoS_2 heterojunctions are anticipated to have power conversion efficiency (PCE) of 0.4–1.5%, whereas photovoltaics based on a graphene/MoS_2 monolayer architecture is anticipated to have PCE of only 0.1–1%, both of which are significantly less efficient than commercial Si-based photovoltaics. However, multilayer stacks may be used to further increase PCE (Bernardi et al. 2013).

3.3 PREPARATION TECHNIQUE OF TRANSITION METAL DICHALCOGENIDES

3.3.1 CHEMICAL VAPOR DEPOSITION

Large-scale production of atomically thin 2D TMD for device applications using chemical vapor deposition (CVD) is very efficient. The coevaporation of metal oxide and chalcogen precursors that drive the vapor phase reaction, followed by the formation of stable 2D TMD over a suitable substrate, is the most essential type of CVD for creating 2D TMD (Choi et al. 2017). The CVD method for the synthesis of monolayer 2D materials for the development of graphene was originally disclosed in 2007 (Blake et al. 2007). Since then, the CVD method has advanced as a result of its cost-effective and scalable production. It is now the most used method for creating 2D materials due to its ability to control the growth location, number of layers, grain size, and doping impurities in addition to large-scale growth (Das et al. 2015, Kim et al. 2016, Hallam et al. 2017, Wang et al. 2017, Yu et al. 2017, Xu, Zhou, et al. 2018). The first study on the large-area development of MoS_2 atomic layers based on a direct chemical vapor phase reaction between MoO_3 and S powder was conducted by Li's group (Lee et al. 2012, 2013). During MoS_2 growth, MoO_3 in the vapor phase experiences two-step reactions, the first of which involves the creation of MoO_{3-x}, which subsequently combines with the sulfur vapor to grow MoS_2 layers. Figure 3.4 depicts a schematic of a CVD setup for the growth of TMD. This approach, which allows for the direct synthesis of single crystalline MoS_2 flakes on arbitrary substrates, is commonly utilized for the production of synthetic TMD in single layers. MoS_2 growth is particularly sensitive to pregrowth substrate treatment (Lee et al. 2012). On SiO_2, sapphire, and glass substrates, up to 400 μm^2 single-layer MoS_2 flakes with a triangular shape can be formed. Because the materials-forming process is also impacted by substrate characteristics, temperature, and atomic gas flux, the growth mechanism of the CVD method varies in each synthesis procedure. CVD can also be used to make a variety of TMD heterostructures (Manzeli et al. 2017). CVD has been utilized to construct atomically sharp interfaces between MoS_2 and $MoSe_2$, WSe_2 and $MoSe_2$, MoS_2 and WS_2, and WSe_2 and MoS_2 monolayers (Manzeli et al. 2017). The surfaces are atomically sharp, and there are no transfer residues, which are advantages of this

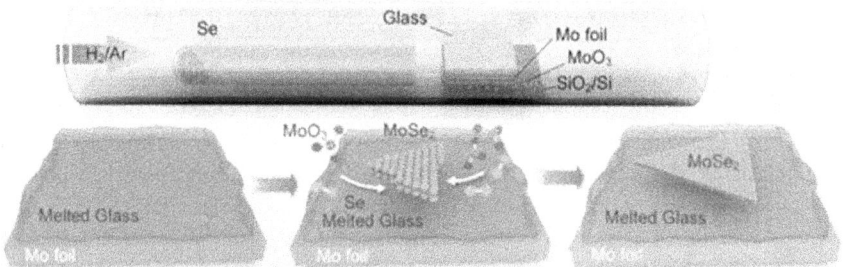

FIGURE 3.4   Schematic diagram of CVD setup for TMD growth.

Source: Chen et al. 2017.

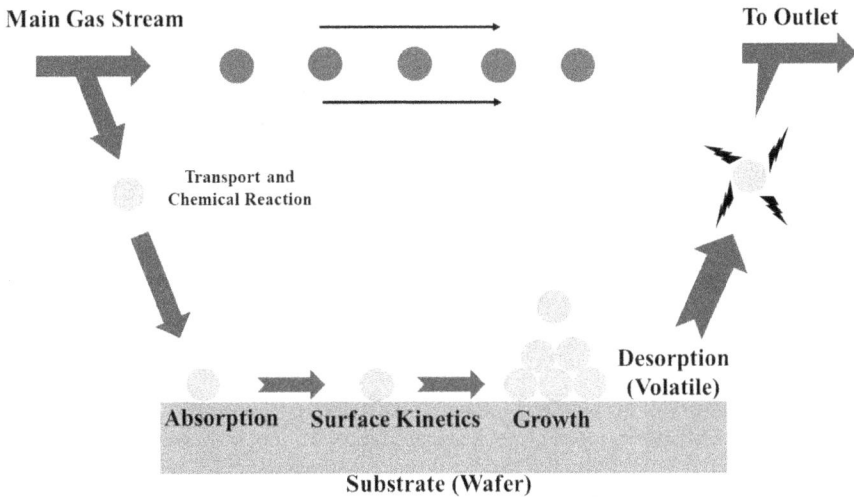

FIGURE 3.5 Growing active layers on the substrate during deposition on the surface and surface processes in MOCVD.

Source: Choi et al. 2017.

technology over manual assembly via transfer procedures. Theoretical modeling predicts exotic behavior for such heterostructures, such as effective charge separation and valley filtering inside a single material (Manzeli et al. 2017).

3.3.2 METAL-ORGANIC CHEMICAL VAPOR DEPOSITION

Metal-organic chemical vapor deposition (MOCVD) has only recently been used to grow 2D TMD. The benefit of MOCVD is that it enables uniform, large-scale synthesis of 2D TMD and fine control over both metal and chalcogen precursors, permitting manipulation of the composition and shape of 2D TMD (Choi et al. 2017). It is identical to a conventional CVD method. With the help of the MOCVD technique, TMD is grown by introducing gases of organic molecules containing a transition metal and chalcogen over a substrate and allowing them to break down through the use of heat energy to deposit a TMD thin layer on the substrate (Kang et al. 2020). The MOCVD method, in particular, can precisely control the partial pressure of the precursors. As a result, this technique enables uniform TMD deposition on large substrates (Kang et al. 2020). MoS_2 and WS_2 were recently synthesized at the wafer scale on a variety of substrates, and the samples that resulted showed good electrical characteristics (Kang et al. 2015). Figure 3.5 is a representative schematic of the MOCVD process, illustrating the numerous steps required in the synthesis of 2D materials. Eichfeld et al. published the first report on the large-area development of mono- and few-layer WSe_2 through MOCVD utilizing $W(CO)_6$ and $(CH_3)_2Se$ precursors (Eichfeld et al. 2015). They demonstrated that the morphology of WSe_2 films is significantly influenced by factors such as temperature, pressure, Se:W ratio, and substrate selection. On various substrates, such as epitaxial graphene, CVD graphene, sapphire, and BN, WSe_2 has a distinctive morphology. On graphene, WSe_2 expanded with a high nucleation density, but sapphire showed the largest domain size of about 5–8 mm. MoS_2 and WS_2 were deposited in a monolayer and a few layers by Kang et al. (2015). $Mo(CO)_6$, $W(CO)_6$, $(C_2H_5)_2S$, Ar, and H_2 are used to create a SiO_2 substrate. The MoS_2 film exhibited electron mobility of 30 cm^2 V^{-1} s^{-1} at room temperature conditions and 114 cm^2 V^{-1} s^{-1} at 90 K temperature. The H_2 increases grain size and crystalline quality while the Ar has been used as a carrier gas. $Mo(NtBu)_2(dpamd)_2$ and $W(NtBu)_2(dpamd)_2$ were successfully employed as precursors by S. Cwik et al. to generate $2H-MoS_2$ and $2H-WS_2$ films on Si, quartz glass, and FTO (Cwik et al. 2018). Andrzejewski D. et al. prepared a MoS_2 film via MOCVD on the (0001) sapphire substrate using $Mo(CO)_6$ and di-tert-butyl sulfide (DTBS) as precursors (Andrzejewski et al. 2018). This technology overcomes the annealing treatment issue that plagues the traditional CVD method and produces great substrate surface coverage over a wide temperature range. MOCVD can be used to grow additional TMDCs, including $MoSe_2$, WS_2, WSe_2, ReS_2, ReS_2, $MoTe_2$, and WTe_2 (Chang et al. 2014, Wang, Gong, et al. 2014, Xia et al. 2014, Zhou et al. 2017). However, more research is necessary to fully characterize the electrical characteristics of the large-area films made of these materials.

3.3.3 LIQUID PHASE EXFOLIATION

A highly helpful technique for producing 2D materials in large quantities and at a low cost for a variety of applications is liquid phase exfoliation (LPE). Furthermore, because the procedure can be carried out at a low temperature, 2D material can be deposited on a variety of substrates (Kang et al. 2020). Large amounts of TMD nanosheets are needed to fully utilize the exceptional potential of layered materials. Solution processing would be more suitable to produce large quantities of single- or few-layer TMD nanosheets. Many studies into the process of exfoliating sheets of TMD have been influenced by the initial report on the liquid phase exfoliation of sheets of clay materials in the early 1960s (Dines 1974, Sasaki et al. 1996, Feng et al. 2011, Golberg 2011). According to Jianfeng Shen et al., the LPE process is split into three subsequent steps: soaking 2D materials in solvents, inserting 2D materials, and stabilizing exfoliated 2D materials (Figure 3.6) (Shen et al. 2015). When the dispersive London interactions are the dominant source of potential energy between neighboring layers, sufficient liquid immersion is one of the easiest and most efficient ways to weaken the van der Waals attractions (Halim et al. 2013). The more compatible they should be, the closer the dispersive components of surface tension between solvents and 2D materials are. Only if the solvents and 2D materials have comparable surface tension components can the immersion of solvents to 2D materials take place effectively in the subsequent stage and prior to exfoliation.

The vdW attractions between the adjacent layers must therefore be further overcome for successful exfoliation of 2D materials. High vapor pressure will typically be reached when the solvent is exposed to strong ultrasound, producing high-intensity vacuum bubbles. During a high-pressure cycle, bubbles grow to a specific size before abruptly collapsing. Liquid jets are created at high temperatures, pressures, and velocities. Agglomerates in solution are capable of disintegrating due to the resulting

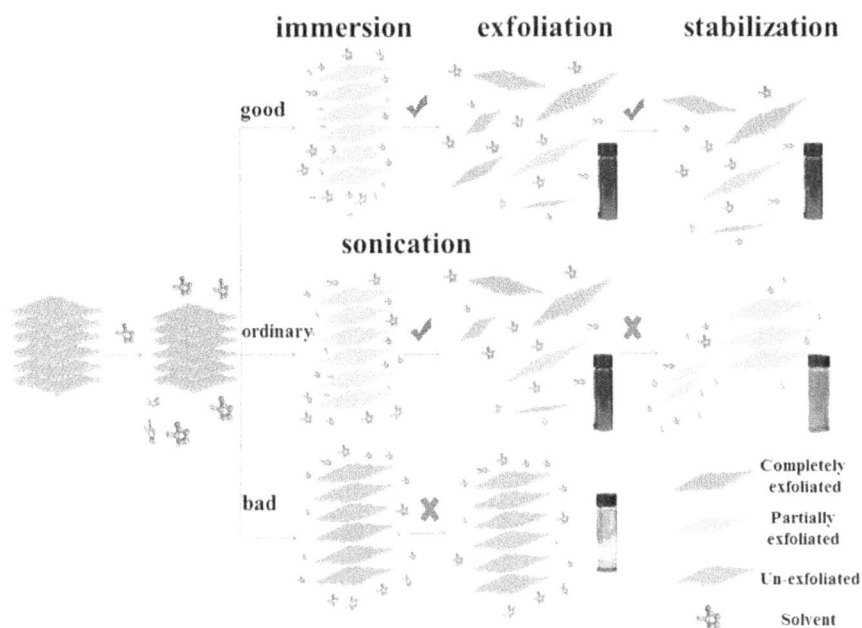

FIGURE 3.6 The graphic illustration of the solvent-based exfoliation method.

Source: Shen et al. 2015.

hydrodynamic forces (Štengl et al. 2014). When an improper solvent is utilized with a surface tension that differs from the surface tension of the molecules of the 2D materials, the balance of the "stabilisation and aggregation" process is disrupted. This is owing to the continuing Brownian motion of the exfoliated 2D materials and solvent molecules after exfoliation. Thus, the entire LPE process will be impacted by the surface tension components matching (Shen et al. 2015).

The best-known solvents are IPA/water, acetone/water, and THF/water based on the solvent requirements. Depending on the 2D material, the ideal volumetric solvent-to-water ratio will vary (Kang et al. 2020). For graphene, hBN, WS_2, and $MoSe_2$, Shen et al. recommended a 1:1 IPA/water ratio, and for MoS_2, they recommended a 7:3 ratio. Depending on the 2D material, the ideal volumetric solvent-to-water ratio will vary (Shen et al. 2015). Ramakrishna Matte et al. reported using n-butyllithium (n-Buli) in hexane as the intercalation agent to insert MoS_2 and WS_2 with lithium, followed by exfoliation in water using ultrasonication to produce single-layer materials (Ramakrishna Matte et al. 2010). The n-Buli solution in hexane has been widely used for lithium intercalation. The Li^+ ion intercalates for the charge balance as the n-Bu⁻ transfers an electron to the TMD layers. Lithium intercalation efficiency has been increased using ultrasonication or microwaves. By hydrolyzing and sonicating, the lithium-ion intercalated TMD bulk crystal is exfoliated (Ramakrishna Matte et al. 2010). High-yield, single-layer TMD nanosheets were created using a controllable electrochemical lithiation technique developed by Zeng et al. (2011).

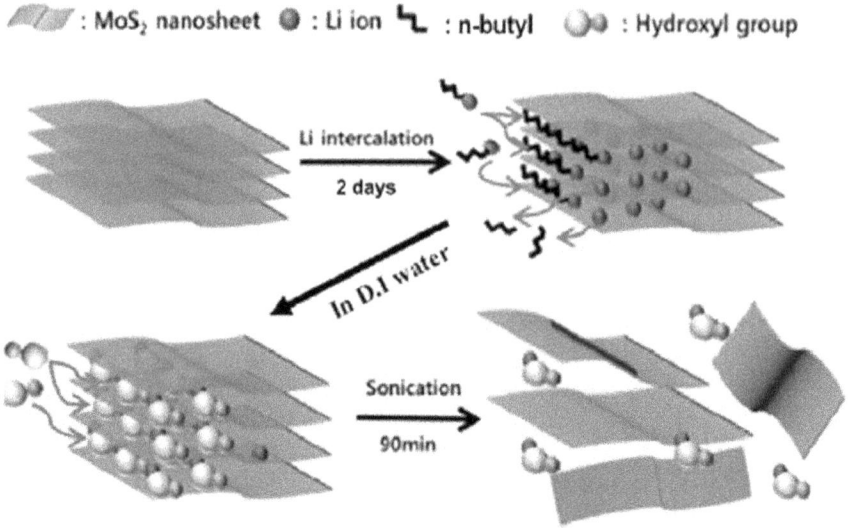

FIGURE 3.7   Lithium intercalation and exfoliation process.

Source: Lee, Jang, et al. 2014.

The lithium intercalation in these materials can be observed and precisely controlled during the discharge process by using the bulk layered TMD materials, such as MoS_2, WS_2, TiS_2, and ZrS_2, as the cathode in an electrochemical cell. Figure 3.7 shows the Lithium intercalation and exfoliation process. To produce high-grade TMD single-layer materials in large quantities, the obtained intercalated compounds can be ultra-sonically exfoliated in water or ethanol.

3.3.4   Atomic Layer Deposition

Atomic layer deposition (ALD) is a gas phase process that deposits atomically thin films of several materials layer by layer by reacting them with the substrate. The procedure is distinguished by the introduction of the precursor gas and the reaction gas into the reaction chamber on a regular and alternate basis, and the formation of a growth film via a chemical reaction process. Because this type of deposition method allows for exact control of film thickness, wafer-level single-layer or multi-layer 2D TMD films can be easily created. Despite the fact that ALD has been widely employed for oxide materials, various research groups have successfully examined several binary sulfide materials. Examples include TiS_2, WS_2, MoS_2, tin(II) sulfide (SnS), and lithium sulfide (Li_2S) (Choi et al. 2017, Xiong et al. 2020). Tan et al. demonstrated precise control over the thickness of the MoS_2 film formed by the self-limiting reactions of $MoCl_5$, which was then flushed through N_2 and introduced into HS_2 for reaction (Figure 3.8) (Tan et al. 2014). Finally, N_2 was utilized to thoroughly clean the chamber. MoS_2 films can be successfully deposited during this deposition cycle. However, high-temperature annealing at 800°C was employed to create massive

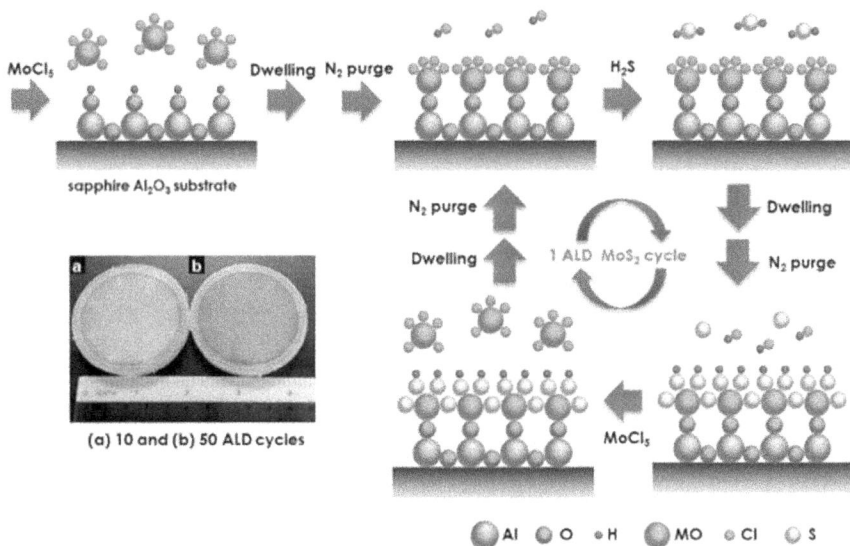

FIGURE 3.8 Growth cycle of an ALD MoS_2 film: (a) 10 and (b) 50 cycles of the ALD MoS_2 film grown.

Source: Tan et al. 2014.

2 m triangular MoS_2 crystals. Song et al. demonstrated wafer-scale development of WS_2 via ALD synthesis of WO_3 and conversion via H_2S annealing. The stages of ALD development for the fabrication of WS_2 nanosheets are shown in Figure 3.9(a) (Song et al. 2013). It is feasible to efficiently manage the number of MoS_2 layers by altering the number of ALD cycles for MoO_3 growth. Figure 3.9(a) shows a camera image of a 13 cm long large-area mono-, bi-, and tetralayer WS_2 nanosheets on a SiO_2 substrate. As seen in Figure 3.9(b), top-gate monolayer WS_2 FETs demonstrated n-type conduction with electron mobility of 3.9 $cm^2\,V^{-1}\,s^{-1}$. When $MoCl_5$ and H_2S were used as reaction gases, Lee et al. explored the deposition of MoS_2 films on a 2 in. (0001) single-sided polished sapphire at low temperatures (about 300°C) and low pressures (approximately 1×10^{-2} torr). Under 20, 10, and 5 cycles, respectively, MoS_2 films having thicknesses of around 3.2 nm, 1.7 nm, and 0.6 nm were produced (Tan et al. 2014). Additionally, it was found that annealing can enhance the crystallinity of the MoS_2 films produced by the ALD process.

3.3.5 MOLECULAR BEAM EPITAXY

The synthesis of TMD has also been carried out using MBE, one of the most sophisticated and controllable growth techniques. The utmost attainable purity is produced by its ultrahigh vacuum environment. As seen in Figure 3.10, a conventional solid source MBE system uses effusion cells for metal evaporation; however, due to the high melting temperatures of transition metals like Mo and W, electron-beam

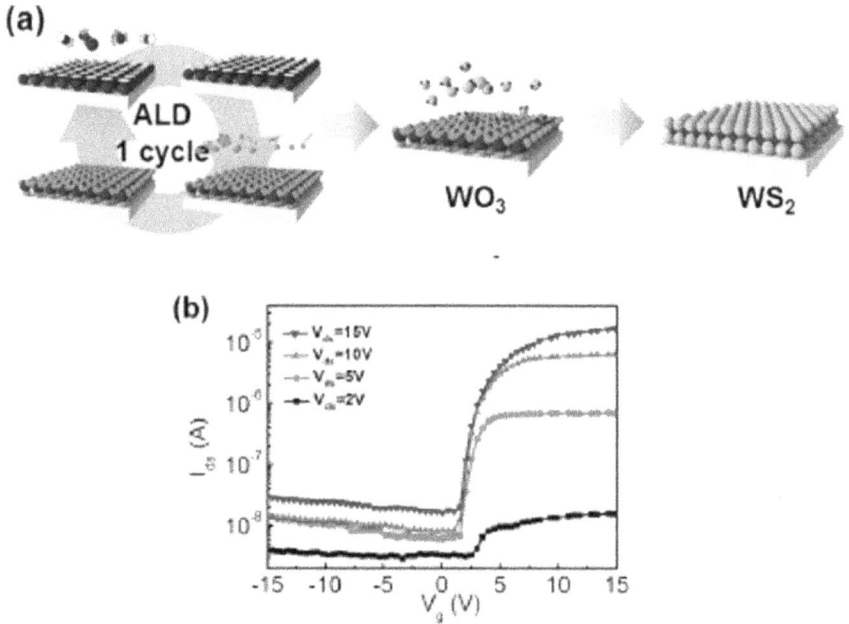

FIGURE 3.9 (a) ALD process for the synthesis of large-area and thickness-controlled WS_2 films and (b) transfer characteristics of a single-layer WS_2 FET with n-type conduction behavior.

Source: Song et al. 2013.

evaporation is required. Additionally, in ultrahigh vacuum circumstances, the sticking coefficient of S (and Se to some extent) on the growth surface is low, which has an impact on growth stoichiometry (Choudhury et al. 2020), especially at high growth temperatures that are advantageous for promoting metal surface diffusion. As a result, compared to MOCVD, TMD domain sizes are frequently smaller. To solve this issue, van der Waals materials like mica and epitaxial graphene are frequently used in the MBE formation of TMD because they promote improved surface diffusion (Vishwanath et al. 2015). The Knudsen cell-evaporated Mo and Se source with a Mo:Se flux ratio of 1:8 produces high-quality $MoSe_2$. ARPES directly detects the change from the indirect to direct band gap in the $MoSe_2$ thin films produced by MBE with precise layer thickness. It may regulate the film's deficit by adjusting the flux ratio. The Se deficit in the MBE-produced $MoSe_2$ films have been shown to exhibit a heavy network along the twin mirror grain boundary.

One of the earliest scalable techniques for fabricating TMD monolayers is MBE. Beginning in the 1980s, Koma et al. produced a monolayer of $MoSe_2$ on a CaF_2 (111) substrate. The addition of an additional molecular beam source makes it possible to dope TMD; therefore MBE has the potential to fabricate heterostructures with doped layers. Fu et al. successfully created a monolayer of MoS_2 on hBN. The continuous MoS_2 monolayer formed on an hBN/sapphire wafer. This approach can produce TMD

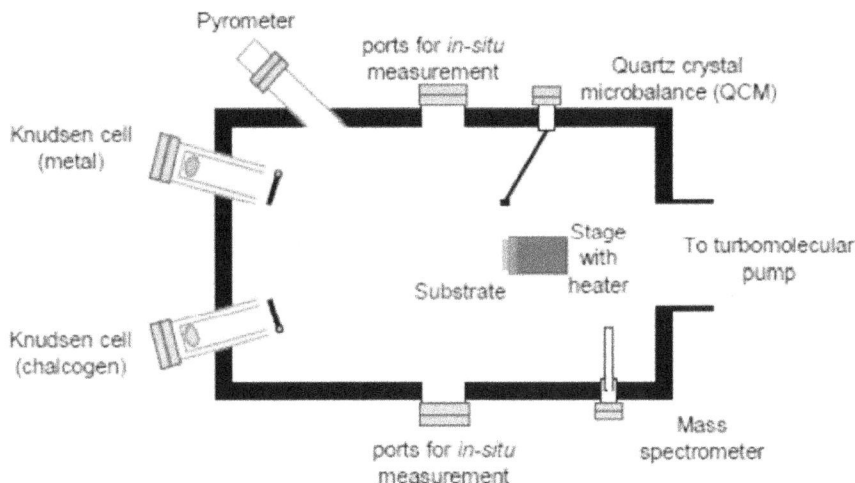

FIGURE 3.10 MBE source supply and substrate heating combinations are shown in schematic drawings of vapor phase deposition procedures for TMD.

Source: Choudhury et al. 2020.

monolayers on wafers, but it requires pricey effusion equipment and requires around 10 hours to create MoS_2 monolayers on 2-in. wafers. Due to its epitaxial nature, the MBE has also been employed to create a vdW heterostructure. The $MoSe_2$ and WSe_2 are aligned similarly to the graphene template.

3.4 DOPING OF TRANSITION METAL DICHALCOGENIDES

The development of both n- and p-type FETs is necessary to use TMD in low-power, high-performance complementary logic, gas, and sensor applications. Doping is therefore one of the most important technologies in TMD. Novel approaches that allow stable doping against changes in external parameters like temperature and humidity are required. Maintaining the doping stable and highly efficient for practical applications remains a difficulty (Song et al. 2017). Due to its wide bandgap and high surface-to-volume ratio, MoS_2 has demonstrated remarkable use in nanoelectronics devices. MoS_2 has been the subject of extensive research aimed at developing sensitive sensors with minimal power consumption and quick reaction times (Zettl 2000, Kalantar-zadeh and Fry 2008). According to Wang et al. analysis of the electrical performance of MoS_2 doped with transition metal (TM), the properties of doped MoS_2 are distinct from those of pure MoS_2 (Wang et al. 2019). DFT was utilized by Y. Wang et al. to establish that Fe-doped MoS_2 might function as a spintronic gas sensor to identify molecules of NO gas (Wang et al. 2015). Additionally, transition metal–doped MoS_2 can be employed as a spintronic gas sensor to detect CO gas molecules, according to research by Y.-H. Zhang et al. (2017), who utilized DFT. Linga Xu et al. investigated the gas sensing capabilities of Pt_n-doped WSe_2 nanosheet to SF_6 breakdown products and found that adding a Pt atom significantly enhances the sensing capabilities of WSe$_2$ nanosheet (Xu et al. 2021). The overall findings highlight MoS$_2$ novelty following appropriate doping. Because the doped TM atoms in MoS$_2$ have strong interactions with one another, the material is stable. Additionally, it offers several free electrons, increasing the electrical conductivity of monolayer MoS$_2$ (Karki et al. 2021, Sharma et al. 2021). Doping reportedly has a significant impact on the overall profile of nanomaterials and is essential for modifying their electrical and magnetic properties. Recent research has also revealed that certain 2D TMD patients exhibit ambipolar behavior (Podzorov et al. 2004, Zhang et al. 2012, Bao et al. 2013). To achieve spatial modification of the doping profile using solid gate dielectrics, local control of carrier types and densities of such devices can be electrostatically controlled by both globally and locally designed gate electrodes (Baugher et al. 2013, Pospischil et al. 2014, Ross et al. 2014). As depicted in Figure 3.11(a–b), it is simple to create lateral P-N, P-P, and N-N junctions by altering the gate voltages of two separately patterned back gate electrodes (Baugher et al. 2013). The actual application in electronics is constrained by the need for additional voltage for local gate configurations. However, even without sophisticated lateral PN junction construction by CVD or selective doping, such gate configurations can still be seen as a useful experimental platform to investigate electrical transport and optoelectronic aspects of 2D materials. To adjust the carrier transfer properties of 2D TMD, surface charge transfer doping (SCTD), which relies on the interfacial charge transfer between dopants and materials, has been thoroughly explored. It has been an alternate approach for surface transfer doping to provide controllable channel doping using molecular/chemical doping, which can address the issue of Schottky barrier (SB)/ contact resistance (Rc), and to obtain increased performance FET. However, because of the irreversible and nonprogrammable nature of conventional SCTD technology,

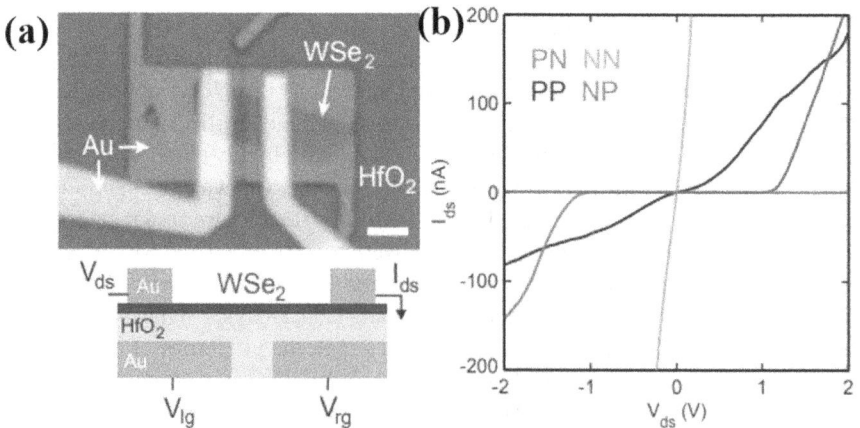

FIGURE 3.11   (a) A monolayer WSe$_2$ device controlled by two local gates is shown in an optical micrograph. Bottom: the device's schematic side view. (b) Current-voltage (I_{ds}-V_{ds}) curves for four different doping configurations.

Source: Baugher et al. 2013.

the doping effect of doped material cannot be freely written and erased. Additionally, space-controlled doping using photoresist and multistep lithography via standard SCTD is unavoidable, which will complicate production and impair the quality of the device. Ion gating is another simple and efficient method for modifying the electrical properties of 2D-TMD. The limit of solid dielectric gating (10^{13} cm^{-2}) is substantially greater than the high capacitance caused by the creation of an electric double layer (EDL) on the material surface, which can tune the carrier density up to 10^{15} cm^{-2} with a tiny gate bias (Song et al. 2017). In Figure 3.11(c–d), a typical experiment is depicted (Ye et al. 2012). Even in an intrinsic n-type TMD FET (such as MoS_2), the local Fermi level modulation can be extended from the conduction band to the valance band and exhibit ambipolar behavior due to its higher control capability over the carrier density. The surface electric dipole layer next to the electrodes can improve gate control in addition to reducing contact resistance by narrowing the Schottky barrier width (Ye et al. 2012).

3.5 NOBLE TRANSITION METAL DICHALCOGENIDES (NTMDS)

Emerging classes of 2D NTMD have distinguished themselves recently for their distinctive structure and new physical characteristics. Due to their distinct interlayer vibrational behaviors and largely programmable electronic structures, 2D NTMDs are anticipated to be more desirable than the majority of TMD semiconductors. Electronics, optoelectronics, catalysis, and sensing applications have all been interested in the new features of 2D NTMD (Pi et al. 2019, Chettri, Sharma, et al. 2022). Group-10 NTMD has just been introduced as a new 2D material with a variety of fascinating features, such as tunable bandgap, moderate carrier mobility, anisotropy, and better air stability (Li et al. 2016, Zhao et al. 2016, 2017, Cui et al. 2020). NTMD d-orbitals are pretty much entirely occupied, in contrast to most typical TMD, which have fewer d-electrons, and the corresponding pz orbital of interlayer chalcogen atoms is substantially hybridized, making robust layer-dependent characteristics and interlayer interactions (Zhao et al. 2016, Yang et al. 2017). For instance, PtS_2 is anticipated to fill the gap between graphene and most TMD with a large bandgap spanning from 0.25 to 1.6 eV (Zhao et al. 2016). Furthermore, the estimated mobility based on PtS_2, $PtSe_2$, and $PdSe_2$ FETs is as high as ~200 cm^2 V^{-1} s^{-1}, which is greater than that of most other TMD (Zhao et al. 2016, 2017, Oyedele et al. 2017). Furthermore, NTMD has excellent air stability, with the performance of the $PtSe_2$ FET remaining unaffected after 5 months of air exposure (Zhao et al. 2017). Pd atoms work with four Se atoms to create the square backbone network in the peculiar pentagonal structure of the monolayer $PdSe_2$ material (see Figure 3.12) (Zhao et al. 2020). Akinola D. Oyedele et al. successfully exfoliated monolayer $PdSe_2$ for the first time, opening up new avenues for research into pentagonal 2D materials (Oyedele et al. 2017). The few-layered $PdSe_2$ behaves as an ambipolar semiconductor with high electron-apparent field-effect mobility. Furthermore, the extraordinary electronic structures of $PdSe_2$ are layer dependent. The indirect band gap of monolayer $PdSe_2$ is about 1.43 eV, while the band gap of bulk $PdSe_2$ is 0.03 eV (Oyedele et al. 2017, Deng et al. 2018). Dan Qin et al. also demonstrated the capable thermoelectric performance of $PdSe_2$ using DFT and the Boltzmann transport

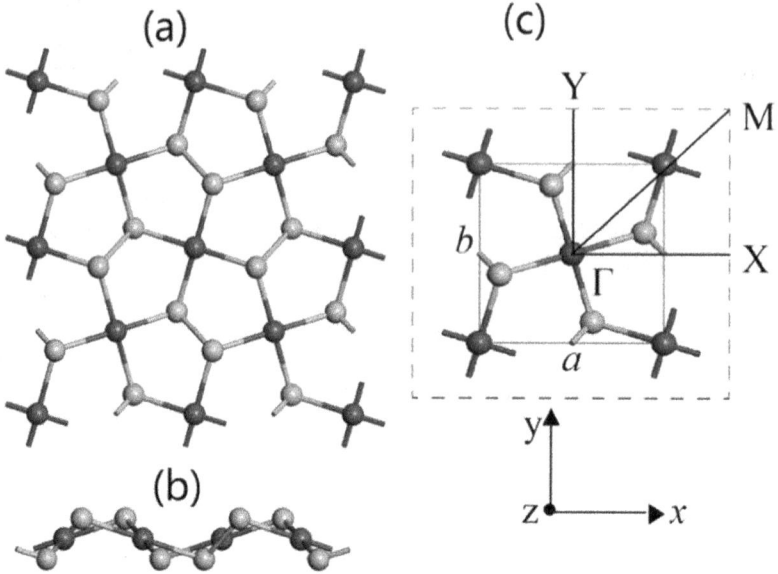

FIGURE 3.12 (a) Top view; (b) side view of PdSe$_2$ monolayer; and (c) unit cell and the Brillouin zone path.

Source: Qin et al. 2018.

equation (Qin et al. 2018). Zhao et al. used DFT to explore the electronic and optical properties of Se and Pd defect PdSe$_2$ monolayers. It was determined that Se defect PdSe$_2$ is a more energetically favorable type than the Pd defect (Zhao et al. 2020). Furthermore, after the introduction of the defect, it was observed that the introduction of midgap in the band structure of PdSe$_2$ (see Figure 3.13) played an important role in boosting the electronic and optical properties of PdSe$_2$ for the application of nanoelectronics devices.

Overall, the discovery of pentagonal PdSe$_2$ allows for the development of physics related to such low-symmetry structures. Sharma et al. explore how external strain can be used to tune the pentagonal PdSe$_2$ monolayer's electrical characteristics. By applying uniaxial and biaxial strains, the PdSe$_2$ monolayer's electronic structure can be flexibly altered (Sharma and Singh 2022). With CBM at the M point and VBM at a k-point along the C-X path, the bandgap remains indirect under the tensile tension. Under uniaxial (biaxial) tensile strains of 8%, the bandgap monotonically drops from 1.34 to 0.83 eV (0.34 eV). At k-points along the C-X path, the bandgap switches from indirect to quasidirect under compressive strain, with both CBM and VBM. When compared to the corresponding value of tensile strains, the change in bandgap under compressive strain is not substantial. Moreover, since PtTe$_2$ and PdTe$_2$ are type-II Dirac fermions, they provide an excellent platform for investigating novel transport related to topological phase transition and chiral anomaly (Noh et al. 2017, Yan et al. 2017). 2D NTMD is becoming increasingly interesting in 2D materials research.

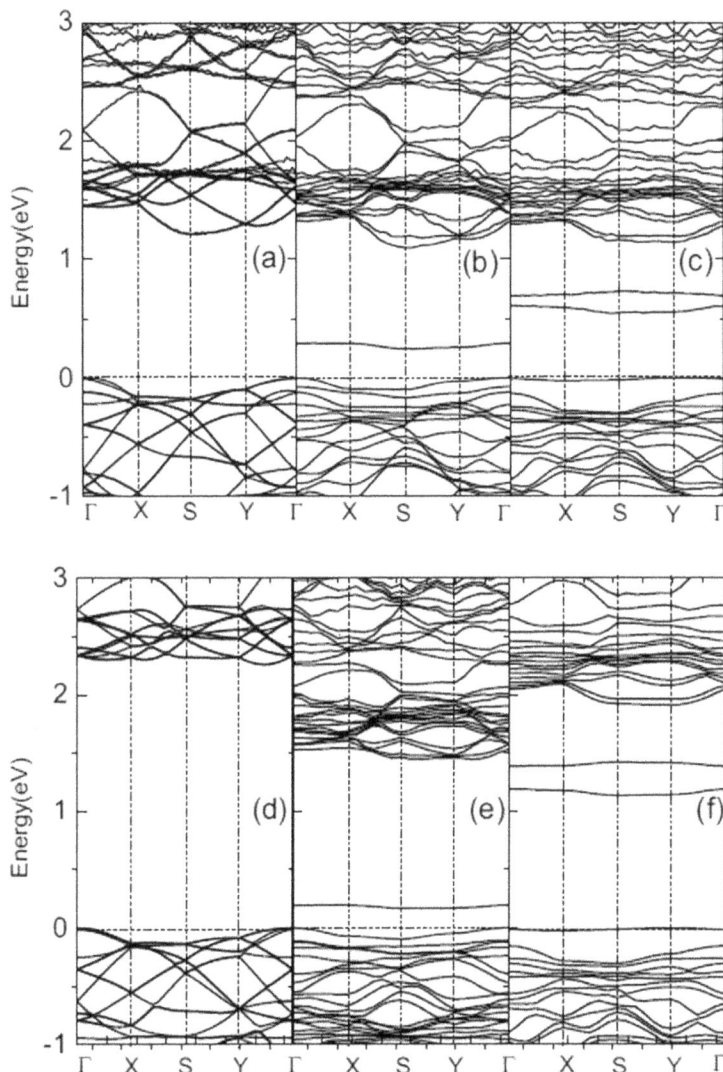

FIGURE 3.13 PBE band structures for pristine $PdSe_2$ (a), $PdSe_2$ with Pd defect (b), and Se defect (c). (d–f) Cases with HSE06.

Source: Zhao et al. 2020.

3.6 APPLICATION OF TRANSITION METAL DICHALCOGENIDES

The development of synthetic methods like CVD and LPE, as well as fabrication techniques like exfoliation, has made it possible to produce ultrathin monolayer TMD. Electronics, photonics, and sensing are just a few of the applications for which these 2D TMD are thought to be attractive. These uses are motivated by the special characteristics of layered materials, such as their thin atomic profile, which is ideal for maximum mechanical strength, tunable electronic structure, optical transparency, and sensor sensitivity (Kang et al. 2014, Huang et al. 2020). These materials hold promise for digital electronics and optoelectronics due to their direct bandgap in the visible energy range. Flexible nanotechnology is of particular relevance for applications in potentially pervasive electronics and energy devices that might benefit from the wide range of exceptional features offered by 2D materials (Huang et al. 2020).

Single-layer MoS_2 nanosheets are regarded as the most supplementing materials to graphene for the fabrication of low-power electronic devices due to their excellent semiconducting properties with a direct bandgap (Radisavljevic et al. 2011). High-performance 2D TFTs based on synthesized MoS_2 have now been accomplished after much research and development. These TFTs have a high on/off current ratio and current saturation that are typical of high-quality TMD when they are operating at room temperature. In particular, the current density of 250 μA/m and electron mobility of 50 cm^2 V^{-1} s^{-1} have been noted. Notably, at a channel length of 0.5 μm, cut-off frequencies greater than 5 GHz have been realized on flexible plastic substrates (Chang et al. 2016). A buffer layer made of a dielectric material like HfO_2 can significantly increase the carrier mobilities of back-gate transistors made of MoS_2 nanosheets with SiO_2 as the dielectric, which have been experimentally observed to have low carrier mobilities of less than 10 cm^2 V^{-1} s^{-1}. As a result, MoS_2-based devices are capable of having carrier mobility of over 200 cm^2 V^{-1} s^{-1}, a high on/off ratio of 1×10^8, and extremely low standby power dissipation at room temperature (Chang et al. 2016). Yoon et al. conducted a theoretical analysis of the performance limit of MoS_2 transistors with HfO_2 as the dielectric and demonstrated that an on/off current ratio greater than 10^{10} is possible (Yoon et al. 2011). P-n diodes, a significant electronic component, are easily realizable through the combination of particular TMD with various charge carriers (Pezeshki et al. 2016). Due to the TMD ultrathin thickness and excellent compatibility with other 2D materials, tunneling transistors can be created using a straightforward process by sandwiching insulating hBN between TMD layers (Xu, Kozawa, et al. 2018). Field-effect transistor active channel materials can be made from semiconductor TMD, which differ from graphene in that they lack a bandgap (FETs). The application of general TMD in high-frequency devices is constrained by their relatively low carrier mobility, but InSe has been shown to have high carrier mobility exceeding 10^3 cm^2 V^{-1} s^{-1} at room temperature (Bandurin et al. 2017). According to the TMD photoelectric characteristics, phototransistors based on TMD, like $SnSe_2$ and WSe_2, have a high photoresponsivity and a quick response time (Zhou et al. 2015, Guo et al. 2018, Wang et al. 2018). High-speed waveguide photodetectors in the telecommunication spectral range have been made with $MoTe_2$ (Ma et al. 2018). Additionally, different TMDs have been researched for solar applications (Zhao et al. 2018). Perovskite solar cells have been shown to perform better when MoS_2 is used (Kakavelakis et al. 2018). Researchers recently discovered intrinsic magnetism in 2D materials, which will encourage practical applications in storage and spintronic devices (Gong et al. 2017, Huang et al. 2017, Miao et al. 2018). Even at room temperature, monolayer VSe_2 exhibits excellent ferromagnetic ordering despite its paramagnetic bulk (Bonilla et al. 2018). Valleytronics research on monolayer MoS_2 was reported by

Wu et al. (2018). For use in superconductors, some TMDs (like $NbSe_2$, TaS_2, and $TaSe_2$) have also been studied (Tsen et al. 2016, Manzeli et al. 2017). In the visible region of the electromagnetic spectrum, monolayer MoS_2 and WS_2 have a direct bandgap and high binding energies. Strong photoluminescence is also present in these two materials. These characteristics make them promising candidates for use in photodetectors, photovoltaics, and light-emitting devices, among other optoelectronic applications. Recently, it was shown that single-layer MoS_2-based phototransistors can be switched on and off in about 50 ms, which is faster than single-layer graphene-based devices (Xia et al. 2009, Yin et al. 2012). They have an on/off ratio of 103 and carrier mobility of 0.11 cm^2 V^{-1} s^{-1} (Yin et al. 2012). Additionally, the MoS_2 phototransistor's photoresponsivity (7.5 mA W^{-1}) is significantly higher than that of graphene-based devices (1 mA W^{-1}) (Yin et al. 2012).

3.7   TRANSITION METAL DICHALCOGENIDES FOR POTENTIAL APPLICATION AS A GAS SENSOR

The use of 2D TMD nanostructures in biomedicine, electronic devices, hydrogen evolution catalysis, and environmental and biological investigation is widespread. While this is ongoing, the excitement surrounding 2D TMD has spread to their counterparts in other dimensions, such as 1D and 0D TMD nanostructures. Future physical and chemical characteristics of TMD nanostructures continue to be strongly influenced by their dimensions and scale, and recent inventive research into these physical and chemical characteristics has had a significant impact on the sensing potential of TMD nanomaterials (Li, Wang, et al. 2016). Until recently, the integration of analytical instruments and electronics only took place at the macroscale. Molecular and nanoscale sensory components are now being imaginatively merged with similarly sized scaled electrical components, opening up new and intriguing frontiers. These nanoinstruments, which represent the upcoming few quantum leaps in sensor technology, are unexpectedly robust and very sensitive (Garaj et al. 2010, 2013, Nakanishi et al. 2014). Recent research on layer-structured TMD with semiconductive characteristics has opened a new chapter in the development of unique electronic device-based sensors that can identify a variety of gases, including triethylamine, NO_2, NO, NH_3, and NO_2 (Perkins et al. 2013). Through surface-gas analyte interactions and charge transfer mechanisms, the reactive gaseous species deposited on the device's surface altered its resistance. Thin-film transistors with gas sensing TMD were able to track changes in the adsorption characteristics of various gas molecules in terms of their typical transient times, channel conductance, and low-frequency current variations (Samnakay et al. 2015). Prior to now, carbon-based nanomaterials might have attracted a lot of attention for their potential as gas sensors because of factors like their high surface-to-volume ratios and distinctive shapes (Liu et al. 2012, Llobet 2013, Wan et al. 2017). Now, to create a complete gas sensor, further crucial features like semiconductive characteristics and a plentiful supply of reactive sites for redox reactions should be taken into account (Late et al. 2013). More NO_2, NH_3, and even humidity-detecting field-effect transistors (FETs) made of single- and multilayered MoS_2 nanosheets are possible (He, Zeng, et al. 2012, Donarelli et al. 2015). The results of the experiments demonstrated that the MoS_2

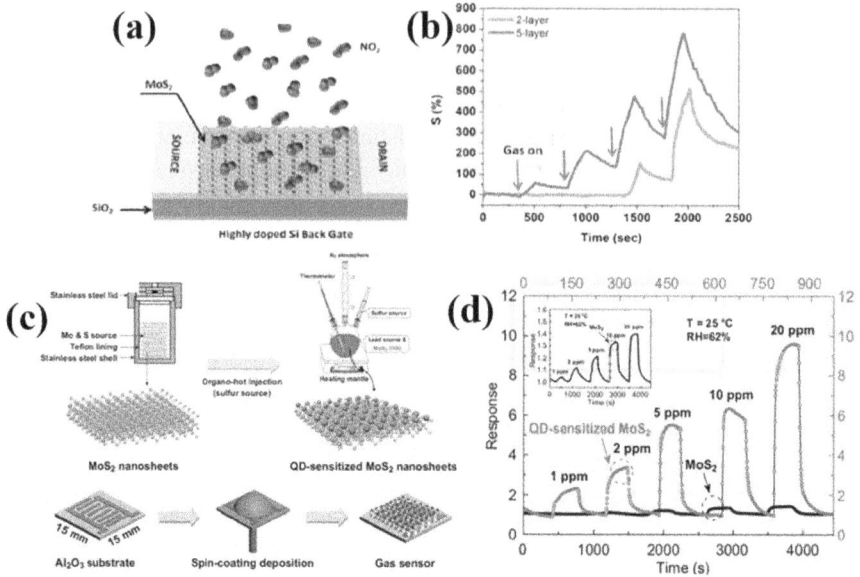

FIGURE 3.14 (a) Representation of the NO_2 gas-sensing device based on MoS_2 transistors. (b) Performance of two- and five-layer MoS_2 cyclic sensing with NO_2. (c) Schematic representation of the formation of MoS_2 nanosheets and QD-sensitized MoS_2 nanosheets. (d) The response of sensors to various NO_2 concentrations.

Sources: Late et al. 2013; Liu et al. 2020.

nanosheet transistors had a high recovery, good sensitivity, and easily controllable features (Figure 3.14(a–b)) (He, Wu, et al. 2012, Late et al. 2013, Chen et al. 2014). MoS_2 alignment had a significant impact on the capacity to detect gases because the edge sites of the material had high d-orbital electron densities that led to robust binding interactions. Therefore, the experimental findings showed that the vertically oriented MoS_2 outperformed the MoS_2 that was produced horizontally in terms of its ability to detect NO_2 (Cho et al. 2015).

Low sensitivity, a slow reaction time and recovery rate, and insufficient selectivity continue to be problems for the individual TMD sensor. Through the development of vdW heterostructures, numerous research attempts have been made to improve the gas-sensing capabilities of TMD materials (Zheng et al. 2021). Quantum dots (QDs), nanoparticles, and nanoclusters are examples of 0D materials, and their boundary sizes typically range from 2 to 100 nm. To create a 0D/2D heterojunction, these tiny particles can be produced or placed onto the surface of 2D TMD. 0D nanomaterials are ideal for creating gas sensors because they have high surface energy, a big surface-to-volume ratio, and strong catalytic activity (Zhang et al. 2014). TMD-based 0D/2D heterojunctions have recently been shown to have better RT gas sensing capabilities. Through surface epitaxial growth or alteration of 0D nanoparticles on TMD surfaces, 0D/2D junctions can be created. The wet chemical technique or stripping of the bulk components in solution is typically used to produce the 2D materials

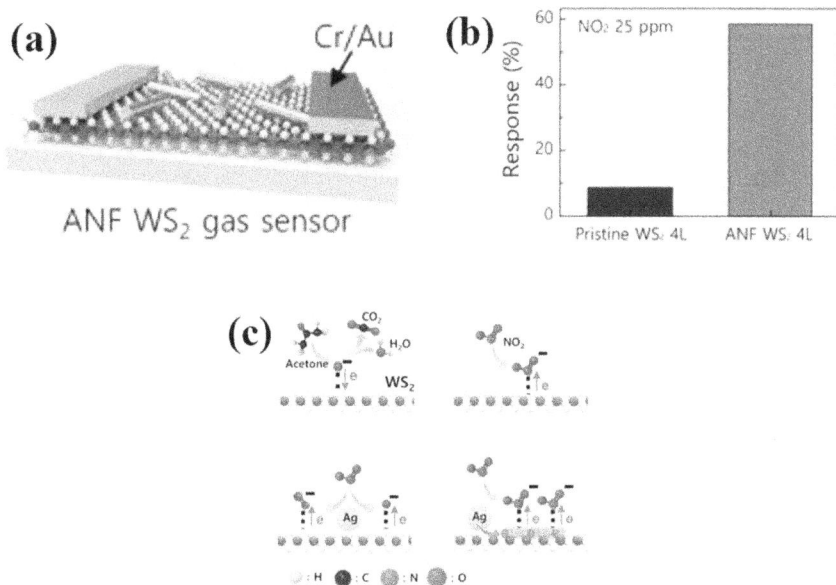

FIGURE 3.15 (a) ANF WS_2 gas sensor schematic image. (b) NO_2 gas sensing performance of ANF WS_2 sensors. (c) Schematic of a gas-sensing mechanism.

Source: Ko et al. 2016.

during synthesis. To create 0D/2D vdW heterojunctions, the 0D nanomaterials are then integrated onto the surface of 2D materials in solution (Zheng et al. 2021). MoS_2 was created via a hydrothermal process by Liu et al. To increase the weak response and incomplete recovery of MoS_2 sensors at RT, they next changed the PbS quantum dots on the surface of MoS_2 using vdW forces. Figure 3.14(c) depicts the precise preparation procedure. When exposed to NO_2 gas, the PbS-sensitized MoS_2 sensor showed a significant improvement that was five times more than that of pristine MoS_2 Devices. In contrast to the MoS_2 sensor, the sensor has a quick reaction and recovery time (15/62 s) (Figure 3.14(d)) (Liu et al. 2020). In order to create the large-area WS_2 nanosheets, Kim et al. sulfurized the WO_3 layer produced using atomic layer deposition (ALD) (Figure 3.15(a)) (Ko et al. 2016). The $CuCl_2$-modified polyol method can produce Ag NWs, which can act as an efficient catalyst. In contrast to the Ag NWs-functionalized ANF WS_2 sensor, which has a strong response of 58% to 25 ppm NO_2 and good recovery performance, the pristine four-layer WS_2 sensor exhibits weak gas sensing performance and partial recovery, as shown in Figure 3.15(b) Due to the Ag catalyst's ability to extract more electrons, the ANF WS_2 gas sensor has great properties. These electrons can produce a middle state between the free and adsorbed NO_2 on WS_2, which will cause NO_2 to break down into NO and O (Figure 3.15(c)). Due to NO_2 ability to bind to WS_2 as an electron acceptor, the Ag NWs also play a role in n-type doping, which can raise the carrier concentration and further improve NO_2 responsiveness (Ko et al. 2016).

3.8 CONCLUSION

We have outlined some of the most important findings in this chapter regarding the characteristics, structure, synthesis techniques, and application of TMDCs. TMD not only has a diverse set of physical and chemical properties, but it is also a prospective candidate for a wide range of applications in low-power electronics, optoelectronics, and spintronics due to its excellent electrical properties. The atomically thin size of the TMD has a significant impact on its electrical characteristics. As we downscale TMD, we notice more and more distinct features. TMD has a flexible bandgap when compared to graphene, which has limited the usage of graphene as an active element in FETs. TMD has also piqued the interest of several researchers for use as an active gas sensor and biosensor. Not only that, but TMD has demonstrated exceptional performance in the drug delivery application to treat various cancers. The emerging classes of 2D NTMD have distinguished themselves through their distinct structure and novel physical properties. NTMDs are expected to be more appealing than most transition metal dichalcogenide semiconductors due to their unique interlayer vibrational behaviors and largely tunable electronic structures. The novel properties of 2D NTMD have piqued the interest of researchers in a variety of fields, including electronics, optoelectronics, catalysis, and sensors. TMD knowledge can benefit the reader's research applications in the field of modern nanoelectronics devices.

ACKNOWLEDGMENT

This work was supported by the All India Council for Technical Education (AICTE), Govt. of India, under the Research Promotion Scheme for North-East Region (RPS-NER) vide ref.: File No. 8-139/RIFD/RPS-NER/Policy-1/2018-19 (PI: Bikash Sharma).

The author (Sanat Kr. Das) acknowledges the TMA Pai University Research Fund-Award of Minor Grant, Sikkim Manipal Institute of Technology, Sikkim Manipal University, Sikkim (Sanction No.: 6100/SMIT/R&D/Project/12/2020, dated on 20th July 2020).

REFERENCES

Ali, M.N., Xiong, J., Flynn, S., Tao, J., Gibson, Q.D., Schoop, L.M., Liang, T., Haldolaarachchige, N., Hirschberger, M., Ong, N.P., and Cava, R.J., 2014. Large, non-saturating magneto-resistance in WTe$_2$. *Nature*, 514 (7521), 205–208.

Andrzejewski, D., Marx, M., Grundmann, A., Pfingsten, O., Kalisch, H., Vescan, A., Heuken, M., Kümmell, T., and Bacher, G., 2018. Improved luminescence properties of MoS$_2$ monolayers grown via MOCVD: Role of pre-treatment and growth parameters. *Nanotechnology*, 29 (29), 295704.

Bandurin, D.A., Tyurnina, A. v., Yu, G.L., Mishchenko, A., Zólyomi, V., Morozov, S. v., Kumar, R.K., Gorbachev, R. v., Kudrynskyi, Z.R., Pezzini, S., Kovalyuk, Z.D., Zeitler, U., Novoselov, K.S., Patanè, A., Eaves, L., Grigorieva, I. v., Fal'Ko, V.I., Geim, A.K., and Cao, Y., 2017. High electron mobility, quantum Hall effect and anomalous optical response in atomically thin InSe. *Nature Nanotechnology*, 12 (3), 223–227.

Bao, W., Cai, X., Kim, D., Sridhara, K., and Fuhrer, M.S., 2013. High mobility ambipolar MoS_2 field-effect transistors: Substrate and dielectric effects. *Applied Physics Letters*, 102 (4), 042104.

Baugher, B.W.H., Churchill, H.O.H., Yafang, Y., and Jarillo-Herrero, P., 2013. Electrically Tunable PN Diodes in a Monolayer Dichalcogenide. *CondMat*.

Bernardi, M., Palummo, M., and Grossman, J.C., 2013. Extraordinary sunlight absorption and one nanometer thick photovoltaics using two-dimensional monolayer materials. *Nano Letters*, 13 (8), 3664–3670.

Blake, P., Hill, E.W., Castro Neto, A.H., Novoselov, K.S., Jiang, D., Yang, R., Booth, T.J., and Geim, A.K., 2007. Making graphene visible. *Applied Physics Letters*, 91 (6), 063124.

Bonilla, M., Kolekar, S., Ma, Y., Diaz, H.C., Kalappattil, V., Das, R., Eggers, T., Gutierrez, H.R., Phan, M.H., and Batzill, M., 2018. Strong room-temperature ferromagnetism in VSe_2 monolayers on van der Waals substrates. *Nature Nanotechnology*, 13 (4), 289–293.

Butler, S.Z., Hollen, S.M., Cao, L., Cui, Y., Gupta, J.A., Gutiérrez, H.R., Heinz, T.F., Hong, S.S., Huang, J., Ismach, A.F., Johnston-Halperin, E., Kuno, M., Plashnitsa, V. v., Robinson, R.D., Ruoff, R.S., Salahuddin, S., Shan, J., Shi, L., Spencer, M.G., Terrones, M., Windl, W., and Goldberger, J.E., 2013. Progress, challenges, and opportunities in two-dimensional materials beyond graphene. *ACS Nano*, 7 (4), 2898–2926.

Castellanos-Gomez, A., Quereda, J., van der Meulen, H.P., Agraït, N., and Rubio-Bollinger, G., 2016. Spatially resolved optical absorption spectroscopy of single- and few-layer MoS_2 by hyperspectral imaging. *Nanotechnology*, 27 (11), 115705.

Chang, H.Y., Yogeesh, M.N., Ghosh, R., Rai, A., Sanne, A., Yang, S., Lu, N., Banerjee, S.K., and Akinwande, D., 2016. Large-Area Monolayer MoS_2 for Flexible Low-Power RF Nanoelectronics in the GHz Regime. *Advanced Materials*, 28 (9), 1818–1823.

Chang, Y.H., Zhang, W., Zhu, Y., Han, Y., Pu, J., Chang, J.K., Hsu, W.T., Huang, J.K., Hsu, C.L., Chiu, M.H., Takenobu, T., Li, H., Wu, C.I., Chang, W.H., Wee, A.T.S., and Li, L.J., 2014. Monolayer $MoSe_2$ grown by chemical vapor deposition for fast photodetection. ACS Nano, 8 (8), 8582–8590.

Chen, J., Zhao, X., Tan, S.J.R., Xu, H., Wu, B., Liu, B., Fu, D., Fu, W., Geng, D., Liu, Y., Liu, W., Tang, W., Li, L., Zhou, W., Sum, T.C., and Loh, K.P., 2017. Chemical vapor deposition of large-size monolayer $MoSe_2$ crystals on molten glass. *Journal of the American Chemical Society*, 139 (3), 1073–1076.

Chen, L., Liu, B., Abbas, A.N., Ma, Y., Fang, X., Liu, Y., and Zhou, C., 2014. Screw-Dislocation-Driven growth of Two-Dimensional few-layer and pyramid-like WSe_2 by sulfur-assisted Chemical Vapor Deposition. *ACS Nano*, 8 (11), 11543–11551.

Cheng, P., Sun, K., and Hu, Y.H., 2016a. Mechanically-induced reverse phase transformation of MoS_2 from stable 2H to metastable 1T and its memristive behavior. *RSC Advances*, 6 (70), 65691–65697.

Cheng, P., Sun, K., and Hu, Y.H., 2016b. Memristive Behavior and Ideal Memristor of 1T Phase MoS_2 Nanosheets. *Nano Letters*, 16 (1), 572–576.

Cheng, Y., Lu, S., Liao, F., Liu, L., Li, Y., and Shao, M., 2017. Rh-MoS_2 Nanocomposite Catalysts with Pt-Like Activity for Hydrogen Evolution Reaction. *Advanced Functional Materials*, 27 (23), 1700359.

Chettri, B., Sharma, A., Das, S.K., and Sharma, B., 2022. First principle study of Rh/Ru doped pentagonal $PdSe_2$ for detection of SO_2 and SO_3 gas. *Materials Today: Proceedings*, 58, 696–701.

Chettri, B., Thapa, A., Das, S., Chettri, P., and Sharma, B., 2022. First principle insight into co-doped MoS_2 for sensing NH_3 and CH_4. *Facta universitatis* – series: Electronics and Energetics, 35 (1), 43–59.

Chettri, B., Thapa, A., Das, S.K., Chettri, P., and Sharma, B., 2021. Computational Study of Adsorption behavior of CH_4N_2O and CH_3OH on Fe decorated MoS_2 monolayer. *Solid State Electronics Letters*, 3, 32–41.

Chhowalla, M., Shin, H.S., Eda, G., Li, L.J., Loh, K.P., and Zhang, H., 2013. The chemistry of two-dimensional layered transition metal dichalcogenide nanosheets. *Nature Chemistry*, 5 (4), 263–275.

Cho, S.Y., Kim, S.J., Lee, Y., Kim, J.S., Jung, W. bin, Yoo, H.W., Kim, J., and Jung, H.T., 2015. Highly Enhanced Gas Adsorption Properties in Vertically Aligned MoS_2 Layers. *ACS Nano*, 9 (9), 9314–9321.

Choe, D.H., Sung, H.J., and Chang, K.J., 2016. Understanding topological phase transition in monolayer transition metal dichalcogenides. *Physical Review B*, 93 (12), 125109.

Choi, W., Choudhary, N., Han, G.H., Park, J., Akinwande, D., and Lee, Y.H., 2017. Recent development of two-dimensional transition metal dichalcogenides and their applications. *Materials Today*. 20 (3), 116–130.

Choudhary, N., Park, J., Hwang, J.Y., Chung, H.S., Dumas, K.H., Khondaker, S.I., Choi, W., and Jung, Y., 2016. Centimeter Scale Patterned Growth of Vertically Stacked Few Layer Only 2D MoS_2/WS_2 van der Waals Heterostructure. *Scientific Reports*, 6 (1), 1–7.

Choudhury, T.H., Zhang, X., al Balushi, Z.Y., Chubarov, M., and Redwing, J.M., 2020. Epitaxial Growth of 2D Layered Transition Metal Dichalcogenides. *Annual Review of Materials Research*, 50, 155–177.

Cui, N., Zhang, F., Zhao, Y., Yao, Y., Wang, Q., Dong, L., Zhang, H., Liu, S., Xu, J., and Zhang, H., 2020. The visible nonlinear optical properties and passively Q-switched laser application of a layered $PtSe_2$ material. *Nanoscale*, 12 (2), 1061–1066.

Cwik, S., Mitoraj, D., Mendoza Reyes, O., Rogalla, D., Peeters, D., Kim, J., Schütz, H.M., Bock, C., Beranek, R., and Devi, A., 2018. Direct Growth of MoS_2 and WS_2 Layers by Metal Organic Chemical Vapor Deposition. *Advanced Materials Interfaces*, 5 (16), 1800140.

Das, S., Chen, H.Y., Penumatcha, A.V., and Appenzeller, J., 2013. High performance multilayer MoS_2 transistors with scandium contacts. *Nano Letters*, 13 (1), 100–105.

Das, S., Demarteau, M., and Roelofs, A., 2015. Nb-doped single crystalline MoS_2 field effect transistor. *Applied Physics Letters*, 106 (17), 173506.

Deng, S., Li, L., and Zhang, Y., 2018. Strain Modulated Electronic, Mechanical, and Optical Properties of the Monolayer PdS_2, $PdSe_2$, and $PtSe_2$ for Tunable Devices. *ACS Applied Nano Materials*, 1 (4), 1932–1939.

Dervin, S., Dionysiou, D.D., and Pillai, S.C., 2016. 2D nanostructures for water purification: Graphene and beyond. *Nanoscale*, 8 (33), 15115–15131.

Dickinson, R.G. and Pauling, L., 1923. The crystal structure of molybdenite. *Journal of the American Chemical Society*, 45 (6), 1466–1471.

Dines, M.B., 1974. Intercalation in layered compounds. *Journal of Chemical Education*, 51 (4), 221.

Donarelli, M., Prezioso, S., Perrozzi, F., Bisti, F., Nardone, M., Giancaterini, L., Cantalini, C., and Ottaviano, L., 2015. Response to NO_2 and other gases of resistive chemically exfoliated MoS_2-based gas sensors. *Sensors and Actuators, B: Chemical, 207*, 602–613.

Eda, G., Yamaguchi, H., Voiry, D., Fujita, T., Chen, M., and Chhowalla, M., 2011. Photoluminescence from chemically exfoliated MoS_2. *Nano Letters*, 11 (12), 5111–5116.

Eichfeld, S.M., Hossain, L., Lin, Y.-C., Piasecki, A.F., Kupp, B., Birdwell, A.G., Burke, R.A., Lu, N., Peng, X., Li, J., Azcatl, A., McDonnell, S., Wallace, R.M., Kim, M.J., Mayer, T.S., Redwing, J.M., and Robinson, J.A., 2015. Highly Scalable, Atomically Thin WSe_2 Grown via Metal–Organic Chemical Vapor Deposition. *ACS Nano*, 9 (2), 2080–2087.

Feng, J., Sun, X., Wu, C., Peng, L., Lin, C., Hu, S., Yang, J., and Xie, Y., 2011. Metallic few-layered VS_2 ultrathin nanosheets: High two-dimensional conductivity for in-plane supercapacitors. *Journal of the American Chemical Society*, 133 (44), 17832–17838.

Frey, G. and Elani, S., 1998. Optical-absorption spectra of inorganic fullerenelike W). *Physical Review B – Condensed Matter and Materials Physics*, 57 (11), 6666–6671.

Garaj, S., Hubbard, W., Reina, A., Kong, J., Branton, D., and Golovchenko, J.A., 2010. Graphene as a subnanometre trans-electrode membrane. *Nature*, 467 (7312), 190–193.

Garaj, S., Liu, S., Golovchenko, J.A., and Branton, D., 2013. Molecule-hugging graphene nanopores. *Proceedings of the National Academy of Sciences of the United States of America*, 110 (30), 12192–12196.

Geim, A.K. and Grigorieva, I. v., 2013. Van der Waals heterostructures. *Nature*, 499 (7459), 419–425.

Gibbon, J.T. and Dhanak, V.R., 2019a. Properties of Transition Metal Dichalcogenides. *Two Dimensional Transition Metal Dichalcogenides*, 69–106.

Gibbon, J.T. and Dhanak, V.R., 2019b. Properties of Transition Metal Dichalcogenides. In: Two Dimensional Transition Metal Dichalcogenides. *Singapore: Springer Singapore*, 69–106.

Golberg, D., 2011. Exfoliating the inorganics. *Nature Nanotechnology*, 6 (4), 200–201.

Gong, C., Li, L., Li, Z., Ji, H., Stern, A., Xia, Y., Cao, T., Bao, W., Wang, C., Wang, Y., Qiu, Z.Q., Cava, R.J., Louie, S.G., Xia, J., and Zhang, X., 2017. Discovery of intrinsic ferro-magnetism in two-dimensional van der Waals crystals. *Nature*, 546 (7657), 265–269.

Guo, J., Liu, Y., Ma, Y., Zhu, E., Lee, S., Lu, Z., Zhao, Z., Xu, C., Lee, S.J., Wu, H., Kovnir, K., Huang, Y., and Duan, X., 2018. Few-Layer GeAs Field-Effect Transistors and Infrared Photodetectors. *Advanced Materials*, 30 (21), 1705934.

Habib, M., Khalil, A., Muhammad, Z., Khan, R., Wang, C., Rehman, Z. ur, Masood, H.T., Xu, W., Liu, H., Gan, W., Wu, C., Chen, H., and Song, L., 2017. WX_2(X=S, Se) Single Crystals: A Highly Stable Material for Supercapacitor Applications. *Electrochimica Acta*, 258, 71–79.

Halim, U., Zheng, C.R., Chen, Y., Lin, Z., Jiang, S., Cheng, R., Huang, Y., and Duan, X., 2013. A rational design of cosolvent exfoliation of layered materials by directly probing liquid-solid interaction. *Nature Communications*, 4, 2213.

Hallam, T., Monaghan, S., Gity, F., Ansari, L., Schmidt, M., Downing, C., Cullen, C.P., Nicolosi, V., Hurley, P.K., and Duesberg, G.S., 2017. Rhenium-doped MoS_2 films. *Applied Physics Letters*, 111 (20).

Han, M.Y., Özyilmaz, B., Zhang, Y., and Kim, P., 2007. Energy band-gap engineering of graphene nanoribbons. *Physical Review Letters*, 98 (20), 206805.

He, Q., Wu, S., Yin, Z., and Zhang, H., 2012. Graphene-based electronic sensors. *Chemical Science, 3(6)*, 1764–1772 .

He, Q., Zeng, Z., Yin, Z., Li, H., Wu, S., Huang, X., and Zhang, H., 2012. Fabrication of flex-ible MoS_2 thin-film transistor arrays for practical gas-sensing applications. *Small*, 8 (19), 2994–2999.

Huang, B., Clark, G., Navarro-Moratalla, E., Klein, D.R., Cheng, R., Seyler, K.L., Zhong, Di., Schmidgall, E., McGuire, M.A., Cobden, D.H., Yao, W., Xiao, D., Jarillo-Herrero, P., and Xu, X., 2017. Layer-dependent ferromagnetism in a van der Waals crystal down to the monolayer limit. *Nature*, 546 (7657), 270–273.

Huang, H.H., Fan, X., Singh, D.J., and Zheng, W.T., 2020. Recent progress of TMD nanomaterials: Phase transitions and applications. *Nanoscale*, 12 (3), 1247–1268.

Islam, M.R., Kang, N., Bhanu, U., Paudel, H.P., Erementchouk, M., Tetard, L., Leuenberger, M.N., and Khondaker, S.I., 2014. Tuning the electrical property via defect engineering of single layer MoS_2 by oxygen plasma. *Nanoscale*, 6 (17), 10033–10039.

Jariwala, D., Sangwan, V.K., Lauhon, L.J., Marks, T.J., and Hersam, M.C., 2014. Emerging device applications for semiconducting two-dimensional transition metal dichalcogenides. *ACS Nano*, 8 (2), 1102–1120.

Kakavelakis, G., Paradisanos, I., Paci, B., Generosi, A., Papachatzakis, M., Maksudov, T., Najafi, L., del Rio Castillo, A.E., Kioseoglou, G., Stratakis, E., Bonaccorso, F., and Kymakis, E., 2018. Extending the Continuous Operating Lifetime of Perovskite Solar Cells with a Molybdenum Disulfide Hole Extraction Interlayer. *Advanced Energy Materials*, 8 (12), 1702287.

Kalantar-zadeh, K. and Fry, B., 2008. Organic Nanotechnology Enabled Sensors. In: *Nanotechnology-Enabled Sensors, Springer US*, 283–370.

Kang, J., Cao, W., Xie, X., Sarkar, D., Liu, W., and Banerjee, K., 2014. Graphene and beyond-graphene 2D crystals for next-generation green electronics. In: *Micro- and Nanotechnology Sensors, Systems, and Applications VI*, 9083, 20–26.

Kang, K., Chen, S., and Yang, E.-H., 2020. Synthesis of transition metal dichalcogenides. In: *Synthesis, Modeling, and Characterization of 2D Materials, and Their Heterostructures*. Elsevier, 247–264.

Kang, K., Xie, S., Huang, L., Han, Y., Huang, P.Y., Mak, K.F., Kim, C.J., Muller, D., and Park, J., 2015. High-mobility three-atom-thick semiconducting films with wafer-scale homogeneity. *Nature*, 520 (7549).

Karki, P., Chettri, B., Thapa, A., Chettri, P., and Sharma, B., 2021. First principle study of MoS_2 adsorbed transition metal for sensing NH_3 and CH_4. In: *Proceedings of 4th International Conference on 2021 Devices for Integrated Circuit, DevIC 2021*, 659–661.

Keum, D.H., Cho, S., Kim, J.H., Choe, D.H., Sung, H.J., Kan, M., Kang, H., Hwang, J.Y., Kim, S.W., Yang, H., Chang, K.J., and Lee, Y.H., 2015. Bandgap opening in few-layered monoclinic $MoTe_2$. *Nature Physics*, 11 (6), 482–486.

Kim, Y., Bark, H., Ryu, G.H., Lee, Z., and Lee, C., 2016. Wafer-scale monolayer MoS_2 grown by chemical vapor deposition using a reaction of MoO_3 and H_2S. *Journal of Physics Condensed Matter*, 28 (18), 184002.

Ko, K.Y., Song, J.G., Kim, Y., Choi, T., Shin, S., Lee, C.W., Lee, K., Koo, J., Lee, H., Kim, J., Lee, T., Park, J., and Kim, H., 2016. Improvement of Gas-Sensing Performance of Large-Area Tungsten Disulfide Nanosheets by Surface Functionalization. *ACS Nano*, 10 (10), 9287–9296.

Kuc, A., Zibouche, N., and Heine, T., 2011. Influence of quantum confinement on the electronic structure of the transition metal sulfide TS_2. *Physical Review B – Condensed Matter and Materials Physics*, 83 (24), 245213.

Late, D.J., Huang, Y.K., Liu, B., Acharya, J., Shirodkar, S.N., Luo, J., Yan, A., Charles, D., Waghmare, U. v., Dravid, V.P., and Rao, C.N.R., 2013. Sensing behavior of atomically thin-layered MoS_2 transistors. *ACS Nano*, 7 (6), 4879–4891.

Lee, C.H., Lee, G.H., van der Zande, A.M., Chen, W., Li, Y., Han, M., Cui, X., Arefe, G., Nuckolls, C., Heinz, T.F., Guo, J., Hone, J., and Kim, P., 2014. Atomically thin p-n junctions with van der Waals heterointerfaces. *Nature Nanotechnology*, 9 (9), 676–681.

Lee, G.H., Cui, X., Kim, Y.D., Arefe, G., Zhang, X., Lee, C.H., Ye, F., Watanabe, K., Taniguchi, T., Kim, P., and Hone, J., 2015. Highly Stable, Dual-Gated MoS_2 Transistors Encapsulated by Hexagonal Boron Nitride with Gate-Controllable Contact, Resistance, and Threshold Voltage. *ACS Nano*, 9 (7), 7019–7026.

Lee, H.S., Kim, M.S., Kim, H., and Lee, Y.H., 2016. Identifying multiexcitons in MoS2 monolayers at room temperature. *Physical Review B*, 93 (14), 140409.

Lee, J., Dak, P., Lee, Y., Park, H., Choi, W., Alam, M.A., and Kim, S., 2014. Two-dimensional layered MoS_2 biosensors enable highly sensitive detection of biomolecules. *Scientific Reports*, 4, 7352.

Lee, J.H., Jang, W.S., Han, S.W., and Baik, H.K., 2014. Efficient hydrogen evolution by mechanically strained MoS$_2$ nanosheets. *Langmuir*, 30 (32), 9866–9873.

Lee, S.Y., Duong, D.L., Vu, Q.A., Jin, Y., Kim, P., and Lee, Y.H., 2015. Chemically Modulated Band Gap in Bilayer Graphene Memory Transistors with High On/Off Ratio. *ACS Nano*, 9 (9), 9034–9042.

Lee, Y.H., Yu, L., Wang, H., Fang, W., Ling, X., Shi, Y., Lin, C. te, Huang, J.K., Chang, M.T., Chang, C.S., Dresselhaus, M., Palacios, T., Li, L.J., and Kong, J., 2013. Synthesis and transfer of single-layer transition metal disulfides on diverse surfaces. *Nano Letters*, 13 (4), 1852–1857.

Lee, Y.-H., Zhang, X.-Q., Zhang, W., Chang, M.-T., Lin, C.-T., Chang, K.-D., Yu, Y.-C., Wang, J.T.-W., Chang, C.-S., Li, L.-J., and Lin, T.-W., 2012. Synthesis of Large-Area MoS$_2$ Atomic Layers with Chemical Vapor Deposition. *Advanced Materials*, 24 (17), 2320–2325.

Li, B.L., Wang, J., Zou, H.L., Garaj, S., Lim, C.T., Xie, J., Li, N.B., and Leong, D.T., 2016. Low-Dimensional Transition Metal Dichalcogenide Nanostructures Based Sensors. Advanced Functional Materials, 26 (39), 7034–7056.

Li, J., Hu, H., Qin, F., Zhang, P., Zou, L., Wang, H., Zhang, K., and Lai, Y., 2017. Flower-like MoSe$_2$/C Composite with Expanded (0 0 2) Planes of Few-layer MoSe$_2$ as the Anode for High-Performance Sodium-Ion Batteries. *Chemistry – A European Journal*, 23 (56), 14004–14010.

Li, Y., Duerloo, K.A.N., Wauson, K., and Reed, E.J., 2016. Structural semiconductor-to-semimetal phase transition in two-dimensional materials induced by electrostatic gating. *Nature Communications*, 7, 10671.

Lin, J.D., Han, C., Wang, F., Wang, R., Xiang, D., Qin, S., Zhang, X.A., Wang, L., Zhang, H., Wee, A.T.S., and Chen, W., 2014. Electron-doping-enhanced trion formation in monolayer molybdenum disulfide functionalized with cesium carbonate. *ACS Nano*, 8 (5), 5323–5329.

Liu, J., Hu, Z., Zhang, Y., Li, H.Y., Gao, N., Tian, Z., Zhou, L., Zhang, B., Tang, J., Zhang, J., Yi, F., and Liu, H., 2020. MoS$_2$ Nanosheets Sensitized with Quantum Dots for Room-Temperature Gas Sensors. *Nano-Micro Letters*, 12 (1), 1–13.

Liu, Y., Dong, X., and Chen, P., 2012. Biological and chemical sensors based on graphene materials. Chemical Society Reviews, 41 (6), 2283–2307.

Llobet, E., 2013. Gas sensors using carbon nanomaterials: A review. *Sensors and Actuators, B: Chemical, 179*, 32–45.

Lui, C.H., Frenzel, A.J., Pilon, D. v., Lee, Y.H., Ling, X., Akselrod, G.M., Kong, J., and Gedik, N., 2014. Trion-induced negative photoconductivity in monolayer MoS$_2$. *Physical Review Letters*, 113 (16), 166801.

Lukowski, M.A., Daniel, A.S., Meng, F., Forticaux, A., Li, L., and Jin, S., 2013. Enhanced hydrogen evolution catalysis from chemically exfoliated metallic MoS$_2$ nanosheets. *Journal of the American Chemical Society*, 135 (28), 10274–10277.

Ma, F., Gao, G., Jiao, Y., Gu, Y., Bilic, A., Zhang, H., Chen, Z., and Du, A., 2016. Predicting a new phase (T″) of two-dimensional transition metal di-chalcogenides and strain-controlled topological phase transition. *Nanoscale*, 8 (9), 4969–4975.

Ma, P., Flöry, N., Salamin, Y., Baeuerle, B., Emboras, A., Josten, A., Taniguchi, T., Watanabe, K., Novotny, L., and Leuthold, J., 2018. Fast MoTe$_2$ Waveguide Photodetector with High Sensitivity at Telecommunication Wavelengths. *ACS Photonics*, 5 (5), 1846–1852.

Ma, Y., Kou, L., Li, X., Dai, Y., and Heine, T., 2016. Two-dimensional transition metal dichalcogenides with a hexagonal lattice: Room-temperature quantum spin Hall insulators. *Physical Review B*, 93 (3), 035442.

Mak, K.F., Lee, C., Hone, J., Shan, J., and Heinz, T.F., 2010. Atomically thin MoS_2: A new direct-gap semiconductor. *Physical Review Letters*, 105 (13), 136805.

Manzeli, S., Ovchinnikov, D., Pasquier, D., Yazyev, O. v., and Kis, A., 2017. 2D transition metal dichalcogenides. *Nature Reviews Materials*, 2 (8), 1–15.

Miao, N., Xu, B., Zhu, L., Zhou, J., and Sun, Z., 2018. 2D Intrinsic Ferromagnets from van der Waals Antiferromagnets. *Journal of the American Chemical Society*, 140 (7), 2417–2420.

Nakanishi, W., Minami, K., Shrestha, L.K., Ji, Q., Hill, J.P., and Ariga, K., 2014. Bioactive nanocarbon assemblies: Nanoarchitectonics and applications. *Nano Today, 9(3)*, 378–394.

Nie, S.M., Song, Z., Weng, H., and Fang, Z., 2015. Quantum spin Hall effect in two-dimensional transition-metal dichalcogenide haeckelites. *Physical Review B – Condensed Matter and Materials Physics*, 91 (23), 235434.

Noh, H.J., Jeong, J., Cho, E.J., Kim, K., Min, B.I., and Park, B.G., 2017. Experimental Realization of Type-II Dirac Fermions in a $PdTe_2$ Superconductor. *Physical Review Letters*, 119 (1), 016401.

Novoselov, K.S., Geim, A.K., Morozov, S. v., Jiang, D., Katsnelson, M.I., Grigorieva, I. v., Dubonos, S. v., and Firsov, A.A., 2005. Two-dimensional gas of massless Dirac fermions in graphene. *Nature*, 438 (7065), 197–200.

Oyedele, A.D., Yang, S., Liang, L., Puretzky, A.A., Wang, K., Zhang, J., Yu, P., Pudasaini, P.R., Ghosh, A.W., Liu, Z., Rouleau, C.M., Sumpter, B.G., Chisholm, M.F., Zhou, W., Rack, P.D., Geohegan, D.B., and Xiao, K., 2017. $PdSe_2$: Pentagonal Two-Dimensional Layers with High Air Stability for Electronics. *Journal of the American Chemical Society*, 139 (40), 14090–14097.

Pan, X.C., Chen, X., Liu, H., Feng, Y., Wei, Z., Zhou, Y., Chi, Z., Pi, L., Yen, F., Song, F., Wan, X., Yang, Z., Wang, B., Wang, G., and Zhang, Y., 2015. Pressure-driven dome-shaped superconductivity and electronic structural evolution in tungsten ditelluride. *Nature Communications*, 6, 7805.

Perkins, F.K., Friedman, A.L., Cobas, E., Campbell, P.M., Jernigan, G.G., and Jonker, B.T., 2013. Chemical vapor sensing with monolayer MoS_2. *Nano Letters*, 13 (2), 668–673.

Pezeshki, A., Shokouh, S.H.H., Nazari, T., Oh, K., and Im, S., 2016. Electric and Photovoltaic Behavior of a Few-Layer α-$MoTe_2$/MoS_2 Dichalcogenide Heterojunction. *Advanced Materials*, 28 (16), 3216–3222.

Pi, L., Li, L., Liu, K., Zhang, Q., Li, H., and Zhai, T., 2019. Recent Progress on 2D Noble-Transition-Metal Dichalcogenides. *Advanced Functional Materials*, 29 (51), 1904932.

Podzorov, V., Gershenson, M.E., Kloc, C., Zeis, R., and Bucher, E., 2004. High-mobility field-effect transistors based on transition metal dichalcogenides. *Applied Physics Letters*, 84 (17), 3301–3303.

Poorahong, S., Izquierdo, R., and Siaj, M., 2017. An efficient porous molybdenum diselenide catalyst for electrochemical hydrogen generation. *Journal of Materials Chemistry A*, 5 (39), 20993–21001.

Pospischil, A., Furchi, M.M., and Mueller, T., 2014. Solar-energy conversion and light emission in an atomic monolayer p-n diode. *Nature Nanotechnology*, 9 (4), 257–261.

Qi, Y., Naumov, P.G., Ali, M.N., Rajamathi, C.R., Schnelle, W., Barkalov, O., Hanfland, M., Wu, S.C., Shekhar, C., Sun, Y., Süß, V., Schmidt, M., Schwarz, U., Pippel, E., Werner, P., Hillebrand, R., Förster, T., Kampert, E., Parkin, S., Cava, R.J., Felser, C., Yan, B., and Medvedev, S.A., 2016. Superconductivity in Weyl semimetal candidate $MoTe_2$. *Nature Communications*, 7 (1), 11038.

Qian, X., Liu, J., Fu, L., and Li, J., 2014. Quantum spin hall effect in two – Dimensional transition metal dichalcogenides. *Science*, 346 (6215), 1344–1347.

Qin, D., Yan, P., Ding, G., Ge, X., Song, H., and Gao, G., 2018. Monolayer $PdSe_2$: A promising two-dimensional thermoelectric material. *Scientific Reports*, 8 (1), 1–8.

Radisavljevic, B., Radenovic, A., Brivio, J., Giacometti, V., and Kis, A., 2011. Single-layer MoS_2 transistors. Nature Nanotechnology, 6 (3), 147–150.

Ramakrishna Matte, H.S.S., Gomathi, A., Manna, A.K., Late, D.J., Datta, R., Pati, S.K., and Rao, C.N.R., 2010. MoS_2 and WS_2 analogues of graphene. *Angewandte Chemie – International Edition*, 49 (24), 4059–4062.

Rasmussen, F.A. and Thygesen, K.S., 2015. Computational 2D Materials Database: Electronic Structure of Transition-Metal Dichalcogenides and Oxides. *Journal of Physical Chemistry C*, 119 (23), 13169–13183.

Reshak, A. and Auluck, S., 2003. Calculated optical properties of $2H\text{-}MoS_2$ intercalated with lithium. *Physical Review B – Condensed Matter and Materials Physics*, 68 (12), 125101.

Ross, J.S., Klement, P., Jones, A.M., Ghimire, N.J., Yan, J., Mandrus, D.G., Taniguchi, T., Watanabe, K., Kitamura, K., Yao, W., Cobden, D.H., and Xu, X., 2014. Electrically tunable excitonic light-emitting diodes based on monolayer WSe2 p-n junctions. *Nature Nanotechnology*, 9 (4), 268–272.

Samnakay, R., Jiang, C., Rumyantsev, S.L., Shur, M.S., and Balandin, A.A., 2015. Selective chemical vapor sensing with few-layer MoS_2 thin-film transistors: Comparison with graphene devices. *Applied Physics Letters*, 106 (2), 023115.

Sasaki, T., Watanabe, M., Hashizume, H., Yamada, H., and Nakazawa, H., 1996. Macromolecule-like aspects for a colloidal suspension of an exfoliated titanate. Pairwise association of nanosheets and dynamic reassembling process initiated from it. Journal of the American Chemical Society, 118 (35), 8329–8335.

Sharma, M. and Singh, R., 2022. Tuning electronic properties of pentagonal PdSe2 monolayer by applying external strain. *Indian Journal of Physics*, 96 (4), 1037–1043.

Sharma, P., Lepcha, M., Chettri, B., Thapa, A., Chettri, P., and Sharma, B., 2021. First principle study of MoS_2 adsorbed transition metal for sensing urea and methanol. In: *Proceedings of 4th International Conference on 2021 Devices for Integrated Circuit, DevIC 2021.* 655–658.

Shen, J., He, Y., Wu, J., Gao, C., Keyshar, K., Zhang, X., Yang, Y., Ye, M., Vajtai, R., Lou, J., and Ajayan, P.M., 2015. Liquid Phase Exfoliation of Two-Dimensional Materials by Directly Probing and Matching Surface Tension Components. *Nano Letters*, 15 (8), 5449–5454.

Sie, E.J., Frenzel, A.J., Lee, Y.H., Kong, J., and Gedik, N., 2015. Intervalley biexcitons and many-body effects in monolayer MoS_2. *Physical Review B – Condensed Matter and Materials Physics*, 92 (12), 125417.

Sie, E.J., Lui, C.H., Lee, Y.H., Kong, J., and Gedik, N., 2016. Observation of Intervalley Biexcitonic Optical Stark Effect in Monolayer WS_2. *Nano Letters*, 16 (12), 7421–7426.

Singh, A., Moody, G., Tran, K., Scott, M.E., Overbeck, V., Berghäuser, G., Schaibley, J., Seifert, E.J., Pleskot, D., Gabor, N.M., Yan, J., Mandrus, D.G., Richter, M., Malic, E., Xu, X., and Li, X., 2016. Trion formation dynamics in monolayer transition metal dichalcogenides. *Physical Review B*, 93 (4), 041401.

Song, J.G., Park, J., Lee, W., Choi, T., Jung, H., Lee, C.W., Hwang, S.H., Myoung, J.M., Jung, J.H., Kim, S.H., Lansalot-Matras, C., and Kim, H., 2013. Layer-controlled, wafer-scale, and conformal synthesis of tungsten disulfide nanosheets using atomic layer deposition. *ACS Nano*, 7 (12), 11333–11340.

Song, X., Guo, Z., Zhang, Q., Zhou, P., Bao, W., and Zhang, D.W., 2017. Progress of Large-Scale Synthesis and Electronic Device Application of Two-Dimensional Transition Metal Dichalcogenides. *Small*, 13(35), 1700098.

Splendiani, A., Sun, L., Zhang, Y., Li, T., Kim, J., Chim, C.Y., Galli, G., and Wang, F., 2010. Emerging photoluminescence in monolayer MoS_2. Nano Letters, 10 (4), 1271–1275.

Štengl, V., Henych, J., Slušná, M., and Ecorchard, P., 2014. Ultrasound exfoliation of inorganic analogues of graphene. *Nanoscale Research Letters*, 9 (1), 1–14.

Tamang, S., Thapa, A., Chettri, K., Datta, B., and Biswas, J., 2022. Analysis of dipyridine dipyrrole based molecules for solar cell application using computational approach. *Journal of Computational Electronics*, 21 (1), 94–105.

Tan, C., Luo, Z., Chaturvedi, A., Cai, Y., Du, Y., Gong, Y., Huang, Y., Lai, Z., Zhang, X., Zheng, L., Qi, X., Goh, M.H., Wang, J., Han, S., Wu, X.J., Gu, L., Kloc, C., and Zhang, H., 2018. Preparation of High-Percentage 1T-Phase Transition Metal Dichalcogenide Nanodots for Electrochemical Hydrogen Evolution. *Advanced Materials*, 30 (9), 1705509.

Tan, L.K., Liu, B., Teng, J.H., Guo, S., Low, H.Y., and Loh, K.P., 2014. Atomic layer deposition of a MoS_2 film. Nanoscale, 6 (18), 10584–10588.

Tsen, A.W., Hunt, B., Kim, Y.D., Yuan, Z.J., Jia, S., Cava, R.J., Hone, J., Kim, P., Dean, C.R., and Pasupathy, A.N., 2016. Nature of the quantum metal in a two-dimensional crystalline superconductor. *Nature Physics*, 12 (3), 208–212.

Varghese, S.S., Varghese, S.H., Swaminathan, S., Singh, K.K., and Mittal, V., 2015. Two-dimensional materials for sensing: Graphene and beyond. *Electronics (Switzerland)*, 4 (3), 651–687.

Vishwanath, S., Liu, X., Rouvimov, S., Mende, P.C., Azcatl, A., McDonnell, S., Wallace, R.M., Feenstra, R.M., Furdyna, J.K., Jena, D., and Xing, H.G., 2015. Comprehensive structural and optical characterization of MBE grown $MoSe_2$ on graphite, CaF_2 and graphene. *2D Materials*, 2 (2), 024007.

Wan, S., Shao, Z., Zhang, H., Yang, Y., Shao, Z., Wan, N., and Sun, L., 2017. Graphene-based gas sensor. Kexue Tongbao/Chinese Science Bulletin, 62 (27), 3121–3133.

Wang, H., Lu, Z., Kong, D., Sun, J., Hymel, T.M., and Cui, Y., 2014. Electrochemical tuning of MoS_2 nanoparticles on three-dimensional substrate for efficient hydrogen evolution. *ACS Nano*, 8 (5), 4940–4947.

Wang, H., Zhang, C., Chan, W., Manolatou, C., Tiwari, S., and Rana, F., 2016. Radiative lifetimes of excitons and trions in monolayers of the metal dichalcogenide MoS2. *Physical Review B*, 93 (4), 045407.

Wang, J., Zhou, Q., Lu, Z., Gui, Y., and Zeng, W., 2019. Adsorption of H_2O molecule on TM (Au, Ag) doped-MoS_2 monolayer: A first-principles study. *Physica E: Low-Dimensional Systems and Nanostructures*, 113, 72–78.

Wang, T., Andrews, K., Bowman, A., Hong, T., Koehler, M., Yan, J., Mandrus, D., Zhou, Z., and Xu, Y.Q., 2018. High-Performance WSe_2 Phototransistors with 2D/2D Ohmic Contacts. *Nano Letters*, 18 (5), 2766–2771.

Wang, X., Gong, Y., Shi, G., Chow, W.L., Keyshar, K., Ye, G., Vajtai, R., Lou, J., Liu, Z., Ringe, E., Tay, B.K., and Ajayan, P.M., 2014. Chemical vapor deposition growth of crystalline monolayer $MoSe_2$. *ACS Nano*, 8 (5), 5125–5131.

Wang, X., Kang, K., Chen, S., Du, R., and Yang, E.H., 2017. Location-specific growth and transfer of arrayed MoS_2 monolayers with controllable size. *2D Materials*, 4 (2), 025093.

Wang, Y., Shang, X., Wang, X., Tong, J., and Xu, J., 2015. Density functional theory calculations of NO molecule adsorption on monolayer MoS_2 doped by Fe atom. *Modern Physics Letters B*, 29 (27), 1550160.

Wang, Z., Li, Q., Besenbacher, F., and Dong, M., 2016. Facile Synthesis of Single Crystal $PtSe_2$ Nanosheets for Nanoscale Electronics. *Advanced Materials*, 28 (46), 10224–10229.

Wilson, J.A. and Yoffe, A.D., 1969. The transition metal dichalcogenides discussion and interpretation of the observed optical, electrical and structural properties. *Advances in Physics*, 18 (73), 193–335.

Wu, Y.J., Shen, C., Tan, Q.H., Shi, J., Liu, X.F., Wu, Z.H., Zhang, J., Tan, P.H., and Zheng, H.Z., 2018. Valley Zeeman splitting of monolayer MoS_2 probed by low-field magnetic circular dichroism spectroscopy at room temperature. *Applied Physics Letters*, 112 (15), 153105.

Xia, F., Mueller, T., Lin, Y.M., Valdes-Garcia, A., and Avouris, P., 2009. Ultrafast graphene photodetector. *Nature Nanotechnology*, 4 (12), 839–843.

Xia, J., Huang, X., Liu, L.Z., Wang, M., Wang, L., Huang, B., Zhu, D.D., Li, J.J., Gu, C.Z., and Meng, X.M., 2014. CVD synthesis of large-area, highly crystalline $MoSe_2$ atomic layers on diverse substrates and application to photodetectors. *Nanoscale*, 6 (15), 8949–8955.

Xiao, S., Chen, J.H., Adam, S., Williams, E.D., and Fuhrer, M.S., 2010. Charged impurity scattering in bilayer graphene. Physical Review B – Condensed Matter and Materials Physics, 82 (4), 041406.

Xiong, L., Wang, K., Li, D., Luo, X., Weng, J., Liu, Z., and Zhang, H., 2020. Research progress on the preparations, characterizations and applications of large scale 2D transition metal dichalcogenides films. *FlatChem*.

Xu, G., Wang, J., Yan, B., and Qi, X.L., 2014. Topological superconductivity at the edge of transition-metal dichalcogenides. *Physical Review B – Condensed Matter and Materials Physics*, 90 (10), 100505.

Xu, H., Zhou, W., Zheng, X., Huang, J., Feng, X., Ye, L., Xu, G., and Lin, F., 2018. Control of the nucleation density of molybdenum disulfide in large-scale synthesis using chemical vapor deposition. *Materials*, 11 (6), 870.

Xu, K., Chen, P., Li, X., Wu, C., Guo, Y., Zhao, J., Wu, X., and Xie, Y., 2013. Ultrathin nanosheets of vanadium diselenide: A metallic two-dimensional material with ferro-magnetic charge-density-wave behavior. *Angewandte Chemie – International Edition*, 52 (40), 10671–10675.

Xu, L., Gui, Y., Li, W., Li, Q., and Chen, X., 2021. Gas-sensing properties of Ptn-doped WSe_2 to SF_6 decomposition products. *Journal of Industrial and Engineering Chemistry*, 97, 452–459.

Xu, W., Kozawa, D., Liu, Y., Sheng, Y., Wei, K., Koman, V.B., Wang, S., Wang, X., Jiang, T., Strano, M.S., and Warner, J.H., 2018. Determining the Optimized Interlayer Separation Distance in Vertical Stacked 2D WS_2:hBN:MoS_2 Heterostructures for Exciton Energy Transfer. *Small*, 14 (13), 1703727.

Yan, M., Huang, H., Zhang, K., Wang, E., Yao, W., Deng, K., Wan, G., Zhang, H., Arita, M., Yang, H., Sun, Z., Yao, H., Wu, Y., Fan, S., Duan, W., and Zhou, S., 2017. Lorentz-violating type-II Dirac fermions in transition metal dichalcogenide $PtTe_2$. *Nature Communications*, 8 (1), 257.

Yang, H., Kim, S.W., Chhowalla, M., and Lee, Y.H., 2017. Structural and quantum-state phase transition in van der Waals layered materials. *Nature Physics*, 13 (10), 931–937.

Ye, J.T., Zhang, Y.J., Akashi, R., Bahramy, M.S., Arita, R., and Iwasa, Y., 2012. Superconducting dome in a gate-tuned band insulator. *Science*, 338 (6111), 1193–1196.

Yin, Z., Li, H., Li, H., Jiang, L., Shi, Y., Sun, Y., Lu, G., Zhang, Q., Chen, X., and Zhang, H., 2012. Single-Layer MoS_2 Phototransistors. ACS Nano, 6 (1), 74–80.

Yoon, Y., Ganapathi, K., and Salahuddin, S., 2011. How good can monolayer MoS2 transistors be? *Nano Letters*, 11 (9), 3768–3773.

You, Y., Zhang, X.-X., Berkelbach, T.C., Hybertsen, M.S., Reichman, D.R., and Heinz, T.F., 2015. Observation of biexcitons in monolayer WSe_2. *Nature Physics*, 11 (6), 477–481.

Yu, H., Liao, M., Zhao, W., Liu, G., Zhou, X.J., Wei, Z., Xu, X., Liu, K., Hu, Z., Deng, K., Zhou, S., Shi, J.A., Gu, L., Shen, C., Zhang, T., Du, L., Xie, L., Zhu, J., Chen, W., Yang, R., Shi, D., and Zhang, G., 2017. Wafer-Scale Growth and Transfer of Highly-Oriented Monolayer MoS_2 Continuous Films. *ACS Nano*, 11 (12), 12001–12007.

Yuan, N.F.Q., Mak, K.F., and Law, K.T., 2014. Possible topological superconducting phases of MoS$_2$. *Physical Review Letters*, 113 (9), 097001.

Zeng, Z., Yin, Z., Huang, X., Li, H., He, Q., Lu, G., Boey, F., and Zhang, H., 2011. Single-layer semiconducting nanosheets: High-yield preparation and device fabrication. *Angewandte Chemie – International Edition*, 50 (47), 11289–11293.

Zettl, A., 2000. Extreme oxygen sensitivity of electronic properties of carbon nanotubes. *Science*, 287 (5459), 1801–1804.

Zhang, D.K., Kidd, D.W., and Varga, K., 2015. Excited Biexcitons in Transition Metal Dichalcogenides. *Nano Letters*, 15 (10), 7002–7005.

Zhang, Y., Duan, L.F., Zhang, Y., Wang, J., Geng, H., and Zhang, Q., 2014. Advances in Conceptual Electronic Nanodevices based on 0D and 1D Nanomaterials. *Nano-Micro Letters*, 6, 1–19

Zhang, Y., Ye, J., Matsuhashi, Y., and Iwasa, Y., 2012. Ambipolar MoS2 thin flake transistors [Supporting information]. *Nano Letters*, 12 (3), 1136–1140.

Zhang, Y.H., Chen, J.L., Yue, L.J., Zhang, H.L., and Li, F., 2017. Tuning CO sensing properties and magnetism of MoS$_2$ monolayer through anchoring transition metal dopants. *Computational and Theoretical Chemistry*, 1104, 12–17.

Zhao, Q., Guo, Y., Zhou, Y., Yao, Z., Ren, Z., Bai, J., and Xu, X., 2018. Band alignments and heterostructures of monolayer transition metal trichalcogenides MX$_3$ (M = Zr, Hf; X = S, Se) and dichalcogenides MX$_2$ (M = Tc, Re; X=S, Se) for solar applications. *Nanoscale*, 10 (7), 3547–3555.

Zhao, X.W., Yang, Z., Guo, J.T., Hu, G.C., Yue, W.W., Yuan, X.B., and Ren, J.F., 2020. Tuning electronic and optical properties of monolayer PdSe$_2$ by introducing defects: first-principles calculations. *Scientific Reports*, 10 (1), 1–8.

Zhao, Y., Qiao, J., Yu, P., Hu, Z., Lin, Z., Lau, S.P., Liu, Z., Ji, W., and Chai, Y., 2016. Extraordinarily Strong Interlayer Interaction in 2D Layered PtS$_2$. *Advanced Materials*, 28 (12), 2399–2407 .

Zhao, Y., Qiao, J., Yu, Z., Yu, P., Xu, K., Lau, S.P., Zhou, W., Liu, Z., Wang, X., Ji, W., and Chai, Y., 2017. High-Electron-Mobility and Air-Stable 2D Layered PtSe2 FETs. *Advanced Materials*, 29 (5), 1604230.

Zheng, W., Liu, X., Xie, J., Lu, G., and Zhang, J., 2021. Emerging van der Waals junctions based on TMDs materials for advanced gas sensors. *Coordination Chemistry Reviews*, 447, 214151.

Zhou, B.T., Yuan, N.F.Q., Jiang, H.L., and Law, K.T., 2016. Ising superconductivity and Majorana fermions in transition-metal dichalcogenides. *Physical Review B*, 93 (18), 180501.

Zhou, J., Liu, F., Lin, J., Huang, X., Xia, J., Zhang, B., Zeng, Q., Wang, H., Zhu, C., Niu, L., Wang, X., Fu, W., Yu, P., Chang, T.R., Hsu, C.H., Wu, D., Jeng, H.T., Huang, Y., Lin, H., Shen, Z., Yang, C., Lu, L., Suenaga, K., Zhou, W., Pantelides, S.T., Liu, G., and Liu, Z., 2017. Large-Area and High-Quality 2D Transition Metal Telluride. *Advanced Materials*, 29 (3), 1603471.

Zhou, X., Gan, L., Tian, W., Zhang, Q., Jin, S., Li, H., Bando, Y., Golberg, D., and Zhai, T., 2015. Ultrathin SnSe$_2$ Flakes Grown by Chemical Vapor Deposition for High-Performance Photodetectors. *Advanced Materials*, 27 (48), 8035–8041.

4 Conducting Polymer Nanocomposites for Electrochemical Supercapacitor

Bhanita Goswami and Debajyoti Mahanta

4.1 INTRODUCTION

In recent years, the demand for energy has surged exponentially. Energy, in its various forms, is considered a fuel for technological, economic, and social development. A huge amount of energy supply is required to meet the increasing demand for energy across the globe. At present, about 80% of this energy demand is met by fossil fuels, although their reserves are exhausting rapidly (Hossain and Hoque 2018). To circumvent this, there is a need for efficient, clean, and sustainable energy storage and conversion systems. Storage and conversion of energy is significant in the conservation and sustainable usage of natural energy sources (Tajik et al. 2020).

Currently, various research activities are going on, which are mainly focused on energy storage systems (namely, batteries and supercapacitors) and energy conversion systems (fuel cells, solar cells, etc.). Supercapacitors have drawn significant research interest among researchers owing to their advantages like improved power density, extensive cycling stability, fast charge/discharge cycle, and environmental friendliness (Dhibar and Malik 2020). An electrode material with good conductivity is a primary requirement in energy storage devices like supercapacitors for maximum power delivery. Conducting polymers (CPs) like polyaniline (PANI), polypyrrole (PPy), and polythiophenes (PThs) and their nanocomposites have been thoroughly investigated by researchers as supercapacitor electrode materials. CPs with their high specific capacitance values are desirable materials for supercapacitors. CPs and their composites show encouraging electrical conductivity and electroactivity.

4.2 SUPERCAPACITORS AS ENERGY STORAGE SYSTEMS

Supercapacitors, also known as ultracapacitors, are one of the popular electric energy storage devices providing high power output (~10 KW/kg) and low energy density (~5–10 Wh/kg). Supercapacitors mostly find applications where power outbursts is required, like satellite launch, hybrid electric vehicles, military vehicles, and powering remote areas where a large amount of electric energy is required in a very short

DOI: 10.1201/9781003323518-5

period of time (Conway 1999a, Donne 2013, Chen 2017). Supercapacitors bridge the gap between conventional dielectric capacitors (high power output and low energy storage) and batteries (high energy storage and low power output) (González et al. 2016). Supercapacitors are developed based on charging and discharging capacity at the electrode-electrolyte interface of high-surface-area materials, highly suited for rapid storage and release of energy. Because of the higher effective surface area, they provide more capacitance by a factor of 10,000. They can be regarded as functioning as rechargeable batteries but are not considered a replacement for the latter. Their advantages over electrochemical batteries include short charging time, longer cycle life, and high efficiency (Wang, Zhang, et al. 2012). Supercapacitors can deliver performance for millions of charging/discharging cycles owing to their charge storage mechanism, which does not involve irreversible chemical reactions (faradaic process) (Conway 1999b, Yan et al. 2014). Table 4.1 highlights the comparison among capacitors, supercapacitors, and batteries.

A supercapacitor cell comprises two electrodes separated by a dielectric electrolyte. Depending on their charge storage mechanism, supercapacitors are classified as (i) electrochemical double-layer capacitors (EDLCs); (ii) pseudocapacitors (PCs); and (iii) hybrid capacitors (HCs) (Mohd Abdah et al. 2020). The classification and working principle of different supercapacitors is shown in Figure 4.1 (Mohd Abdah et al. 2020). The energy storage mechanism of EDLCs is based on the electrostatic interaction of ions between the electrolyte and the electrode surface, which is essentially a non-faradaic process. In EDLCs, the charge is momentarily stored at the interface of the electrode and the electrolyte. The performance of EDLCs depends on the electrochemical activity and kinetic features of the electrode material. Hence, electrode materials with high specific surface area, high porosity, and pore size distribution between 0.5 and 2 nm are required (Simon and Gogotsi 2008). Carbon-based materials such as activated carbon, graphene, carbon aerogels, carbon nanotubes, etc., are considered to be ideal electrode materials for EDLCs as they possess distinct features such as high conductivity, controlled porous structures, large surface area, etc. (Salunkhe et al. 2015, Tang, Salunkhe, et al. 2015).

TABLE 4.1

A Table Comparing the Characteristics of Capacitors, Supercapacitors, and Batteries

Characteristics	Capacitor	Supercapacitor	Battery
Specific energy (Wh/kg)	<0.1	1–10	10–100
Specific power (W/kg)	>>10,000	500–10,000	<1000
Discharge time	10^{-6} to 10^{-3} seconds	Seconds to minutes	0.3–3 h
Charge time	10^{-6} to 10^{-3} seconds	Seconds to minutes	1–5 h
Coulombic efficiency (%)	About 100	85–98	70–85
Cycle life	Almost infinite	>500,000	About 1000

Source: Reproduced with permission from ref. (González et al. 2016). Copyright 2016 Elsevier.

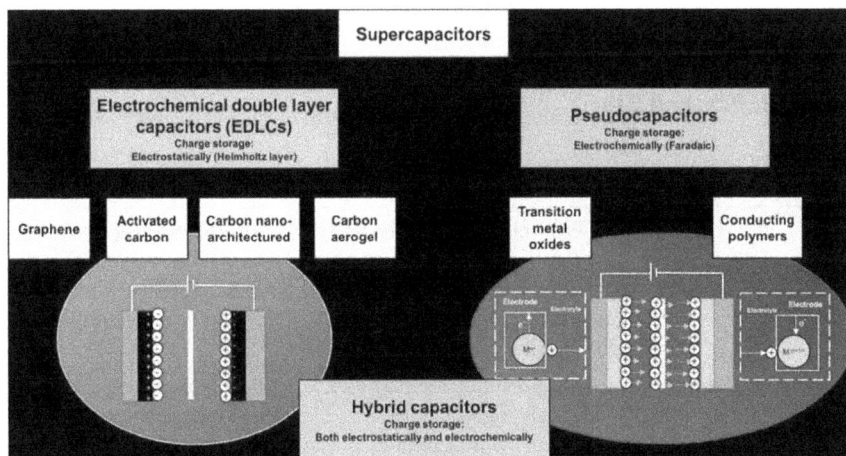

FIGURE 4.1 A figure showing the classification of supercapacitors. Reproduced with permission from ref. Mohd Abdah et al. (2020). Copyright 2020 Elsevier.

In case of PCs, the energy storage mechanism takes place via a quick and reversible faradaic redox reaction between the electrode surface and electrolyte ions. Compared to EDLCs, PCs can store more capacitance per gram material. However, the kinetics of pseudocapacitors is slower in comparison to EDLCs. This is because in pseudocapacitors the energy storage process occurs both on the surface and in bulk of the electrode material, while in EDLCs the charge/discharge process occurs only at the surface of the electrode materials. Metal oxides, especially transition metal oxides and conducting polymers, are mostly suited for pseudocapacitor electrode materials (Snook et al. 2011). Hybrid capacitors are fabricated by combining the advantages of both EDLCs and pseudocapacitors (Mohd Abdah et al. 2020). Hybrid capacitors use both faradaic and non-faradaic processes for charge storage, leading to high specific capacitance and energy density, moderate power density, high cell voltage, and higher cyclic stability (Ehsani et al. 2019).

Conducting polymers have been regarded as more promising pseudocapacitive electrode materials compared to metal oxides or hydroxides due to their unique characteristics like large theoretical specific capacitance, environmental stability, good electrical conductivity, low cost, and easy synthesis. Although CPs has unique properties, they exhibit certain limitations when used as active electrode materials alone. One of the major disadvantages of these conducting polymers for practical application is their poor cyclic stability, caused by continuous swelling/shrinking of the polymer chains during the charge/discharge process (Meng et al. 2017). To overcome such limitations and improve the electrochemical performance and cyclic stabilities of the CPs, researchers have attempted synthesizing binary and even ternary composites of CPs in combination with other active materials (like carbon materials and metal oxides). Fabrication of such CP-based composites is expected to improve their performance as supercapacitor electrode materials via synergistic

TABLE 4.2
A Comparison Table Summarizing the Advantages and Disadvantages of CPs like PANI, PPy, and PTh

Conducting Polymer	Advantages	Disadvantages
PANI	Flexibility, high specific capacitance range, easy fabrication, easy doping/dedoping, high doping level, high theoretical specific capacitance, controllable conductivity	Specific capacitance depends on synthesis conditions, poor cyclic stability, works with proton-type electrolytes
PPy	Flexibility, easy fabrication, high cyclic stability, relatively high specific capacitance, works with neutral electrolytes	Difficulty in doping/dedoping, relatively low specific capacitance, used only as cathode materials
PTh	Flexibility, easy synthesis, favorable environmental stability, and cyclic stability	Poor specific capacitance and poor conductivity

Source: Reproduced with permission from ref. (Meng et al. 2017). Copyright 2017 Elsevier.

effects. Table 4.2 summarizes the advantages and disadvantages of CPs, namely, PANI, PPy, and PTh.

4.3 ELECTROCHEMICAL PROPERTIES OF CONDUCTING POLYMERS

Conducting polymers belong to the class of organic polymers with metallic or semi-conducting features. They are mainly composed of organic monomers with conjugating double bonds. Unlike conventional polymers, they exhibit features like electrical conductivity, higher electron affinity, redox behavior, and optical and magnetic properties (Aydemir et al. 2016). However, CPs retains few characteristics of conventional polymers like facile synthesis, corrosion resistance, and low cost. The advantage of CPs over organic polymers is their modifiable chemical structures which also change their conductivity. Features of CPs like good electrical conductivity, mechanical flexibility, high surface area, low energy optical transition, and high electron affinity as well as lower ionization potential are due to their charge mobility along with p-electron backbones. The physical properties of CPs are governed by parameters like degree of crystallinity, conjugation length, and intra- and interchain interactions (Tajik et al. 2020). PANI, PPy, and PTh are some of the most explored conducting polymers for application in energy storage applications (Hossain and Hoque 2018).

The conduction mechanism in conducting polymers is quite complicated because of their conductivity range, which is approximately 15 orders of magnitude. Many

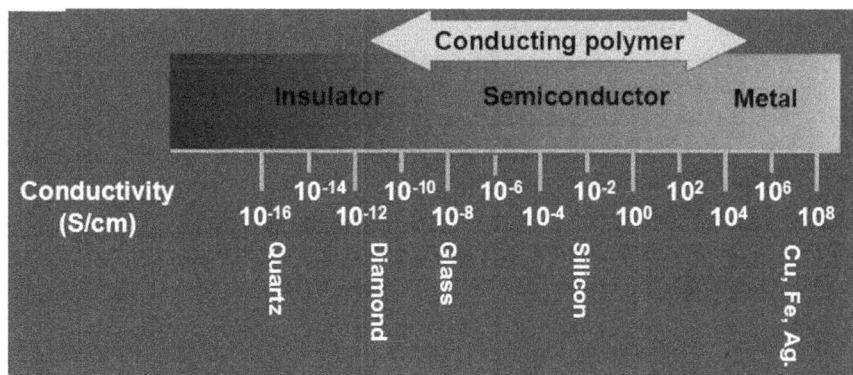

FIGURE 4.2 A scale bar showing the electrical conductivity range of conducting polymers. Reproduced with permission from ref. Wang et al. (2021). Copyright 2021 Elsevier.

of them show diverse mechanisms in distinct order. CPs can be electrochemically doped, creating a p-doped state via oxidation and an n-doped state via reduction. Redox reactions in CPs may affect their conductivity. Features like morphology, thickness, and conductivity of CPs can be regulated via electrochemical preparation techniques. Electrical conductivity in CPs is affected by parameters like polaron length, conjugation length, total chain length, and charge transfer to nearby molecules or charged species. Such charge transfer mechanisms are based on various models. These models are based on intersoliton hopping, intrachain hopping of the bipolarons, hopping between localized states assisted by lattice vibrations, varied range hopping in three sizes, and charged energy restricting the tunneling between the conducting species. Tailoring the features of CPs by functionalization with distinctive materials enhances the mechanical and electrical properties of the polymer chain (Tajik et al. 2020). CPs exhibit a broad range of conductivities, ranging from semiconductor (10^{-11} to 10^{-3} S/cm) to metal (10^{-1} to 10^6 S/cm) behavior, as shown in Figure 4.2 (Wang et al. 2021).

4.4 NANOCOMPOSITES OF CONDUCTING POLYMERS

Nanocomposites are hybrid materials formed by the combination of two or more materials, with one of their constituents being nanosized (dimensions ranging between 1 and 100 nm). Their physical and chemical properties remain separate and distinct on a macroscopic level. Conducting polymer nanocomposites comprise a conducting polymer backbone as matrix and other nanomaterials as inorganic filler. Nanocomposites exhibit synergistic properties with high electrical conductivity and flexibility due to the presence of the polymers and enhanced stability and dielectric properties owing to the presence of the filler materials. A small amount of filler may enhance the property of the nanocomposite to a great extent. The nature and properties of the polymeric backbone and the filler used in the nanocomposites have immense effects on their processability, electrical conductivity, tensile strength, and ionic conductivity, as well as thermal, chemical, and mechanical stability. Among the conducting polymers, PANI, PPy, and PTh are more extensively used for energy storage applications. Fillers or nanofillers are used in combination with conducting polymers for improving the properties of the polymers for their application in energy storage devices. So far, scientists have reported the use of fillers like metals, metal oxides, carbon, carbon nanotubes (CNTs), graphene, titanium nanotubes, ferromagnetic materials, layered silicates, and dendrimers (Hossain and Hoque 2018).

4.5 CONDUCTING POLYMERS NANOCOMPOSITES IN SUPERCAPACITORS

In this chapter, we will highlight the recent progress in the synthesis of various conducting polymer-based nanocomposites for application as supercapacitor electrode materials. To date, researchers have reported a wide range of conducting polymer nanocomposites as supercapacitor electrode materials. Here, we will basically focus on supercapacitor electrodes based on nanocomposites of CPs like PANI, PPy, and PTh.

The performance of CP-based supercapacitor electrode nanomaterials is crucially influenced by the nanostructures of CPs. There are several polymerization techniques for synthesizing CP-based supercapacitor electrode nanomaterials. These techniques include electrochemical polymerization, in situ polymerization, interfacial polymerization, emulsion polymerization, dilute polymerization, etc. Among these, electrochemical polymerization, in situ polymerization, and interfacial polymerization are commonly adopted methods (Meng et al. 2017).

4.5.1 PANI NANOCOMPOSITES

PANI can be synthesized via chemical or electrochemical polymerization of aniline monomer. It possesses many advantages such as easy synthesis, environmental stability, and easy acid/base doping/dedoping chemistry (Meng et al. 2017). It is one of the most promising pseudocapacitor electrode materials whose nanostructured morphology has a crucial influence on its electrochemical properties. Here, we will discuss how nanocomposites of PANI with other materials (like carbon materials, metal oxides, etc.) have shown promising applications in supercapacitors.

Many investigations have demonstrated that CP/carbon-based nanocomposites exhibited better electrochemical performance. A novel PANI/GN (graphene) nanocomposite was prepared by Gómez et al. using the chemical precipitation method, which achieved a specific capacitance of 300–500 F/g at a current density of 0.1 A/g (Gómez et al. 2011, Wang et al. 2021). In another work, Cong et al. fabricated a flexible PANI/GN nanocomposite via the electropolymerization of PANI nanorods on a free-standing GN (graphene) paper. This paper electrode was lightweight with high electrical conductivity and good flexibility. While using this paper directly as a working electrode, it showed an excellent specific capacitance value of 763 F/g and good cyclic stability (82% retention of initial capacitance after 1000 cycles) (Cong et al. 2013).

GO (graphene oxide) is a derivative of GN (graphene) and various investigations on PANI/GO nanocomposites have been carried out for application as supercapacitor electrode materials. Wang et al. reported the preparation of fibrillar PANI doped with GO sheet via a soft chemical route. The as-prepared nanocomposite exhibited high conductivity of 10 S/cm at 22 °C and a specific capacitance of 531 F/g at a current density of 0.2 A/g, while pure PANI exhibited a specific capacitance of 216 F/g, which reflects that the addition of GO sheets has significantly enhanced the performance of PANI (Wang et al. 2009). Free-standing 3D PANI/rGO nanocomposite foam was fabricated by Sun et al. via template-directed in situ polymerization. The composite with a specific capacitance of 701 F/g at current density 1 A/g displayed an enhanced cyclic stability of 92% retention of initial capacitance after 1000 cycles (Sun et al. 2015).

CNTs (carbon nanotubes), with their unique electrical properties and nano size, are used in combination with CPs, forming nanocomposites for deriving high-performance electrode material. A new derivative of PANI, poly(2-methylthioaniline)-coated MWCNTs or,(PMTA@CNT), and its composite with graphene (PMTA@CNT/RGO) were used as supercapacitor by Jena et al. The PMTA@CNT/RGO and PMTA@CNT electrodes showed specific capacitances of 616 F/g and 522 F/g, respectively, in 6 M KOH at a current density of 1 A/g. The PMTA-based composite electrodes showed very good electrochemical function as compared to similar PANI-based composite electrodes. This enhancement of the capacitive behavior of PMTA-based composite could be attributed to the presence of an acceptable electron-donating group (SCH_3) in PMTA. This SCH_3 group made the PMTA molecule electronically rich via resonance effect facilitating the transfer of charges between the pi-electrons of the quinoid rings in PMTA and the benzenoid rings of CNTs or graphene as shown in Figure 4.3 (Jena et al. 2017). Imani and Farzi prepared PANI/multiwalled CNT (MWCNT) nanocomposite with a tubular structure, which exhibited a higher specific capacitance of 552.11 F/g at a current density of 4 mA/cm^2 compared to pure PANI (411.52 F/g) (Imani and Farzi 2015).

Carbon nanofiber is another useful carbon material. Chau et al. fabricated 3D free-standing electrodes using PANI and porous carbon nanofibers. This hybrid electrode showed a specific capacitance of 366 F/g at 100 mV/s (Tran et al. 2015). Nanocomposites of PANI/carbon spheres and PANI/carbon particles as electrode materials have also been investigated by researchers. For example, Shen et al. reported the performance of nano-PANI/hollow carbon sphere composite with a specific capacitance value of 435 F/g at current density 0.5 A/g and 60% cyclic stability after 2000 cycles (Shen et al. 2015). Free-standing composite film of PANI/acid-treated carbon particles as synthesized by Khosrozadeh et al., although showing a low specific capacitance of 272.6 F/g at 0.63 A/g current density, displayed unique advantages like suitable flexibility and thickness along with stable cyclic life (Khosrozadeh et al. 2015).

Recently, research attention has been directed toward the investigation of metal oxides and their nanocomposites with CPs as supercapacitor materials. Among the metal oxides, MnO_2 has gained special attention as it offers higher surface area, favorable cyclic stability, much lower cost, enhanced charge storage capacity, and eco-friendliness (Meng et al. 2017). As investigated by Zheng et al., PANI/intercalated layered MnO_2 was prepared using n-octadecyltrimethylammonium-intercalated MnO_2 via exchange reaction of PANI in N-methyl-2-pyrrolidone. This composite showed excellent stability with 94% retention of initial capacitance value after

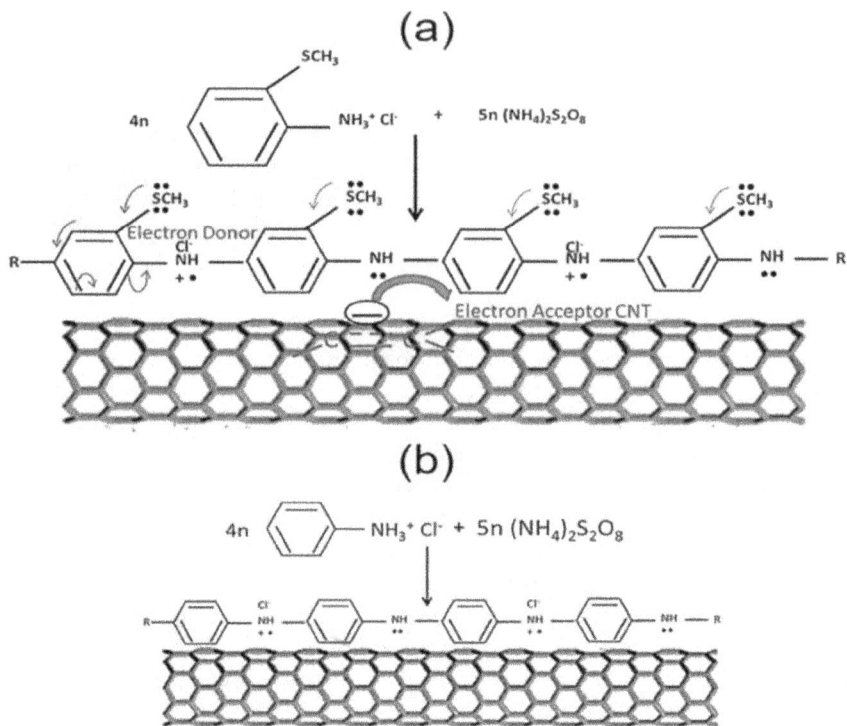

FIGURE 4.3 Schematic representation showing electron conjugation process of the SCH_3 group of (a) PMTA vs (b) PANI followed by donor-acceptor interaction (probable conduction mechanism) (R represents the extended polymer chain). Reproduced with permission from ref. Jena et al. (2017). Copyright 2017 Elsevier.

1000 cycles. Its specific capacitance value stands at 330 F/g at 1 A/g current density (Zhang et al. 2007). Other metal oxide-PANI-based nanocomposites have also gained attention as supercapacitor materials. In this regard, we mention the investigation done by Ates et al. They synthesized films of PANI/CuO, PPy/CuO, and PEDOT/CuO electrochemically on glassy carbon electrodes and studied their electrochemical performance in sulfuric acid solution. The results showed that PANI/CuO electrodes performed better as active electrode material than the other two. The specific capacitance values of PANI/CuO, PPy/CuO, and PEDOT/CuO were found to be 286.35 F/g at 20 mV/s, 198.89 F/g at 5 mV/s, and 20.78 F/g at 5 mV/s, respectively (Ates et al. 2015).

Apart from carbon materials and metal oxides, nanocomposites of PANI with other materials like metal sulfides, metal hydroxides, organic components, and so on were also investigated as supercapacitor materials. For example, cabbage-like PANI/hydroquinone composite and 1D fiber-like PANI/3D flower-like $Ni(OH)_2$ showed good specific capacitance and stable cycle life (Chen, Fan, et al. 2015, Zhang et al. 2015). Hierarchical core-sheath $PANI/MoS_2$ nanocomposite was synthesized by Lei et al. using a hydrothermal process. High specific capacitance of 450 F/g was obtained with 80% cycle stability after 2000 charge/discharge cycles (Lei et al. 2015). As investigated by Zhu et al., PANI/MoS$_2$ architectures were produced from PANI nanoneedle arrays and MoS$_2$ nanosheets. Specific capacitance of 669 F/g was achieved at a potential window +0.6 V to -0.6 V at 1 A/g current density. It also exhibited 91% cyclic stability after 4000 cycles at a current density of 10 A/g (Zhu et al. 2015).

It is interesting to note that many PANI-based ternary composites were also synthesized for investigating their supercapacitive performance. CPs/metal oxides/carbon ternary composites were synthesized. Chen et al. developed a hybrid nanostructured NiMoO$_4$/PANI/CC (carbon cloth) electrode which showed outstanding specific capacitance of 1340 F/g at a current density of 1 mA/cm^2 and 96.7% cyclic stability after 2000 cycles (Chen, Liu, et al. 2015). In another study done by Lin et al., a 3D porous graphene/PANI/Co$_3$O$_4$ ternary hybrid aerogel demonstrated a very high specific capacitance of 1247 F/g at a current density of 1 A/g with almost 100% cyclic stability even after 3500 cycles (Lin et al. 2016). Zhu et al. reported the synthesis of a CuO/PANI/rGO ternary composite which displayed specific capacitance as high as 634.4 F/g with energy density 126.8 Wh/kg and power density 114.2 kW/kg at a current density of 1 A/g. The composite demonstrated extremely high cyclic stability at 97.4% after 10000 cycles (Zhu et al. 2016). A PANI/MoO$_3$/GN nanoplates (GNPs) composite was prepared by Das et al. adopting a simple, inexpensive, and novel in situ polymerization of aniline in the presence of MoO$_3$ and GNP as shown in Figure 4.4.

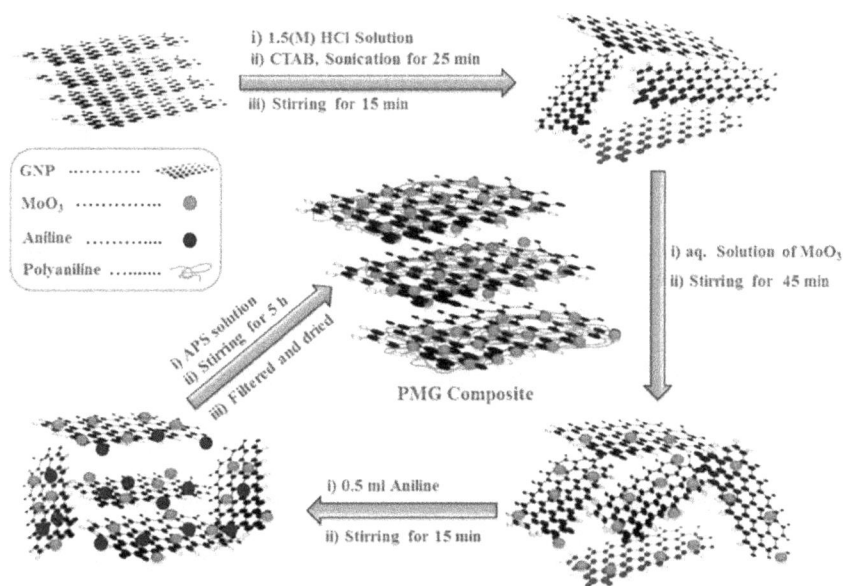

FIGURE 4.4 A schematic illustration showing the preparation steps of PANI/MoO$_3$/GNP ternary composite. Reproduced with permission from ref. Das et al. (2015). Copyright 2015 Elsevier.

Specific capacitance of 734 F/g and 92.4% cyclic stability after 1000 cycles at current density 1 A/g was achieved using this nanocomposite as supercapacitor electrode material (Das et al. 2015). Other examples include graphene/ZrO_2/PANI, rGO/MoO_3/PANI, and graphene/$CoFe_2O_4$/PANI composites with delivering specific capacitances of 1359.99 F/g at 1 mV/s, 553 F/g at 1 mV/s, and 1133.3 F/g at 1 mV/s, respectively (Ehsani et al. 2019).

Some scientists also tried synthesizing ternary nanocomposites by introducing nanoparticles of Au, Ag, and other metals into CP/carbon-based systems. In this case, we can cite the work done by Dhibar et al., where they synthesized PANI/GN/Ag nanocomposite. The nanocomposite showed a high specific capacitance of 591 F/g at a scan rate of 5 mV/s. It also exhibited an excellent energy density of 20.24 Wh/kg at a current density of 0.5 A/g and superior power density of 749.30 W/kg at a current density of 3 A/g. It also retained 96% of its initial capacitance value after 1500 charge/discharge cycles (Dhibar and Das 2015).

4.5.2 PPy Nanocomposites

PPy is another kind of conducting polymer possessing advantages like easy synthesis, high cyclic stability, and relatively high capacitance (Meng et al. 2017). PPy possesses better electrical and electrochemical properties compared to PANI. But it has low conductive and thermal stability. The poor cyclic stability of PPy limits its application. This prompted many researchers to focus on PPy/carbon nanocomposites and other hybrid composites of PPy for supercapacitor applications. Such nanocomposites of PPy exhibited better electrochemical and mechanical properties.

Ji et al. reported the use of a quaternary electrolyte to synthesize conducting polymer-carbon composites via the electrodeposition method to fabricate device-ready electrodes for energy storage. The quaternary electrolyte formulation uses lithium perchlorate (20 mM) as a supporting electrolyte and dopant, with sodium dodecyl-benzenesulfonate at a low concentration of 1.43 mM as surfactant, together with carbon nanomaterials and pyrrole monomers. Polypyrrole (PPy) composites with carbon black, carbon nanotubes (CNTs), and electrochemically exfoliated graphene (EEG) were prepared using this quaternary electrolyte. The specific capacitance of the as-fabricated optimized PPy/EEG composite electrodes was found to be 348.8 F/g at 0.5 mA/cm^2, with a high rate capability of 190.7 F/g at 71 A/g. Supercapacitor devices (both aqueous and solid-state) fabricated with PPy/EEG composite electrodes exhibited ideal capacitive behavior with a high rate capability of up to 500 mV/s. It also showed a high cyclic stability with 94.3% capacity retention after 5000 cycles (Ji et al. 2020).

PPy-based electrode materials are widely synthesized as composites of carbon materials or metal oxides. PPy/MnO_2 nanowire composite was prepared by Shayeh et al. via an electrochemical method. This composite showed a specific capacitance of 203 F/g, while pure PPy showed 109 F/g (Shayeh et al. 2016). Incorporating CNTs as additives or templates to synthesize PPy-based electrode nanocomposites significantly improves the cyclic stability due to the synergistic effect (Meng et al. 2017). A flexible PPy/CNT-based supercapacitor electrode was synthesized by Chen et al. An all-solid-state supercapacitor assembly using this electrode displayed outstanding flexibility and high cyclic stability by retaining 95% initial capacitance value after 10,000 cycles (Chen, Du, et al. 2015). Warren et al. also reported excellent cyclic stability of a hybrid supercapacitor electrode material based on PPy/CNT nanocomposite with 92% capacitance retention over 3000 cycles (Warren et al. 2015).

Graphene (GN) and its derivatives, like GO (graphene oxide) and rGO (reduced graphene oxide), are also considered ideal additives for PPy-based electrodes (Meng et al. 2017). As synthesized by Wu et al., a 3D core-shell GO/PPy nanomaterial exhibited excellent specific capacitance of 370 F/g at a current density of 0.5 A/g with an outstanding cyclic stability of 91.2% capacitance retention after 4000 cycles. The synthesis process involved uniform coating of PPy nanospheres on GO sheets via an in situ surface-initiated polymerization technique (Wu et al. 2015). Many other carbon-based nanocomposites of PPy (such as PPy/carbon cloth, PPy/carbon nanofibers, PPy/graphite sheets, PPy/active carbon, etc.) were also reported as supercapacitor materials. These carbon materials improve the conductivity and cyclic stability of PPy via a synergistic effect (Meng et al. 2017).

Nanocomposites with two electro-active materials like conducting polymers and transition metal oxides exhibit improved electrical, electrochemical, and mechanical properties. Nanocomposites of PPy with transition metal oxides have been extensively investigated for energy storage applications. Here, we can cite the development of a 3D PPy/CoO hybrid nanowire array on nickel foam by Zhou et al. This hybrid array showed an outstanding specific capacitance of 2223 F/g at a current density of 1 mA/cm^2 with a good rate capability and high cyclic stability of 99.8% at a scan rate of 20 mA/cm^2 after 2000 cycles. This hybrid nanocomposite when used as a positive electrode in an aqueous asymmetric supercapacitor device displayed energy density as high as 43.5 Wh/kg and power density of 5500 W/kg at 11.8 Wh/kg. This asymmetric supercapacitor also had an outstanding cycle stability of 20,000 times. Such high specific capacitance value was attributed to the synergistic effect between conductive PPy and CoO nanowires (Zhou et al. 2013). A facile synthesis method was adopted by Ji et al. to fabricate PPy nanotube/MnO$_2$ nanocomposite. This nanocomposite as electrode material showed considerably good specific capacitance (403 F/g at 1 A/g current density) and good cyclic stability (88.6% initial value retention after 800 cycles) (Ji et al. 2015). A unique on-chip supercapacitor with core-shell CuO/PPy nanosheets was fabricated by Qian et al. This supercapacitor electrode was fabricated via direct electrochemical deposition on interdigited electrode. This device showed excellent specific capacitance of 1275.5 F/cm^3 and energy density of 28.35 mWh/cm^3 with an excellent cyclic stability of almost 100% at 2.5 A/cm^3 after 3000 cycles (Qian et al. 2015). In another work, PPy/MoS$_2$ nanocomposite was prepared by Tang et al. via in situ oxide polymerization of PPy monomers on single-layered MoS$_2$ in the presence of ammonium persulfate solution (($NH_4)_2S_2O_8$). The nanocomposite material showed a specific capacitance of 695 F/g at a current density of 0.5 A/g with an excellent cyclic stability of 85% initial value retention after 4000 cycles (Tang, Wang, et al. 2015).

Thus, PPy/metal oxide nanocomposite-based supercapacitor electrode materials showed high electrochemical performance due to the synergic effect of the two components. They exhibited high specific capacitance, high energy density and power density, and outstanding rate capability and cyclic stability.

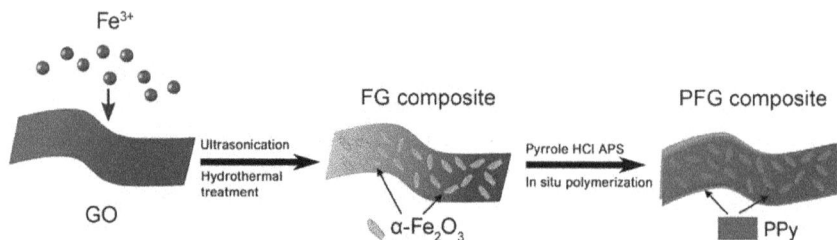

FIGURE 4.5 A schematic illustration showing the preparation of ternary $PPy/Fe_2O_3/rGO$ nanocomposite. Reproduced with permission from ref. Moyseowicz et al. (2017). Copyright 2017 Elsevier.

PPy-based ternary nanocomposites as supercapacitor electrode materials have also drawn a lot of interest from researchers as they are expected to improve the electrochemical performance. For example, a 3D $CNT/PPy/MnO_2$ composite electrode was prepared by Zhou et al., which showed a specific capacitance of 529.3 F/g at a current density of 0.1 A/g and cyclic stability of 98.5% at a current density of 5 A/g after 1000 cycles (Zhou et al. 2015). A flexible $PPy/GO/MnO_x$ supercapacitor was fabricated by Ng et al., which displayed high cyclic stability of 96.58% retention of initial specific capacitance after 1000 cycles (Ng et al. 2015). Besides these, other ternary systems have also been reported, such as $Ni/PPy/MnO_2$ and Ag/PPy/GN nanocomposites exhibiting excellent electrochemical performance (Meng et al. 2017). Asen et al. utilized the electrochemical deposition method to fabricate a ternary nanocomposite of $PPy/rGO/Cu_2O$-$Cu(OH)_2$ on Ni foam as a supercapacitor electrode material. The nanocomposite exhibited excellent specific capacitance of 997 F/g at 10 A/g current density with energy density 20 Wh/kg at a power density of 8000 W/kg, and power density as high as 19,998.5 W/kg at an energy density of 5.8 Wh/kg. The nanocomposite also exhibited 90% cyclic stability after 2000 cycles (Asen and Shahrokhian 2017).

Graphene/SnO_2/PPy composite exhibited a specific capacitance of 616 F/g at 1 mV/s as reported by Wang et al. This ternary electrode material also exhibited specific power density of 9973.26 W/kg and specific energy density of 19.4 Wh/kg (Wang, Hao, et al. 2012). In another study by Moyseowicz et al. a ternary composite of $PPy/Fe_2O_3/rGO$ was prepared via a two-step method. At first, the binary composite was fabricated using hydrothermal treatment of GO in the presence of iron precursor, which was followed by oxidative polymerization of pyrrole on the binary composite. The synthesis procedure of this ternary nanocomposite has been represented in Figure 4.5. This composite displayed a specific capacitance of 140 F/g with excellent capacitance retention of 93% after 5000 consecutive cycles (Moyseowicz et al. 2017).

4.5.3 PTh Nanocomposites

PTh, along with its derivatives or composites, has attracted considerable attention for being used as a supercapacitor electrode material because of its high environmental stability, high electrical conductivity, and long wavelength absorption (Meng et al. 2017).

Ambade et al. reported the preparation of PTh/TiO_2 flexible and wire-shaped all-solid-state symmetric supercapacitor. Here, PTh was electrochemically deposited onto TiO_2 nanowires. This supercapacitor electrode showed outstanding specific capacitance and cyclic stability of 1357.31 mF/g and 97% initial capacitance retention after 3000 cycles, respectively (Ambade et al. 2016). Another study by Nejati showed that deposition of an ultrathin PTh film on an active carbon electrode increased its specific capacitance by 50%. It also exhibited retention of 90% initial capacitance after 5000 cycles (Nejati et al. 2014). The electrochemical performance of PTh depends on various factors like synthesis method, presence of substrate, and PTh morphology, etc. Although the electrochemical performance of PTh has improved a lot, it still shows inferior electrochemical performance compared to PANI and PPy. It has certain limitations like lower specific capacitance and faster loss of power density (Meng et al. 2017).

A lot of research work has been done and reported on nanocomposites of PTh and carbon materials exhibiting considerably good electrochemical performance. For instance, Fu et al. reported the synthesis of PTh/MWCNT composite via electrochemical polymerization in ionic liquid solution. The composite electrode exhibited a specific capacitance of 110 F/g at a scan rate of 60 mV/s, with 90% retention of initial capacitance after 1000 consecutive cycles (Fu et al. 2012). Wang et al. in their work tried synthesizing a series of electrode materials via the electrochemical deposition of CPs on monolithic coral-like porous carbon (MC). They reported that the specific capacitance of the PTh/MC composite electrode was 720 F/g at a current density of 0.5 A/g with 79.4% cyclic stability at 1 A/g current density after 1000 cycles (Wang et al. 2013). PTh/carbon nanomaterials show superior electrochemical performance and cyclic stability compared to pristine PTh. The properties of these PTh-carbon composites depend on factors like deposition amount of PTh, synthesis method, and type of carbon material (Meng et al. 2017).

PTh and PTh-based nanocomposites showed poorer electrochemical performance compared to PANI and PPy. Recently, few works on PTh/metal oxide composites have been reported. Till now, not much work has been reported on PTh-based ternary nanocomposite as a supercapacitor electrode material (Meng et al. 2017).

4.5.4 OTHER CP-BASED NANOCOMPOSITES

In an effort to develop a high-performance supercapacitor based on nanostructured poly(3,4-ethylene-dioxythiophene) (PEDOT), Wang et al. reported a simple, rapid, and flexible method for fabricating nanostructured PEDOT paper. The composite material exhibited a larger surface area of 137 m^2/g with a lower sheet resistance of 1.4 ohm/cm and high active mass loading (7.3 mg/cm^2). These symmetric flexible supercapacitors based on PEDOT paper exhibited higher specific capacitance of 90 F/g, 920 mF/cm^2, and 54 F/cm^3 in 0.1 M H_2SO_4. The cyclic stability of the supercapacitors was very good, showing 93% capacity retention after 15,000 cycles at 30 mA/cm^2. These supercapacitors also retained their electrochemical function even at different bending angles, which demonstrated their flexibility (Li et al. 2008).

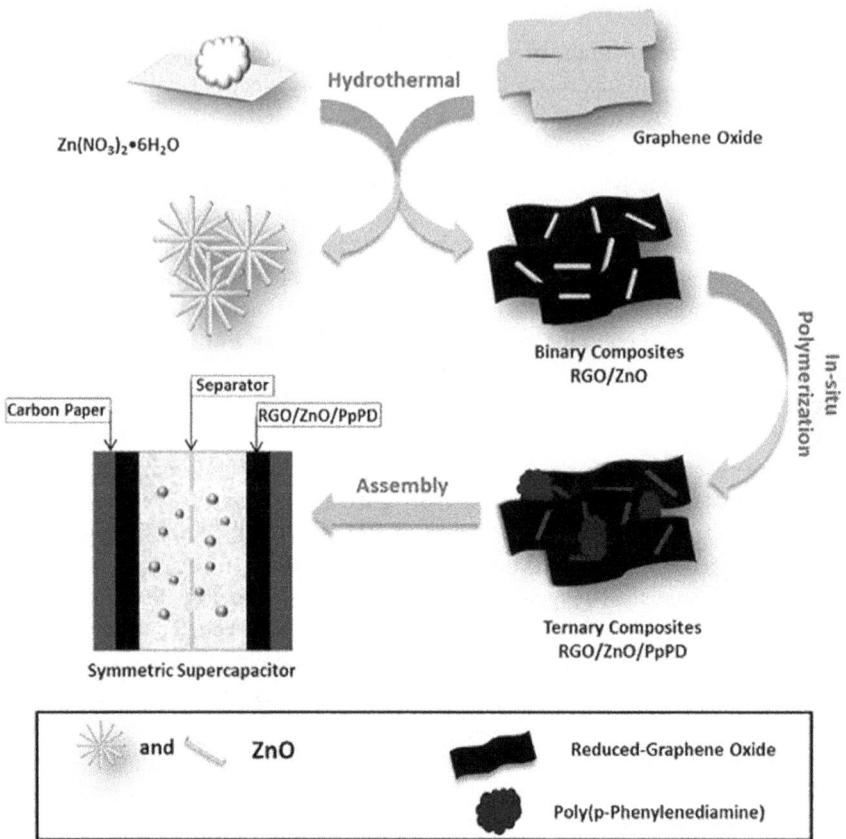

FIGURE 4.6 Schematic illustration representing synthesis steps of ternary composite rGO/ZnO/PpPD. Reproduced with permission from ref. Li et al. (2017). Copyright 2017 Elsevier.

In another study, a ternary hybrid electrode of rGO/ZnO/poly(p-phenylenediamine) was fabricated by Li et al. A two-step preparation approach was adopted. Initially, the reduction of GO and synthesis of ZnO nanorods was done simultaneously via a hydrothermal process, which was followed by in situ polymerization of p-phenylenediamine. The preparation steps have been demonstrated in Figure 4.6. The presence of rGO and poly(p-phenylenediamine) has suppressed the volume expansion of ZnO throughout the charge/discharge cycles. The as-prepared composite displayed quite good specific capacitance of 320 F/g at 5 mV/s and also demonstrated energy density and power density as high as 18.14 Wh/kg and 10 kW/kg, respectively (Li et al. 2017).

PPy-based fiber-shaped supercapacitors (FSSC) are gaining popularity as energy storage devices for flexible and wearable electronic devices. Teng et al. constructed a high-performance flexible FSSC, which is hierarchically interconnected porous poly(3,4-ethylenedioxythiophene):poly(4-styrenesulfonate) (PEDOT:PSS)/PPy composite fiber. Here, in situ polymerization of pyrrole was done on PEDOT:PSS hydrogel support. The hybrid fiber showed superb volumetric/areal/length specific capacitance of 393.8 F/cm^3, 770.6 mF/cm^2, and 18.9 mF/cm, respectively, at a current density of 2 A/cm^3. The hybrid fiber also displayed a high energy density of 8.3 mWh/cm^3 at a high-power density of 389.1 mW/cm^3. It exhibited cyclic stability of 52.8% even at an ultrahigh current density of 50 A/cm^3 (Teng et al. 2020). In a new study, a composite of poly-ortho-aminophenol (POAP) and lignin-derived carbon (LDC) was reported by Ehsani et al. The POAP/LDC composite demonstrated a specific capacitance of 485 F/g at a current density of 1 A/g and 86% cyclic stability after 3000 cycles (Ehsani et al. 2021). In another work by Ehsani et al., a POAP/chitosan/MOF ternary nanocomposite was synthesized via chemical routes that exhibited 3150 mF/cm^2 specific capacitance at a current density of 1 mA/cm^2. The nanocomposite also showed high cyclic stability of 93% over 1000 cycles at 10 mA/cm^2 current density (Ehsani et al. 2020). Li et al. reported a high-performance rGO/PEDOT nanocomposite. PEDOT was deposited on rGO nanosheets via the hydrothermal method. The composite revealed a high specific capacitance of about 202.7 F/g with 90% initial capacitance retention after 9000 consecutive cycles. Moreover, this nanocomposite was deposited on cotton fabric, which exhibited excellent flexibility and 98% initial capacitance retention after 3000 free bendings (Li et al. 2019).

4.6 CONCLUSION

Experts in the energy storage field are giving special attention to studies on developing a novel technology for the generation and storage of electrical energy. With progressive research in the energy storage field scientists are more inclined towards designing of sustainable, eco-friendly energy storage and conversion systems, such as supercapacitors, solar cells, lithium-ion batteries, and fuel cells. Conducting polymers and their nanocomposites, with features like metallic conductivity and reversible electrochemical doping/dedoping capability, are promising substances for energy conversion and storage. The charge transfer mechanism of conducting polymers has been an important feature of their application in energy storage and energy conversion systems. The existence of p-conjugation in molecular backbones of conducting polymer serves as a carrier for electron transfer from the redox-active center to the electrode surface. The conductivity of this CPs can be adjusted between the range of 10^{-10} and 10^4 S/cm by regulating their doping levels. CPs also retains some beneficial features of conventional polymers like low cost, facile synthesis, good durability, and flexibility, along with affinity for other substances forming composites. CPs like PANI, PPy, and PEDOT, in particular, and their nanocomposites can be considered to be potential active electrode substances for electrochemical applications. However, a lot of theoretical and experimental research needs to be done for optimizing the synthesis process of conducting polymer nanocomposites, in order to infuse the desired characteristics into them. Synthesis of binary and ternary conducting polymer nanocomposites with controlled morphology and desired features is the key to fabricating high energy density and high power density supercapacitor electrode materials with enhanced cyclic stability. The use of such nanocomposites is expected to realize the dream of replacing fossil fuels for environmental sustainability and securing the future energy demand.

REFERENCES

Ambade, R.B., Ambade, S.B., Salunkhe, R.R., Malgras, V., Jin, S.H., Yamauchi, Y., and Lee, S.H., 2016. Flexible-wire shaped all-solid-state supercapacitors based on facile electropolymerization of polythiophene with ultra-high energy density. *Journal of Materials Chemistry A*, 4 (19), 7406–7415.

Asen, P. and Shahrokhian, S., 2017. A High Performance Supercapacitor Based on Graphene/Polypyrrole/Cu2O-Cu(OH)2 Ternary Nanocomposite Coated on Nickel Foam. *Journal of Physical Chemistry C*, 121 (12), 6508–6519.

Ates, M., Serin, M.A., Ekmen, I., and Ertas, Y.N., 2015. Supercapacitor behaviors of polyaniline/CuO, polypyrrole/CuO and PEDOT/CuO nanocomposites. *Polymer Bulletin*, 72 (10), 2573–2589.

Aydemir, N., Malmström, J., and Travas-Sejdic, J., 2016. Conducting polymer based electrochemical biosensors. *Physical Chemistry Chemical Physics,* 18 (12), 8264–8277.

Chen, C., Fan, W., Zhang, Q., Ma, T., Fu, X., and Wang, Z., 2015. In situ synthesis of cabbage like polyaniline@hydroquinone nanocomposites and electrochemical capacitance investigations. *Journal of Applied Polymer Science*, 132 (29), 42290.

Chen, G.Z., 2017. Supercapacitor and supercapattery as emerging electrochemical energy stores. *International Materials Reviews*, 62 (4), 173–202.

Chen, Y., Du, L., Yang, P., Sun, P., Yu, X., and Mai, W., 2015. Significantly enhanced robustness and electrochemical performance of flexible carbon nanotube-based supercapacitors by electrodepositing polypyrrole. *Journal of Power Sources*, 287, 68–74.

Chen, Y., Liu, B., Liu, Q., Wang, J., Liu, J., Zhang, H., Hu, S., and Jing, X., 2015. Flexible all-solid-state asymmetric supercapacitor assembled using coaxial NiMoO4 nanowire arrays with chemically integrated conductive coating. *Electrochimica Acta*, 178, 429–438.

Cong, H.P., Ren, X.C., Wang, P., and Yu, S.H., 2013. Flexible graphene-polyaniline composite paper for high-performance supercapacitor. *Energy and Environmental Science*, 6 (4), 1185.

Conway, B.E., 1999a. Electrochemical Capacitors Based on Pseudocapacitance. *In: Electrochemical Supercapacitors*, Springer, Boston, MA, 221–257.

Conway, B.E., 1999b. Similarities and Differences between Supercapacitors and Batteries for Storing Electrical Energy. *In: Electrochemical Supercapacitors*, Springer, Boston, MA, 11–31.

Das, A.K., Karan, S.K., and Khatua, B.B., 2015. High energy density ternary composite electrode material based on polyaniline (PANI), molybdenum trioxide (MoO₃) and graphene nanoplatelets (GNP) prepared by sono-chemical method and their synergistic contributions in superior supercapacitive performance. *Electrochimica Acta*, 180, 1–15.

Dhibar, S. and Das, C.K., 2015. Electrochemical performances of silver nanoparticles decorated polyaniline/graphene nanocomposite in different electrolytes. *Journal of Alloys and Compounds*, 653, 486–497.

Dhibar, S. and Malik, S., 2020. Morphological Modulation of Conducting Polymer Nanocomposites with Nickel Cobaltite/Reduced Graphene Oxide and Their Subtle Effects on the Capacitive Behaviors. *ACS Applied Materials and Interfaces*, 12 (48), 54053–54067.

Donne, S.W., 2013. General Principles of Electrochemistry. *In: Supercapacitors: Materials, Systems, and Applications*, Wiley-VCH Verlag GmbH & Co. KGaA, 1–64.

Ehsani, A., Bigdeloo, M., Assefi, F., Kiamehr, M., and Alizadeh, R., 2020. Ternary nanocomposite of conductive polymer/chitosan biopolymer/metal organic framework: Synthesis, characterization and electrochemical performance as effective electrode materials in pseudocapacitors. *Inorganic Chemistry Communications*, 115, 107885.

Ehsani, A., Heidari, A.A., and Shiri, H.M., 2019. Electrochemical Pseudocapacitors Based on Ternary Nanocomposite of Conductive Polymer/Graphene/Metal Oxide: An Introduction and Review to it in Recent Studies. *Chemical Record*, 19 (5), 908–926.

Ehsani, A., Moftakhar, M.K., and Karimi, F., 2021. Lignin-derived carbon as a high efficient active material for enhancing pseudocapacitance performance of p-type conductive polymer. *Journal of Energy Storage*, 35, 102291.

Fu, C., Zhou, H., Liu, R., Huang, Z., Chen, J., and Kuang, Y., 2012. Supercapacitor based on electropolymerized polythiophene and multi-walled carbon nanotubes composites. *Materials Chemistry and Physics*, 132 (2–3), 596–600.

Gómez, H., Ram, M.K., Alvi, F., Villalba, P., Stefanakos, E., and Kumar, A., 2011. Graphene-conducting polymer nanocomposite as novel electrode for supercapacitors. *Journal of Power Sources*, 196 (8), 4102–4108.

González, A., Goikolea, E., Barrena, J.A., and Mysyk, R., 2016. Review on supercapacitors: Technologies and materials. *Renewable and Sustainable Energy Reviews*, 58, 1189–1206.

Hossain, S.K.S. and Hoque, M.E., 2018. Polymer nanocomposite materials in energy storage: Properties and applications. *In: Polymer-based Nanocomposites for Energy and Environmental Applications: A volume in Woodhead Publishing Series in Composites Science and Engineering*, 239–282.

Imani, A. and Farzi, G., 2015. Facile route for multi-walled carbon nanotube coating with polyaniline: tubular morphology nanocomposites for supercapacitor applications. *Journal of Materials Science: Materials in Electronics*, 26 (10), 7438–7444.

Jena, R.K., Yue, C.Y., Sk, M.M., and Ghosh, K., 2017. A novel high performance poly (2-methyl thioaniline) based composite electrode for supercapacitors application. *Carbon*, 115, 175–187.

Ji, J., Zhang, X., Liu, J., Peng, L., Chen, C., Huang, Z., Li, L., Yu, X., and Shang, S., 2015. Assembly of polypyrrole nanotube@MnO2 composites with an improved electrochemical capacitance. *Materials Science and Engineering B: Solid-State Materials for Advanced Technology*, 198, 51–56.

Ji, S., Yang, J., Cao, J., Zhao, X., Mohammed, M.A., He, P., Dryfe, R.A.W., and Kinloch, I.A., 2020. A Universal Electrolyte Formulation for the Electrodeposition of Pristine Carbon and Polypyrrole Composites for Supercapacitors. *ACS Applied Materials and Interfaces*, 12 (11), 13386–13399.

Khosrozadeh, A., Xing, M., and Wang, Q., 2015. A high-capacitance solid-state supercapacitor based on free-standing film of polyaniline and carbon particles. *Applied Energy*, 153, 87–93.

Lei, J., Jiang, Z., Lu, X., Nie, G., and Wang, C., 2015. Synthesis of Few-Layer MoS2 Nanosheets-Wrapped Polyaniline Hierarchical Nanostructures for Enhanced Electrochemical Capacitance Performance. *Electrochimica Acta*, 176, 149–155.

Li, J., Sun, Y., Li, D., Yang, H., Zhang, X., and Lin, B., 2017. Novel ternary composites reduced-graphene oxide/zine oxide/poly(p-phenylenediamine) for supercapacitor: Synthesis and properties. *Journal of Alloys and Compounds*, 708, 787–795.

Li, S., Chen, Y., He, X., Mao, X., Zhou, Y., Xu, J., and Yang, Y., 2019. Modifying Reduced Graphene Oxide by Conducting Polymer Through a Hydrothermal Polymerization Method and its Application as Energy Storage Electrodes. *Nanoscale Research Letters*, 14 (226), 1–12.

Li, S.H., Xu, Z., Yang, G., Ma, L., and Yang, Y., 2008. Solution-processed poly(3-hexylthiophene) vertical organic transistor. *Applied Physics Letters*, 93 (21).

Lin, H., Huang, Q., Wang, J., Jiang, J., Liu, F., Chen, Y., Wang, C., Lu, D., and Han, S., 2016. Self-Assembled Graphene/Polyaniline/Co3O4 Ternary Hybrid Aerogels for Supercapacitors. *Electrochimica Acta*, 191, 444–451.

Meng, Q., Cai, K., Chen, Y., and Chen, L., 2017. Research progress on conducting polymer based supercapacitor electrode materials. *Nano Energy*, 36, 268–285.

Mohd Abdah, M.A.A., Azman, N.H.N., Kulandaivalu, S., and Sulaiman, Y., 2020. Review of the use of transition-metal-oxide and conducting polymer-based fibres for high-performance supercapacitors. *Materials and Design*, 186, 108199.

Moyseowicz, A., Śliwak, A., Miniach, E., and Gryglewicz, G., 2017. Polypyrrole/iron oxide/reduced graphene oxide ternary composite as a binderless electrode material with high cyclic stability for supercapacitors. *Composites Part B: Engineering*, 109, 23–29.

Nejati, S., Minford, T.E., Smolin, Y.Y., and Lau, K.K.S., 2014. Enhanced charge storage of ultrathin polythiophene films within porous nanostructures. *ACS Nano*, 8 (6), 5413–5422.

Ng, C.H., Lim, H.N., Lim, Y.S., Chee, W.K., and Huang, N.M., 2015. Fabrication of flexible polypyrrole/graphene oxide/manganese oxide supercapacitor. *International Journal of Energy Research*, 39 (3), 344–355.

Qian, T., Zhou, J., Xu, N., Yang, T., Shen, X., Liu, X., Wu, S., and Yan, C., 2015. On-chip supercapacitors with ultrahigh volumetric performance based on electrochemically co-deposited CuO/polypyrrole nanosheet arrays. *Nanotechnology*, 26 (42), 425402.

Salunkhe, R.R., Lin, J., Malgras, V., Dou, S.X., Kim, J.H., and Yamauchi, Y., 2015. Large-scale synthesis of coaxial carbon nanotube/Ni(OH)2 composites for asymmetric supercapacitor application. *Nano Energy*, 11, 211–218.

Shayeh, J.S., Siadat, S.O.R., Sadeghinia, M., Niknam, K., Rezaei, M., and Aghamohammadi, N., 2016. Advanced studies of coupled conductive polymer/metal oxide nano wire composite as an efficient supercapacitor by common and fast Fourier electrochemical methods. *Journal of Molecular Liquids*, 220, 489–494.

Shen, K., Ran, F., Zhang, X., Liu, C., Wang, N., Niu, X., Liu, Y., Zhang, D., Kong, L., Kang, L., and Chen, S., 2015. Supercapacitor electrodes based on nano-polyaniline deposited on hollow carbon spheres derived from cross-linked co-polymers. *Synthetic Metals*, 209, 369–376.

Simon, P. and Gogotsi, Y., 2008. Materials for electrochemical capacitors. *Nature Materials*, 7 (11), 845–854.

Snook, G.A., Kao, P., and Best, A.S., 2011. Conducting-polymer-based supercapacitor devices and electrodes. *Journal of Power Sources*, 196 (1), 1–12.

Sun, H., She, P., Xu, K., Shang, Y., Yin, S., and Liu, Z., 2015. A self-standing nanocomposite foam of polyaniline@reduced graphene oxide for flexible super-capacitors. *Synthetic Metals*, 209, 68–73.

Tajik, S., Beitollahi, H., Nejad, F.G., Shoaie, I.S., Khalilzadeh, M.A., Asl, M.S., van Le, Q., Zhang, K., Jang, H.W., and Shokouhimehr, M., 2020. Recent developments in conducting polymers: Applications for electrochemistry. *RSC Advances*, 10 (62), 37834–37856.

Tang, H., Wang, J., Yin, H., Zhao, H., Wang, D., and Tang, Z., 2015. Growth of polypyrrole ultrathin films on mos2 monolayers as high-performance supercapacitor electrodes. *Advanced Materials*, 27 (6), 1117–1123.

Tang, J., Salunkhe, R.R., Liu, J., Torad, N.L., Imura, M., Furukawa, S., and Yamauchi, Y., 2015. Thermal conversion of core-shell metal-organic frameworks: A new method for selectively functionalized nanoporous hybrid carbon. *Journal of the American Chemical Society*, 137 (4), 1572–1580.

Teng, W., Zhou, Q., Wang, X., Che, H., Hu, P., Li, H., and Wang, J., 2020. Hierarchically interconnected conducting polymer hybrid fiber with high specific capacitance for flexible fiber-shaped supercapacitor. *Chemical Engineering Journal*, 390, 124569.

Tran, C., Singhal, R., Lawrence, D., and Kalra, V., 2015. Polyaniline-coated freestanding porous carbon nanofibers as efficient hybrid electrodes for supercapacitors. *Journal of Power Sources*, 293, 373–379.

Wang, G., Zhang, L., and Zhang, J., 2012. A review of electrode materials for electrochemical supercapacitors. *Chemical Society Reviews*, 41 (2), 797–828.

Wang, H., Hao, Q., Yang, X., Lu, L., and Wang, X., 2009. Graphene oxide doped polyaniline for supercapacitors. *Electrochemistry Communications*, 11 (6), 1158–1161.

Wang, W., Hao, Q., Lei, W., Xia, X., and Wang, X., 2012. Graphene/SnO_2/polypyrrole ternary nanocomposites as supercapacitor electrode materials. *RSC Advances*, 2 (27), 10268.

Wang, X.X., Yu, G.F., Zhang, J., Yu, M., Ramakrishna, S., and Long, Y.Z., 2021. Conductive polymer ultrafine fibers via electrospinning: Preparation, physical properties and applications. *Progress in Materials Science*, 115, 100704.

Wang, Y., Tao, S., An, Y., Wu, S., and Meng, C., 2013. Bio-inspired high performance electrochemical supercapacitors based on conducting polymer modified coral-like monolithic carbon. *Journal of Materials Chemistry A*, 1 (31), 8876.

Warren, R., Sammoura, F., Teh, K.S., Kozinda, A., Zang, X., and Lin, L., 2015. Electrochemically synthesized and vertically aligned carbon nanotube-polypyrrole nanolayers for high energy storage devices. *Sensors and Actuators, A: Physical*, 231, 65–73.

Wu, W., Yang, L., Chen, S., Shao, Y., Jing, L., Zhao, G., and Wei, H., 2015. Core-shell nanospherical polypyrrole/graphene oxide composites for high performance supercapacitors. *RSC Advances*, 5 (111), 91645–91653.

Yan, J., Wang, Q., Wei, T., and Fan, Z., 2014. Recent advances in design and fabrication of electrochemical supercapacitors with high energy densities. *Advanced Energy Materials*, 4 (4), 1300816.

Zhang, J., Shi, L., Liu, H., Deng, Z., Huang, L., Mai, W., Tan, S., and Cai, X., 2015. Utilizing polyaniline to dominate the crystal phase of Ni(OH)2 and its effect on the electrochemical property of polyaniline/Ni(OH)2 composite. *Journal of Alloys and Compounds*, 651, 126–134.

Zhang, X., Ji, L., Zhang, S., and Yang, W., 2007. Synthesis of a novel polyaniline-intercalated layered manganese oxide nanocomposite as electrode material for electrochemical capacitor. *Journal of Power Sources*, 173 (2), 1017–1023.

Zhou, C., Zhang, Y., Li, Y., and Liu, J., 2013. Construction of high-capacitance 3D CoO@ Polypyrrole nanowire array electrode for aqueous asymmetric supercapacitor. *Nano Letters*, 13 (5), 2078–2085.

Zhou, J., Zhao, H., Mu, X., Chen, J., Zhang, P., Wang, Y., He, Y., Zhang, Z., Pan, X., and Xie, E., 2015. Importance of polypyrrole in constructing 3D hierarchical carbon nanotube@ MnO_2 perfect core-shell nanostructures for high-performance flexible supercapacitors. *Nanoscale*, 7 (35), 14697–14706.

Zhu, J., Sun, W., Yang, D., Zhang, Y., Hoon, H.H., Zhang, H., and Yan, Q., 2015. Multifunctional Architectures Constructing of PANI Nanoneedle Arrays on MoS2 Thin Nanosheets for High-Energy Supercapacitors. *Small*, 11 (33), 4123–4129.

Zhu, S., Wu, M., Ge, M.H., Zhang, H., Li, S.K., and Li, C.H., 2016. Design and construction of three-dimensional CuO/polyaniline/rGO ternary hierarchical architectures for high performance supercapacitors. *Journal of Power Sources*, 306, 593–601.

5 Properties and Supercapacitor Applications of Graphene-Based Materials

Suveksha Tamang, Sadhna Rai, Rabina Bhujel,
Nayan Kamal Bhattacharyya, and Joydeep Biswas

5.1 INTRODUCTION

In a pursuit to replace conventional, nonrenewable energy sources and storage devices, research has been piqued for renewable sources and cleaner approaches for high-efficiency devices (Chettri *et al.*, 2021, 2022; Rai *et al.*, 2018, 2019; Tamang *et al.*, 2021). Supercapacitors have emerged as new energy storage devices for a large amount of energy storage for instantaneous applications. The electrode materials for these devices should possess high electrical conductance, mechanical strength, structural flexibility, high surface area, and tunable electrochemical properties. Graphene oxide (GO) and reduced graphene oxides (rGO) fit these criteria. These materials have been studied for their excellent inherent properties like high specific surface area (SSA) and better electrochemical properties (Mandal *et al.*, 2019). The oxygen-containing functional groups (OFs) in the GO layers provide these materials with several active sites for binding active species. Removing these OFs by reducing these layers to rGO increases electrical conductance. It elevates the electrochemical properties for application in supercapacitor devices.

The synthesis procedure of these materials involves graphite powder as a precursor. These tunable properties and the abundance of graphite in nature have opened vast research for many applications for graphene-based materials. Several methods have been developed over the years to synthesize GO and rGO. However, a green process reducing GO to rGO has picked up pace in recent decades as pollution and its effect has become a concern and a severe global threat to the environment. Plant extract-reduced and biomass-derived rGO can be studied as supercapacitor electrode materials for renewable sources of materials for energy storage devices (Rai *et al.*, 2021). The phytochemicals in the plant extract have been reported to act as capping and stabilizing agents for these nanomaterials.

DOI: 10.1201/9781003323518-6

The electrochemical properties of GO and rGO can be tuned by forming composites with metal oxides to achieve high values of specific capacitance. Yang *et al.* electrochemically synthesized rGO (ErGO) for application in supercapacitor electrodes (J. Yang & Gunasekaran, 2013). The specific capacitance was revealed to be 223.6 F/g at a scan rate of 5 mV/s. Rai *et al.* synthesized rGO using ginger extract with a specific capacitance of 79.85 F/g (Rai *et al.*, 2021). Navarro-Suárez *et al.* synthesized a composite of titanium carbide MXene ($Ti_3C_2T_x$) and rGO for a supercapacitor electrode (Navarro-Suárez *et al.*, 2018). They reported an increase in the specific capacitance of the composites as a result of the synergistic effect between the materials. Madhuri *et al.* synthesized a composite of rGO and nickel phthalocyanine (NiPc) with a specific capacitance value of 23.28 F/g at 1 A/g current density (Madhuri & John, 2018). Wang *et al.* synthesized a composite of nickel cobalt oxide and rGO ($NiCo_2O_4$-rGO) with a specific capacitance value of 1222 F/g at a current density of 0.5 A/g. Khodary *et al.* synthesized rGO, exposing GO to microwave radiations resulting in a high specific capacitance of 302 F/g at a scan rate of 1 mV/s (El-Khodary *et al.*, 2014). These materials have a high electrical conductivity that can be improved by altering the surface. Thus, the specific surface area (SSA) can affect the conductivity of these materials. Supercapacitor electrodes can use these tunable properties to their advantage.

5.2 GRAPHENE

The structure of graphite was discovered in 1924 by Bernal (Freise, 1962). The hexagonal array of carbon atoms in graphite is present in the plane so that the first and the third layers exactly fall over the other. The forces of attraction holding these layers together are the weak van der Waals forces of attraction. Unlike these weak forces, the carbon atoms in the hexagonal rings are sp^2 hybridized, and π-electron conjugation is very strong between these carbon atoms. Later in 2004, the discovery of graphene brought about a new revolutionary field of research in nanomaterials and nanotechnology (Novoselov *et al.*, 2004). Graphene is a single-layered hexagonal array of sp^2 hybridized carbon atoms as shown in Figure 5.1. The electron conjugation in the graphene sheets renders this material with superior electronic properties.

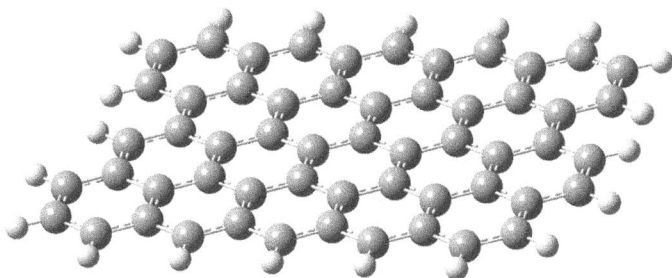

FIGURE 5.1 Structure of graphene.

Source: Novoselov *et al.*, 2004.

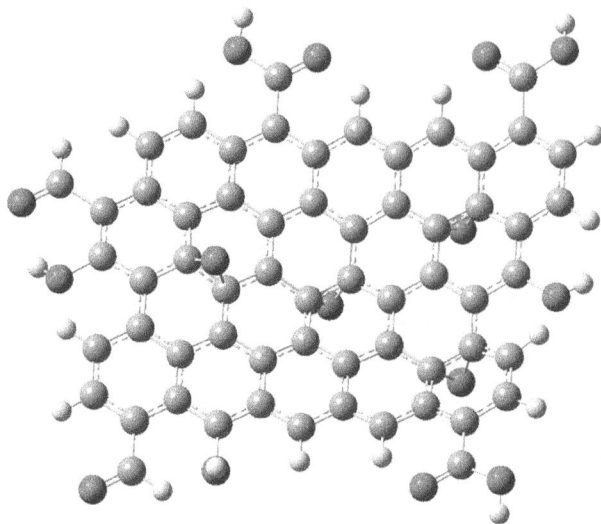

FIGURE 5.2   Structure of GO.

Source: Paranthaman *et al.*, 2018.

5.3   GRAPHENE OXIDE AND REDUCED GRAPHENE OXIDE

GO is an oxidized graphene derivative containing OFs like carbonyl, carboxy, hydroxy, epoxy, etc. Figure 5.2 and Figure 5.3 shows the structure of GO and rGO respectively. The band gap energy of GO falls almost in the insulating region (1–3.5 eV), thus rendering the material almost nonconducting (Jin *et al.*, 2020). The conductivity and other inherent optical and electrical properties of GO can be tuned. The tuning can be carried out by varying the concentration of functional groups in GO. The reduction of these OFs in GO results in the formation of the reduced derivative of GO, known as rGO. The band gap energy of this rGO material lies between 0.2 and 2 eV and thus is conducting in nature (Jin *et al.*, 2020). GO and rGO possess high mechanical strength and SSA, and most importantly, they have a low surface-to-volume ratio. (Mohan *et al.*, 2020; Ray *et al.*, 2015). Specific surface of graphene-based materials (SSA) is influenced by the morphology of GO and rGO. The interlayer spacing between the layers, the orientation of the layers, pore size distribution, and the surface defect influence the SSA (Paranthaman *et al.*, 2018).

5.4   COMPOSITES OF GO AND rGO WITH METAL OXIDES

GO and rGO have been explored in catalysis, sensors, energy storage devices, and many more (Alam *et al.*, 2017; Bhujel *et al.*, 2019; Paranthaman *et al.*, 2018). The properties of these materials, specifically for energy storage materials such as supercapacitor devices, required enhancement. The electrochemical properties of

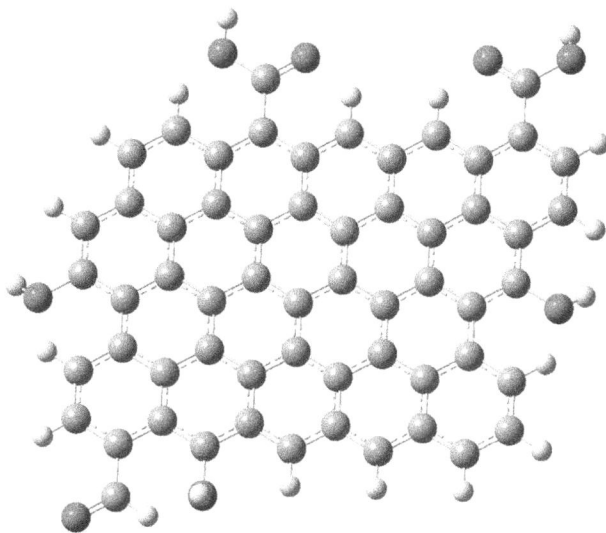

FIGURE 5.3   Structure of rGO.

Source: Paranthaman *et al.*, 2018.

these materials, along with surface modification, enhancement of porosity of layers, and other properties, have been reported in the literature by synthesizing composites of these materials with metal oxides of suitable properties. Agglomerations of metal oxides due to small size alone in metal oxide nanoparticles have also been checked in these composites. The reactivity of metal oxides combined synergistically with high specific surface area, tunable band gap energy and superior mechanical strength, and electrochemical properties of GO and rGO renders these composites better applicability in many research fields (Bhujel *et al.*, 2019; Rai *et al.*, 2019).

5.5   SYNTHESIS OF GO

The chemical synthesis of GO includes various synthesis methods involving the use of several chemical reagents. Some of these reagents involved in the synthesis of GO are reportedly hazardous. The first known chemical synthesis of GO was reported in 1859 from powder graphite, followed by Staudenmaier in 1898 (Staudenmaier, 1898). Both methods involved the liberation of explosive ClO_2 gas. Later, Hummers and Offeman synthesized GO from powdered graphite (Hummers & Offeman, 1958). They were successful in synthesizing GO with higher OFs concentration. However, this method released toxic NO_2 gas. A modified Hummers' method has been reported to eradicate the toxic by-product formation during the chemical synthesis of GO from powder graphite (Rodríguez-Pastor *et al.*, 2022). This method avoids toxic and explosive reagents, provided the oxidation extent and the inherent properties are kept intact. In the modified Hummers' method, sodium nitrate ($NaNO_3$) is avoided to avoid releasing toxic gases.

5.6   SYNTHESIS OF rGO

The reduction of GO to rGO can be procured via chemical, mechanical, thermal, green reduction, *etc*. Out of this, the recent research focus has shifted toward green synthesis techniques. This ensures a nonhazardous approach toward the synthesis of rGO for various applications. Green synthesis involves reducing GO using several plant extracts and biomass-derived rGO (Agarwal & Zetterlund, 2021). Plants extracts of green tea (Akhavan *et al.*, 2012), chrysanthemum (Hou *et al.*, 2016), sugarcane bagasse (B. Li *et al.*, 2018), *Colocasia esculenta* (S. Thakur & Karak, 2012), eucalyptus leaf (C. Li *et al.*, 2017), and many other plant extracts containing phytochemicals have been reported by many researchers for green synthesis of rGO. Chemical reduction involves the use of reducing agents such as hydrazine hydrate (N_2H_4), ascorbic acid (Habte & Ayele, 2019), and sodium borohydride ($NaBH_4$) (Muda *et al.*, 2017). The reduction of GO using ascorbic acid is reported to be chemically more stable (Habte & Ayele, 2019). The toxicity of some of these reagents like N_2H_4 is the major disadvantage of this synthesis procedure. However, the reduction following the chemical route results in higher yield, better tunable inherent properties, and a low time-consuming process. Mechanical processes, even though time consuming with the tedious process involved, result in high-quality materials. Similarly, the thermal procedures are energy consuming and are thus not a very efficient process for synthesizing these nanomaterials.

5.7   PROPERTIES OF GO AND rGO

5.7.1   OPTICAL PROPERTIES

The UV-visible spectrum provides a preliminary characterization to detect the formation of GO and rGO. This plot of absorbance vs wavelength for GO shows two characteristic peaks around 230 and 290 nm (Gurunathan *et al.*, 2012). These two peaks correspond to the π–π* and n–π* transition of electrons in the GO surface. The π–π* transition results due to the electronic conjugation between sp^2 carbon atoms placed in a hexagonal array. The n–π* transitions correspond to the electronic transition from the nonbonding (n) electrons of the oxygen-containing functional groups to the π* orbitals. The reduction of the GO into rGO results in the removal of most OFs. The plot of absorbance vs wavelength of rGO does not show any characteristic peak for the n–π* transition (Sinha *et al.*, 2021). The plot is a single peak at around 270–290 nm. Figure 5.4 illustrates the plot of absorbance vs wavelength for GO and rGO.

5.7.2   VIBRATIONAL PROPERTIES

The functional groups in GO and rGO can be detected using FTIR spectroscopy. The functional groups generally present in GO are hydroxy (-OH), epoxy (-O-), and carbonyl (C=O). These functional groups are recorded at characteristic wavenumber between 4000 and 400 cm^{-1}. The transmittance % vs wavenumber plot for GO exhibits a strong band at around 1620 cm^{-1}, corresponding to the carbonyl functional groups. Similarly, the peaks around 3400 cm^{-1} and 1400 cm^{-1} correspond to carboxylic acid

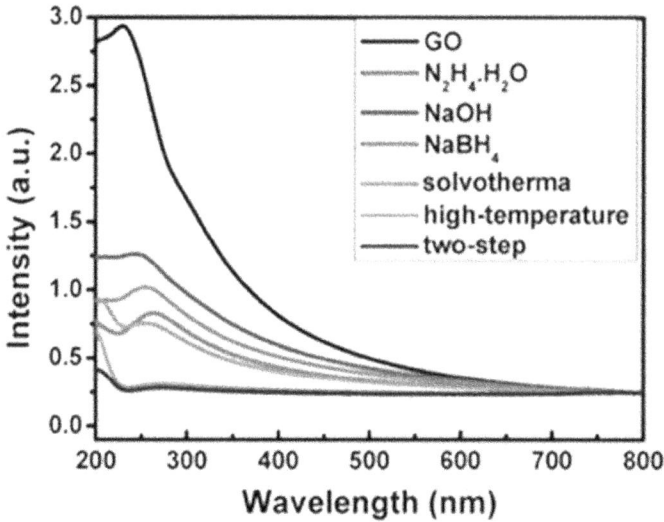

FIGURE 5.4 The plot of absorbance vs wavelength for GO and rGO. Adapted with permission from Luo *et al.* (2011). Copyright 2011 American Chemical Society.

groups and hydroxy group vibrations respectively in the GO molecule, (Alam *et al.*, 2017; A. Thakur *et al.*, 2015). These bands corresponding to carbonyl and acid functional groups decrease in intensity in rGO after GO reduction.

The vibrational and rotational bonds in the nanomaterials, GO and rGO, can also be analyzed using Raman spectroscopy (Alam *et al.*, 2017). GO exhibits two bands, D and G, around 1320 and 1590 cm⁻¹, respectively. The structural defects in the layers during the graphitic exfoliation give rise to the D band in these materials, and the G band arises due to the vibration of sp² carbon. These bands for rGO shift to slightly higher values around 1347 and 1599 cm⁻¹.

5.7.3 STRUCTURAL PROPERTIES

The X-ray diffraction (XRD) spectrum of GO and rGO exhibits characteristic peaks at certain values of 2θ. GO exhibits two peaks at 2θ values of around 11–12° and around 45° (Sinha *et al.*, 2021). The Miller indices corresponding to these angles are (002) and (110), respectively. The 2θ value of rGO lies around 2θ = 24–25° (Alam *et al.*, 2017; Sinha *et al.*, 2021; Stobinski *et al.*, 2014). The Miller coefficient corresponding to this angle is (002). The distance between two consecutive planes (d) can also be determined by the XRD data using equation (1) (Bhujel *et al.*, 2019; Rai *et al.*, 2019).

$$d = \frac{n\lambda}{2\sin\theta} \tag{5.1}$$

where d is the interlayer distance between two consecutive layers, λ denotes the wavelength of X-rays, θ is the angle of incidence, and n is a constant.

The OFs in GO result in a higher value of d than rGO. The d value for GO has been reported to be around 0.76 nm, and that for rGO has been reported to be around 0.34 nm. XRD provides an analysis of the oxidation of graphite to GO and the reduction of GO to rGO (Rai *et al.*, 2019). The value of interplanar distance for graphite has been reported to be around 0.33 nm as OFs are absent in the graphitic layers, which are to be introduced only after oxidation (Bhujel *et al.*, 2019).

5.7.4   BONDING PROPERTIES

5.7.4.1   X-Ray Photoelectron Spectroscopy (XPS)

The full scan XPS spectra of GO and rGO materials aided by the deconvolution plot provide information about the elemental composition present in the material as depicted in Figure 5.5. Along with the composition, the deconvolution curves also provide information about the bonding of the core orbitals. The full scan spectra for GO and rGO reveal characteristic peaks, respectively around binding energy values of 285 and 532 eV (Quan *et al.*, 2016; D. Yang & Bock, 2017). On deconvolution of the core orbitals, analysis becomes more prominent. The deconvolution of core orbital C(1s) for GO reveals characteristic peaks of carbonyl, sp^2 and sp^3 C-C bonds, and carboxy acid groups around 287, 284, and 285 eV, respectively (Sinha *et al.*, 2021; D. Yang & Bock, 2017). The deconvolution plot of C(1s) for rGO exhibits lowering intensities in these peaks of the OFs but a significant increase in the intensity of the peak at 284 eV corresponding to the sp^2 C=C bond. This implies the extent of restoration of sp^2 C=C in rGO upon reduction of GO. Similarly, the deconvolution of O(1s) reveals peaks around 532- 533 eV corresponding to OFs like epoxy and hydroxyl groups that appear lower in intensity in rGO as compared to GO (Quan *et al.*, 2016).

5.7.4.2   Morphological Properties

Morphological analysis of GO and rGO can be provided using scanning electron microscopy (SEM) and transmission electron microscopy (TEM) characterization.

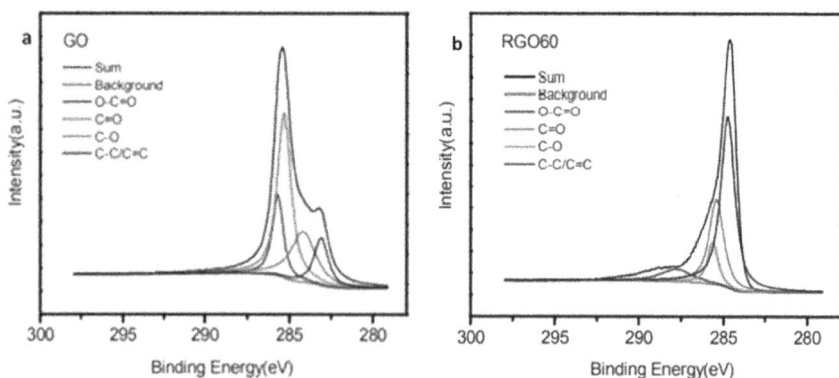

FIGURE 5.5 XPS spectra of GO and rGO. Adapted with permission from Zhang *et al.* (2014). Copyright 2014 American Chemical Society.

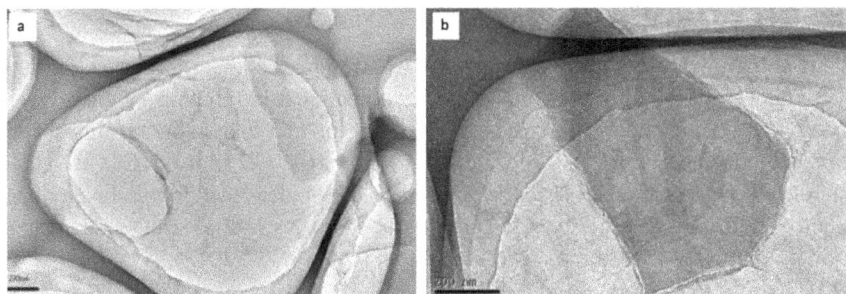

FIGURE 5.6   TEM images of GO and rGO. Adapted with permission from Shen *et al.* (2009). Copyright 2009 American Chemical Society.

Figure 5.6 shows the TEM image of GO and rGO. These electron microscopes can provide images up to the nanoscale range, down to a single atom. These microscopies have enabled the morphological analysis of nanosized particles (Inkson, 2016). The SEM micrographs of GO at different resolutions reveal porous transparent layers of GO with wrinkles. The exfoliation of the layers during the chemical synthesis of GO adds crumples and folds in the layers (Rai *et al.*, 2019). The reduction of GO to rGO resulting in the removal of OFs results in the appearance of more wrinkles and folds around the edge (Alam *et al.*, 2017; A. Thakur *et al.*, 2015). The energy dispersive spectroscopy (EDS) spectra provide information on removing OFs and reducing GO to rGO. The atomic and weight percentages of oxygen decrease in rGO in comparison to GO, whereas the percentage of C increases. This analysis also provides an overview of the extent of oxidation and reduction in these nanomaterials. The TEM provides two-dimensional structures of wrinkled sheets of GO and rGO (Stobinski *et al.*, 2014). In the case of crystalline materials, the lattice planes can be analyzed along with determining lattice spacing (d).

5.8   ELECTROCHEMICAL PROPERTIES FOR SUPERCAPACITOR APPLICATION

GO is electrically nonconducting; the band gap energy value of GO has been reported to be around 3 eV. The insulating OFs and the disruption in the π-conjugation system in GO layers obstruct the flow of electrons in the layers. Removing these OFs and restoring the C=C after reduction renders electrical conductivity to rGO layers. The band gap energy of rGO has been reported to be around 0.2–2 eV. Electrochemical impedance is the net total obstruction to the flow of current by electrical components of a circuit. The Nyquist plot is a plot of real impedance *vs* imaginary impedance. The plots after curve fitting provide values of charge transfer resistance (R_{CT}) (Casero *et al.*, 2012). Analysis of R_{CT} values and the equivalent circuit reveal the electrical conductivity of these materials and their applicability as a supercapacitor material. The R_{CT} values increase in rGO layers as the reduction extent increases in GO (Bhujel *et al.*, 2019), indicating the increase in the extent of restoration of the π-electron conjugation in sp^2 C=C in the rGO layers on the removal of the OFs.

5.9 CYCLIC VOLTAMMETRY

Cyclic voltammetry (CV) provides analysis of the redox behavior of materials within a potential window. This plot of current vs potential is a function of energy scan (Kissinger & Heineman, 1983). The CV curves for GO and rGO are almost rectangular type (Quan *et al.*, 2016). Figure 5.7 shows the CV plot of rGO at various scan rates. The rectangular curves are characteristic of electrical double-layered capacitance (EDLC) type. The specific capacitance of these materials can be determined using Equation 5.2 for a three-electrode system (Chen *et al.*, 2013).

$$C_{sp} = \frac{\int_{V_1}^{V_2} idv}{v*(V_2 - V_1)*m} \tag{5.2}$$

In Equation 5.2, C_{sp}, m, v, V_1, and V_2 represent the specific capacitance (F/g), the active mass of electrode material, scan rate (V/s), and low and high potential (volts), respectively. The term $\int_{V_1}^{V_2} idv$ represents the area under the curve. The specific capacitance value increases as the reduction extent of GO layers increases. As

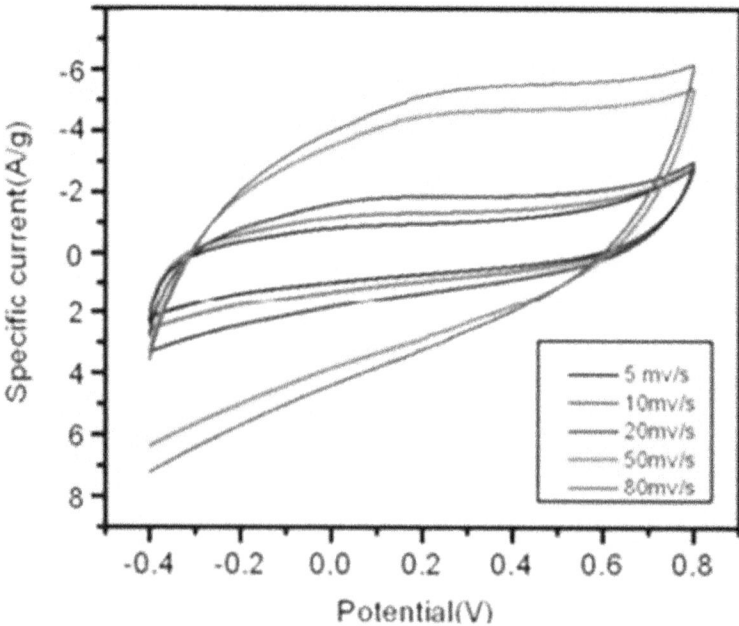

FIGURE 5.7 CV plot of rGO at various scan rates. Adapted with permission from Zhang *et al.* (2014). Copyright 2014 American Chemical Society.

the sp^2 carbon bonds are restored in rGO due to reduction, the π-electron conjugation increases, thus increasing the electrical conjugation in rGO, rendering rGO to have better supercapacitive properties.

For a two-electrode system, the specific capacitance of materials from CV plots can be determined using the Equation 5.3 (Bhujel *et al.*, 2019; Chen *et al.*, 2013):

$$C_{sp} = \frac{2\int_{V_1}^{V_2} idv}{m \times v \times (V_2 - V_1)} \tag{5.3}$$

The energy density (E) and power density (P) are the other two parameters to examine materials for supercapacitor application. EDLC-type supercapacitors exhibit higher P and low E. These parameters for the two-electrode system can be determined using Equations 5.4 and 5.5, and for the three-electrode system, the Equations are 5.6 and 5.7 (Chen *et al.*, 2013).

$$E = \frac{C_{sp} \times \Delta V}{2} \times 3.6 \tag{5.4}$$

$$P = \frac{E \times 3600}{\Delta t} \tag{5.5}$$

$$E = \frac{1}{8} C_{sp} \times \Delta V^2 \tag{5.6}$$

$$P = \frac{E}{\Delta t} \tag{5.7}$$

In these equations, E and P denote the energy and power density in Wh/kg and W/kg, respectively. Other symbols have their usual meaning and Δt denotes the discharge time.

5.10 GALVANOSTATIC CHARGE DISCHARGE (GCD)

The specific capacitance and cyclic retention of material can be analyzed through GCD cycles. The GCD curves are plotted at a constant potential window at varying current densities for specific capacitance. The cyclic retention is determined for many cycles at a constant current density of both GO and rGO. For two-electrode systems, the C_{sp} can be determined using the Equation 5.8 (Chen *et al.*, 2013).

$$C_{sp} = \frac{4I}{m \times dV/dT} \tag{5.8}$$

TABLE 5.1
Specific Capacitance Values of Several rGO and rGO Composites

Entry	Material	Year	Specific capacitance (F/g)	Scan rate (mV/s)	Current density (mA/cm²)	Current density (A/g)	References
1	ErGO	2013	223.6	5	-	-	Yang-2013
2	Gelatin-Mg-Zn	2013	284		-	1	Chen-2013
3	MrGO	2014	302	1	-	-	Khodary-2014
4	α-Fe2O3/rGO	2016	903		-	1	Quan-2016
5	NiPc fiber-rGO	2017	223.28	-	-	1	Madhuri-2017
6	rGO:Ti$_3$C$_2$T$_x$	2018	254	2	-		Navarro- Suárez-2018
7	RGO/Fe2O3	2019	50	0.1			Bhujel-2019
8	CuO-rGO	2019	45.86				Rai-2019

In Equation 5.8, I is the applied current and $\dfrac{dV}{dT}$ is the discharging slope. Other

symbols have their usual significance.

Similarly, for a three-electrode system, Equation 5.9 determines the value of C_{sp}:

$$C_{sp} = \frac{I \times \Delta t}{M \times (\Delta V)} \tag{5.9}$$

The EDLC materials GO and rGO both exhibit triangle-shaped curves (Bhujel *et al.*, 2019). The higher symmetrical plots represent cyclic stability and better discharge time for rGO than GO (Rai *et al.*, 2019). These materials and their composites reveal excellent electrochemical properties exploitable in supercapacitor electrodes for high values of specific capacitance and better energy and power density values. Table 5.1 depicts the reported values of specific capacitance of rGO and rGO composites with various materials, determined using CV plots at different scan rates or GCD cycles at different current density values.

5.11 CONCLUSION

Based on the analysis of these materials' various features, we can infer that GO and rGO-based materials reveal superior applicability in the field of supercapacitor electrodes. For the synthesis of GO, methods have been developed that avoid the use of toxic reagents. In the case of rGO, green reduction techniques are gaining pace for safer approaches in synthesizing this nanomaterial. These green processes also focus on tuning surface and electrochemical properties for better electrical conductance and reactivity. Several studies have reported high values of specific capacitance (C_{sp}), energy density (E), and power density (P). The low charge transfer resistance (R_{CT}) value in the case of rGO reveals the high electrical conductance of this material. These properties can be exploited in several fields, including supercapacitor electrodes.

ACKNOWLEDGMENT

ST thanks Sikkim Manipal University for providing PhD Research Scholarship. RB thanks Sikkim Alpine University for its help and support. All other authors thank Sikkim Manipal Institute of Technology for help and support.

REFERENCES

Agarwal, V., & Zetterlund, P. B. (2021). Strategies for reduction of graphene oxide – A comprehensive review. *Chemical Engineering Journal*, *405*, 127018. https://doi.org/10.1016/j.cej.2020.127018

Akhavan, O., Kalaee, M., Alavi, Z. S., Ghiasi, S. M. A., & Esfandiar, A. (2012). Increasing the antioxidant activity of green tea polyphenols in the presence of iron for the reduction of graphene oxide. *Carbon, 50*(8), 3015–3025. https://doi.org/10.1016/j.car bon.2012.02.087

Alam, S. N., Sharma, N., & Kumar, L. (2017). Synthesis of Graphene Oxide (GO) by Modified Hummers Method and Its Thermal Reduction to Obtain Reduced Graphene Oxide (rGO)*. *Graphene, 06*(01), 1–18. https://doi.org/10.4236/graphene.2017.61001

Bhujel, R., Rai, S., Deka, U., & Swain, B. P. (2019). Electrochemical, bonding network and electrical properties of reduced graphene oxide-Fe2O3 nanocomposite for supercapacitor electrodes applications. *Journal of Alloys and Compounds, 792*, 250–259. https://doi.org/10.1016/j.jallcom.2019.04.004

Casero, E., Parra-Alfambra, A. M., Petit-Domínguez, M. D., Pariente, F., Lorenzo, E., & Alonso, C. (2012). Differentiation between graphene oxide and reduced graphene by electrochemical impedance spectroscopy (EIS). *Electrochemistry Communications, 20*(1), 63–66. https://doi.org/10.1016/j.elecom.2012.04.002

Chen, X. Y., Chen, C., Zhang, Z. J., & Xie, D. H. (2013). Gelatin-derived nitrogen-doped porous carbon via a dual-template carbonization method for high performance supercapacitors. *Journal of Materials Chemistry A, 1*(36), 10903. https://doi.org/10.1039/c3ta12328f

Chettri, B., Thapa, A., Das, S., Chettri, P., & Sharma, B. (2022). First principle insight into co-doped MoS2 for sensing NH_3 and CH_4. *Facta Universitatis – Series: Electronics and Energetics, 35*(1), 43–59. https://doi.org/10.2298/FUEE2201043C

Chettri, B., Thapa, A., Das, S. K., Chettri, P., & Sharma, B. (2021). Computational Study of Adsorption behavior of CH_4N_2O and CH_3OH on Fe decorated MoS2 monolayer. *Solid State Electronics Letters, 3*, 32–41. https://doi.org/10.1016/j.ssel.2021.12.002

El-Khodary, S. A., El-Enany, G. M., El-Okr, M., & Ibrahim, M. (2014). Preparation and Characterization of Microwave Reduced Graphite Oxide for High-Performance Supercapacitors. *Electrochimica Acta, 150*, 269–278. https://doi.org/10.1016/j.electa cta.2014.10.134

Freise, E. J. (1962). Structure of Graphite. *Nature, 193*(4816), 671–672. https://doi.org/10.1038/193671a0

Gurunathan, S., Woong Han, J., Abdal Daye, A., Eppakayala, V., & Kim, J. (2012). Oxidative stress-mediated antibacterial activity of graphene oxide and reduced graphene oxide in Pseudomonas aeruginosa. *International Journal of Nanomedicine, 7*, 5901. https://doi.org/10.2147/IJN.S37397

Habte, A. T., & Ayele, D. W. (2019). Synthesis and Characterization of Reduced Graphene Oxide (rGO) Started from Graphene Oxide (GO) Using the Tour Method with Different Parameters. *Advances in Materials Science and Engineering, 2019*, 1–9. https://doi.org/10.1155/2019/5058163

Hou, D., Liu, Q., Cheng, H., Li, K., Wang, D., & Zhang, H. (2016). Chrysanthemum extract assisted green reduction of graphene oxide. *Materials Chemistry and Physics, 183*, 76–82. https://doi.org/10.1016/j.matchemphys.2016.08.004

Hummers, W. S., & Offeman, R. E. (1958). Preparation of Graphitic Oxide. *Journal of the American Chemical Society, 80*(6), 1339–1339. https://doi.org/10.1021/ja01539a017

Inkson, B. J. (2016). Scanning electron microscopy (SEM) and transmission electron microscopy (TEM) for materials characterization. In Gerhard Hübschen, Iris Altpeter, Ralf Tschuncky, and Hans-Georg Herrmann, eds. *Materials Characterization Using Nondestructive Evaluation (NDE) Methods* (pp. 17–43). Woodhead publishing. https://doi.org/10.1016/B978-0-08-100040-3.00002-X

Jin, Y., Zheng, Y., Podkolzin, S. G., & Lee, W. (2020). Band gap of reduced graphene oxide tuned by controlling functional groups. *Journal of Materials Chemistry C, 8*, 4885. https://doi.org/10.1039/c9tc07063j

Kissinger, P. T., & Heineman, W. R. (1983). Cyclic voltammetry. *Journal of Chemical Education*, *60*(9), 702. https://doi.org/10.1021/ed060p702

Li, B., Jin, X., Lin, J., & Chen, Z. (2018). Green reduction of graphene oxide by sugarcane bagasse extract and its application for the removal of cadmium in aqueous solution. *Journal of Cleaner Production*, *189*, 128–134. https://doi.org/10.1016/j.jclepro.2018.04.018

Li, C., Zhuang, Z., Jin, X., & Chen, Z. (2017). A facile and green preparation of reduced graphene oxide using Eucalyptus leaf extract. *Applied Surface Science*, *422*, 469–474. https://doi.org/10.1016/j.apsusc.2017.06.032

Luo, D., Zhang, G., Liu, J., & Sun, X. (2011). Evaluation Criteria for Reduced Graphene Oxide. *Journal of Physical Chemistry C*, *115*(23), 11327–11335. https://doi.org/10.1021/jp110001y

Madhuri, K. P., & John, N. S. (2018). Supercapacitor application of nickel phthalocyanine nanofibres and its composite with reduced graphene oxide. *Applied Surface Science*, *449*, 528–536. https://doi.org/10.1016/j.apsusc.2017.12.021

Mandal, S. K., Dutta, K., Pal, S., Mandal, S., Naskar, A., Pal, P. K., Bhattacharya, T. S., Singha, A., Saikh, R., De, S., & Jana, D. (2019). Engineering of ZnO/rGO nanocomposite photocatalyst towards rapid degradation of toxic dyes. *Materials Chemistry and Physics*, *223*, 456–465. https://doi.org/10.1016/J.MATCHEMPHYS.2018.11.002

Mohan, V. B., Jayaraman, K., & Bhattacharyya, D. (2020). Brunauer–Emmett–Teller (BET) specific surface area analysis of different graphene materials: A comparison to their structural regularity and electrical properties. *Solid State Communications*, *320*, 114004. https://doi.org/10.1016/j.ssc.2020.114004

Muda, M. R., Ramli, M. M., Isa, S. S. M., Jamlos, M. F., Murad, S. A. Z., Norhanisah, Z., Isa, M. M., Kasjoo, S. R., Ahmad, N., Nor, N. I. M., & Khalid, N. (2017). Fundamental study of reduction graphene oxide by sodium borohydride for gas sensor application. *AIP Conference Proceedings*, *1808*, 020034. https://doi.org/10.1063/1.4975267

Navarro-Suárez, A. M., Maleski, K., Makaryan, T., Yan, J., Anasori, B., & Gogotsi, Y. (2018). 2D Titanium Carbide/Reduced Graphene Oxide Heterostructures for Supercapacitor Applications. *Batteries & Supercaps*, *1*(1), 33–38. https://doi.org/10.1002/batt.201800014

Novoselov, K. S., Geim, A. K., Morozov, S. v., Jiang, D., Zhang, Y., Dubonos, S. v., Grigorieva, I. v., & Firsov, A. A. (2004). Electric Field Effect in Atomically Thin Carbon Films. *Science*, *306*(5696), 666–669. https://doi.org/10.1126/science.1102896

Paranthaman, V., Sundaramoorthy, K., Chandra, B., Muthu, S. P., Alagarsamy, P., & Perumalsamy, R. (2018). Investigation on the Performance of Reduced Graphene Oxide as Counter Electrode in Dye Sensitized Solar Cell Applications. *Physica Status Solidi (a)*, *215*(18), 1800298. https://doi.org/10.1002/pssa.201800298

Quan, H., Cheng, B., Xiao, Y., & Lei, S. (2016). One-pot synthesis of α-Fe_2O_3 nanoplates-reduced graphene oxide composites for supercapacitor application. *Chemical Engineering Journal*, *286*, 165–173. https://doi.org/10.1016/j.cej.2015.10.068

Rai, S., Bhujel, R., Biswas, J., & Swain, B. P. (2019). Effect of electrolyte on the supercapacitive behaviour of copper oxide/reduced graphene oxide nanocomposite. *Ceramics International*, *45*(11), 14136–14145. https://doi.org/10.1016/j.ceramint.2019.04.114

Rai, S., Bhujel, R., Biswas, J., & Swain, B. P. (2021). Biocompatible synthesis of rGO from ginger extract as a green reducing agent and its supercapacitor application. *Bulletin of Materials Science*, *44*(1), 40. https://doi.org/10.1007/s12034-020-02318-w

Rai, S., Bhujel, R., & Swain, B. P. (2018). Electrochemical Analysis of Graphene Oxide and Reduced Graphene Oxide for Super Capacitor Applications. *Proceedings of International*

Conference on 2018 IEEE Electron Device Kolkata Conference, EDKCON 2018. https:// doi.org/10.1109/EDKCON.2018.8770433

Ray, S. C. (2015). Application and Uses of Graphene Oxide and Reduced Graphene Oxide. In *Applications of Graphene and Graphene-Oxide Based Nanomaterials*, 6(8), (pp. 39– 55). Elsevier. https://doi.org/10.1016/B978-0-323-37521-4.00002-9

Rodríguez-Pastor, I., López-Pérez, A., Romero-Sánchez, M. D., Pérez, J. M., Fernández, I., & Martin-Gullon, I. (2022). Effective Method for a Graphene Oxide with Impressive Selectivity in Carboxyl Groups. *Nanomaterials*, *12*(18), 3112. https://doi.org/10.3390/ nano12183112

Shen, J., Hu, Y., Shi, M., Lu, X., Qin, C., Li, C., & Ye, M. (2009). Fast and Facile Preparation of Graphene Oxide and Reduced Graphene Oxide Nanoplatelets. *Chemistry of Materials*, *21*(15), 3514–3520. https://doi.org/10.1021/cm901247t

Sinha, A., Ranjan, P., & Thakur, A. D. (2021). Effect of characterization probes on the properties of graphene oxide and reduced graphene oxide. *Applied Physics A*, *127*(8), 585. https://doi.org/10.1007/s00339-021-04734-z

Staudenmaier, L. (1898). Verfahren zur Darstellung der Graphitsäure. *Berichte Der Deutschen Chemischen Gesellschaft*, *31*(2), 1481–1487. https://doi.org/10.1002/cber.18980310237

Stobinski, L., Lesiak, B., Malolepszy, A., Mazurkiewicz, M., Mierzwa, B., Zemek, J., Jiricek, P., & Bieloshapka, I. (2014). Graphene oxide and reduced graphene oxide studied by the XRD, TEM and electron spectroscopy methods. *Journal of Electron Spectroscopy and Related Phenomena*, *195*, 145–154. https://doi.org/10.1016/j.elspec.2014.07.003

Tamang, S., Thapa, A., Chettri, K., Datta, B., & Biswas, J. (2021). Analysis of dipyridine dipyrrole based molecules for solar cell application using computational approach. *Journal of Computational Electronics*, *21*, 94–105. https://doi.org/10.1007/s10 825-021-01822-4

Thakur, A., Kumar, S., & Rangra, V. S. (2015). Synthesis of reduced graphene oxide (rGO) via chemical reduction. *AIP Conference Proceedings*, *1661*, 080032. https://doi.org/ 10.1063/1.4915423

Thakur, S., & Karak, N. (2012). Green reduction of graphene oxide by aqueous phytoextracts. *Carbon*, *50*(14), 5331–5339. https://doi.org/10.1016/j.carbon.2012.07.023

Yang, D., & Bock, C. (2017). Laser reduced graphene for supercapacitor applications. *Journal of Power Sources*, *337*, 73–81. https://doi.org/10.1016/j.jpowsour.2016.10.108

Yang, J., & Gunasekaran, S. (2013). Electrochemically reduced graphene oxide sheets for use in high performance supercapacitors. *Carbon*, *51*(1), 36–44. https://doi.org/10.1016/ j.carbon.2012.08.003

Zhang, W., Zhang, Y., Tian, Y., Yang, Z., Xiao, Q., Guo, X., Jing, L., Zhao, Y., Yan, Y., Feng, J., & Sun, K. (2014). Insight into the Capacitive Properties of Reduced Graphene Oxide. *ACS Applied Materials & Interfaces*, *6*(4), 2248–2254. https://doi.org/10.1021/ am4057562

Part II

Advanced Nanomaterials
Bio-Medical Applications

6 First Principles Approach Toward Electrically Doped Nanodevices

Debarati Dey Roy, Pradipta Roy, and Debashis De

6.1 BACKGROUND

This chapter presents the theoretical background for the analysis of nanoscale molecular electronic devices. The first principles approach is generally a conjugated form of density functional theory and nonequilibrium Green's function. The major areas of discussion in this chapter are as follows:

6.2 DENSITY FUNCTIONAL THEORY

Density functional theory (DFT) is an extremely useful tool to design nanoscale electronic devices. For the last 50 years, DFT has had the capability to dominate the quantum-mechanical and quantum-ballistic simulation of periodic systems. It is also used to calculate the surface energy of the nanoscale molecular devices. In this chapter, we introduce and outline the basic concepts and features of DFT. Therefore, modern progress in exchange correlation functionals is introduced in this chapter. The main formula for DFT is derived from the solution of the Schrödinger Equation. The energy can be calculated by solving the Schrödinger equation in a time-independent, nonrelativistic Born-Oppenheimer approximation platform in Equation 6.1.

$$\mathbf{H}\psi\left(r_1 r_2 \ldots \ldots r_n\right) = E\psi\left(r_1 r_2 \ldots \ldots r_n\right) \tag{6.1}$$

The Hamiltonian operator (H) is a summation of three individuals, for example, the kinetic energy, the interactive part of the external potential (V_{ext}), and the interaction between two electrons (V_{ee}). The external potential, which is the force that is acting between atomic nuclei and electron, plays an important role in the simulation of materials which is shown in Equation 6.2.

$$V_{ext} = \sum_{\alpha}^{N_{at}} \left| Z_\alpha \middle/ r_i - R_\alpha \right| \tag{6.2}$$

DOI: 10.1201/9781003323518-8

In Equation 6.2, $\mathbf{r}i$ is the electron coordinate for i and R_α is the charge on the nucleus at Z_α. Equation 6.1 is solved for a set of external potential which subject to the restraint that the Ψ are antisymmetric and the sign of electrons are opposite if their coordinates are interchanged. The lowest energy eigenvalue, E0, is the ground state energy and the probability density of finding an electron with any particular set of coordinates $\{\mathbf{r}i\}$ is $|\Psi\ 0|^2$. The average energy can be formulated as shown in Equation 6.3.

$$E(\Psi) = \int \Psi^* H\Psi \, dr \qquad (6.3)$$

Ψ describes the wave function of the functionals. Ψ_0 is the ground state wavefunction which is related to energy as stated in Equation 6.4.

$$E(\Psi) \geq E_0 \qquad (6.4)$$

Hartree-Fock theory can also be implemented for Ψ, which is assumed to be a product of Φ_I which is obtained for a single electron coordinate system, which is described in Equation 6.5.

$$\Psi_{HF} = \frac{1}{\sqrt{N!}} \det\left[\varnothing_1 \varnothing_2 \ldots \varnothing_N\right] \qquad (6.5)$$

The Hartree-Fock equations explain electrons that are noninteracting under the pressure of an average field potential in the circumstances of classical Coulomb potential and a nonlocal exchange potential (Callaway and March 1984, Parr and Yang 1989, March 1999, N. M. Harrison 2003). The year-wise evolution of DFT is shown in Figure 6.1.

Moreover, it can be formulated for many-body complex electronic structure calculations where nuclei of the treated molecules are assumed as fixed with external static potential V and the wave function is Ψ. Therefore, after solving the time-independent Schrodinger equation, formalisms of DFT are implemented. DFT is often described as an ab initio method by the researchers. Therefore, for an N-electron system DFT can be formulated as shown in Equation 6.6. In this equation H is Hamiltonian, U is the energy between two electrons, and E is the total energy for kinetic energy T and potential energy V, which are received from the external field due to positively charged nuclei.

$$H\Psi = [T + V + U]\Psi = \left[\sum_i^N \left(\frac{\hbar^2}{2m_i}\Delta_i^2\right) + \sum_i^N V(r_i) + \sum_{i<j}^N V(r_i r_j)\right]\Psi = E\Psi \qquad (6.6)$$

DFT provides an appealing alternative to implement the key variable, that is, electron density, which is represented as n(r). Therefore normalized Ψ is given by Equation 6.7:

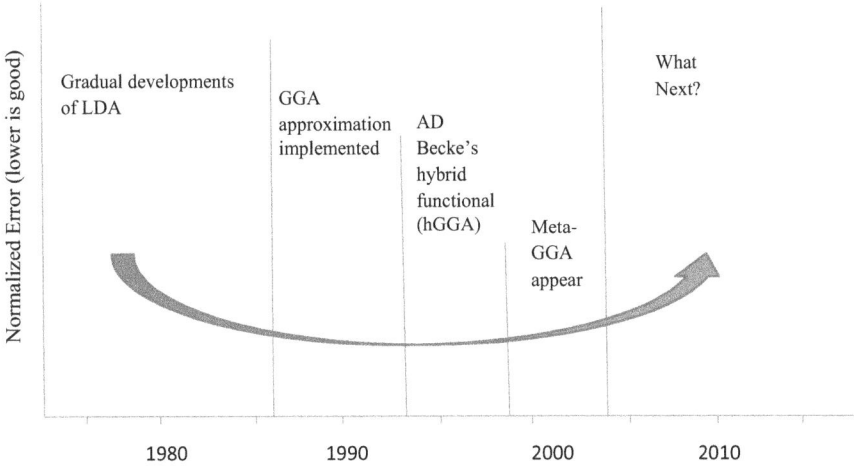

FIGURE 6.1　DFT evolution year wise as per electron density.

$$n(r) = N \int d^3 r_2 \ldots \ldots \int d^3 r_N \Psi^* \left(r r_2 \ldots \ldots r_N \right) \Psi(r r_2 \ldots \ldots r_N). \qquad (6.7)$$

where Ψ is a unique function of n_0 as shown in Equation 6.8.

$$\Psi_0 = \Psi\left[n_0 \right] \qquad (6.8)$$

In the context of computational material science and chemistry, DFT plays a crucial role to allow the prediction and calculation of material behavior on the basis of quantum mechanical considerations. It does not require higher-order fundamental material properties as its parameters, whereas it is the technique to calculate many-body electronic systems and also evaluate potentials that act on the system. The origin of DFT is the Hohenberg-Kohn theorem which was formulated by Walter Kohn and Pierre Hohenberg in the era of the 70s (Hohenberg and Kohn 1964). Figure 6.2 shows the flowchart of solving Schrodinger equation in solids.

Generally, DFT has its application in the field of molecular chemistry and materials science to determine and predict complex system behavior on the atomic scale. To be specific, DFT-based approaches are used to synthesize many-body complex systems. For example, the effect of dopants on phase transformation behavior in oxides and magnetic behavior in dilute magnetic semiconductor materials and study and investigation of the magnetic and electromechanical properties for semiconductor materials, etc. are easier with this computational method. (Michelini et al. 1998, Segall et al. 2002, Rastegar et al. 2013, 2014, Music et al. 2016). Figure 6.3 shows the complete DFT in a glance.

Concept of solving Schrödingers equation in solids

Relativistic treatment of electron	Form of potential	Exchange & corelation potential
Non relativistic Semi relativistic Fully relativistic	Non-self consistent Muffin-tin (MT) Atomic Sphere Approximation (ASA) Full potential (FP) Pseudo potential (PP)	Hartree-Fock (+correlation) Density Functional Theory (DFT) Local Density Approximation (LDA) Generalized Gradient Approximation (GGA) Beyond LDA (ex: LDA+U)

$$\left[-\frac{1}{2}\nabla^2 + V(r)\right]\varphi_i^k = \varepsilon_i^k \varphi_i^k$$

Schrödinger equation
(Kohn-Sham equation)

Representation of Solid	Basis Function	Treatment of Spin
Non periodic (Cluster) Periodic (Unit cell)	Plain waves (PW) Augmented Plain Waves (APW) Atomic orbitals (ex: Slater (STO), Gaussians (GTO), Numerical basis	Non spin polarized Spin polarized (with certain magnetic order)

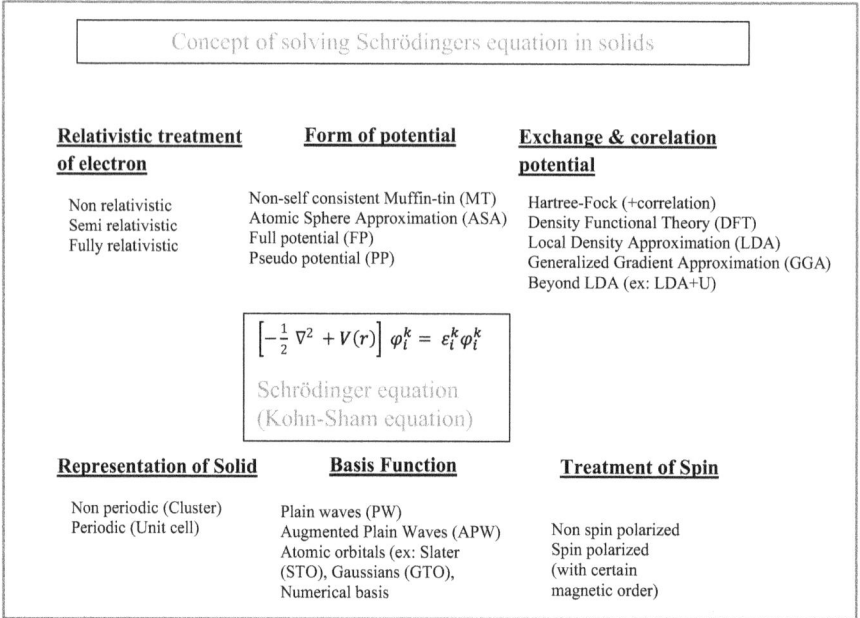

FIGURE 6.2　Flowchart of solving Schrodinger equation (for example, in solid).

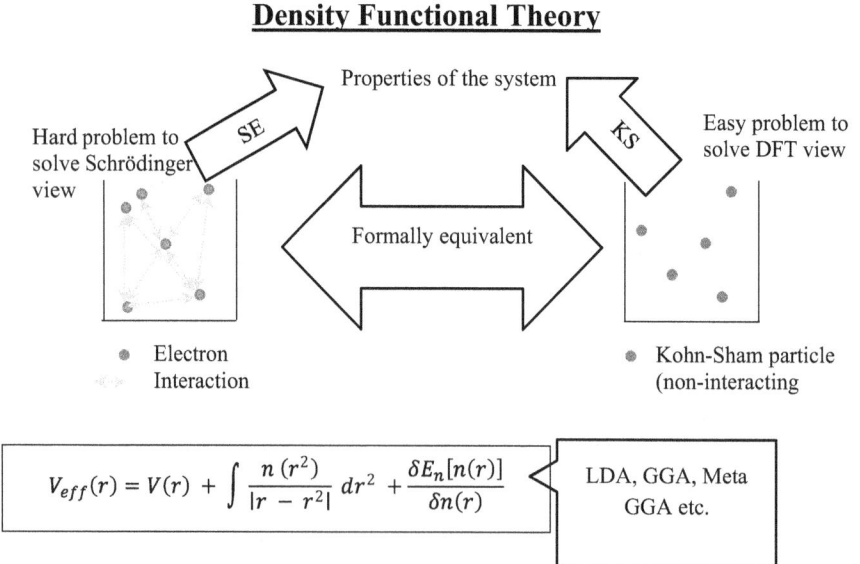

Density Functional Theory

Properties of the system

Hard problem to solve Schrödinger view　SE

KS　Easy problem to solve DFT view

Formally equivalent

Electron Interaction

Kohn-Sham particle (non-interacting

$$V_{eff}(r) = V(r) + \int \frac{n(r^2)}{|r - r^2|} dr^2 + \frac{\delta E_n[n(r)]}{\delta n(r)}$$

LDA, GGA, Meta GGA etc.

FIGURE 6.3　DFT at a glance.

6.3 NONEQUILIBRIUM GREEN'S FUNCTION

The quantum-ballistic transport phenomenon can be satisfactorily explained with the help of nonequilibrium Green's function (NEGF) as shown in Figure 6.4. It is well described for mesoscopic and nanoscale systems. These nanoscale and mesoscopic systems are the thrust area for researchers nowadays. This low-dimension quantum-ballistic transmission phenomenon is to be divided into two parts: stationary and time-dependent occurrence. NEGF is also known as Keldysh formalism, which is widely used to describe quantum transport phenomena in the nanoscale regime. NEGF helps to investigate and predict the interaction of particles within a time-dependent many-body system. Time-dependent current-voltage (I-V) characteristics can be eventually solved by using NEGF formalisms which are divided into two basic phases: (a) static and (b) time-dependent electronic transport in mesoscopic systems. The main difference between ordinary equilibrium theory and NEGF formalisms is that all time-dependent functions are defined for time arguments on a contour, which is called the Keldysh contour. To create equilibrium perturbation theory it is to be considered that the complex system should return to its first state as t → +∞, but in NEGF theory this may not work as the initial state at t = −∞ can be dissimilar from the final state t = +∞. Therefore, in the NEGF formalisms any reference to large times should be avoided.

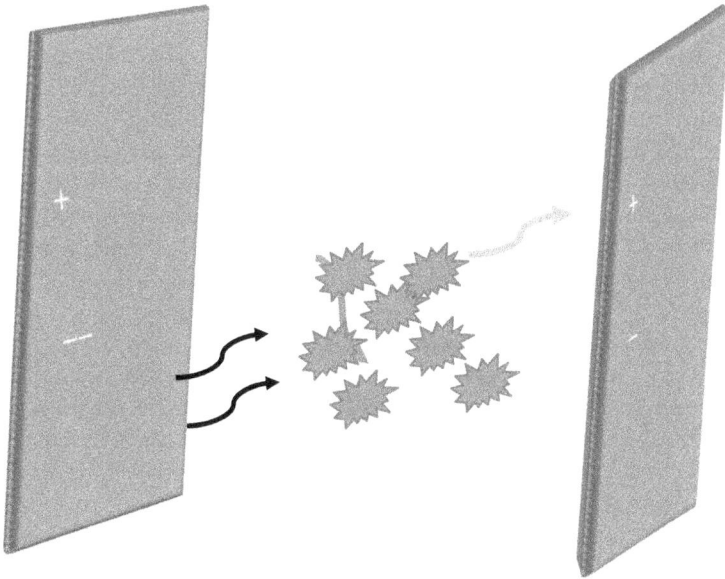

FIGURE 6.4 NEGF-based fundamental molecular modeling with a two-electrode experimental set.

6.4 MOLECULAR MODELING OF INORGANIC NANODEVICES

Shrinking of conformist electronic devices is the maximum developing research area nowadays. There are numerous tactics that lead to inspiring investigators to explore their research work and also investigate the nature of nanoscale devices. The greatest significant method is to model a design and simulate the analytical nanoassemblies (Serge L. Rudaz 1998, Yamamori et al. 1998, Gao and Kahn 2003, Kahn et al. 2003, 2006, An et al. 2012, Ling et al. 2015, Yu et al. 2015, Dey, Roy, De, et al. 2017). Frequently, it is noteworthy that the molecular nanoscale devices are designed using this simulation process and the gained results analyzed. Depending on the consequences, the scientists can adapt the numerous simulation limitations in addition to the diverse features of the nanoscale investigative prototypical. Among these simulation approaches, the first principle style is the most actual and prevalent procedure. Transformation of the features of nanoscale molecular devices inspires scientists to revolutionize conformist devices in an adapted form. For instance, traditional semiconductor devices can be intended using biomolecules. In the case of biomolecules usually the nucleobases such as cytosine, guanine, adenine, and thymine are measured that are recognized as the elementary structural building slabs of DNA. It is extremely reciprocated to develop conservative inorganic semiconductor devices in the area of nanoscale technology. Though, it is tough to design and model organic nanoscale electronic devices, which are mostly using bioinspired materials. These semiconductors are termed doped semiconductor, which is generally depending on the doping characteristics. If the semiconductor does not have any contamination doping, then it is termed an intrinsic or pure semiconductor. Instead, if the semiconductor is doped with external atoms or molecules then it is known as an extrinsic or impure semiconductor (Chen et al. 2011, Deng et al. 2012, A. Shah and Rashid Dar 2014, Biswal et al. 2016, Dey, Roy, and De 2017a, Anasthasiya et al. 2018, Brahma et al. 2018). In this work, instead of the conventional doping process, another interesting doping process has been introduced. This doping process is known as electrical doping. This procedure has several advantages over the conventional doping process. For example, the basic hazards of conventional doping like disrupters of ionic or covalent bonding and heat generation have been avoided using the electrical doping process. In this work, we have dealt with the theoretical electrical doping process and applied to design several nanoelectronic devices and achieved successful results in almost every case. Therefore, these nanoelectronic devices have been made with the electrical doping process, where the doping concentration has been changed by using some mathematical formulas, where the length, width, and height of the nanodevice play an important role. The doping concentration can be varied by changing the potential at the two ends of the nanodevices.

6.5 ELECTRICAL DOPING FOR THE ORGANIC AND INORGANIC NANODEVICES

The manner by which a potential change has been shaped amid the two ends of the nanodevice. This theoretic simulation is known as the electrical doping technique. Using this property we have decided to provide the different polarity with similar valued voltage at the two ends of the nanodevice through two probe electrodes. The graphic illustration for this hypothetical process is already shown in Figure 6.1. This notional tactic is intricate to create extremely doped positive (p+) and negative (n+) regions, which are significant to project nanosemiconductor devices both for organic and inorganic resources. This doping process is unconditionally dissimilar from the conservative doping process. In addition to this conformist progressive doping method, the semiconductor material is doped with external dopants or foreign contamination. This procedure is mainly temperature-dependent process. It is reported that there is a probability of coincidental breaking of bonds that may occur throughout this large hike in temperature doping development. The ionization technique is also accepted to instigate this doping technique. Furthermore, electrical doping development is not associated with contaminations at all. Thus, in this technique, opposite potential charges are introduced at the two ends of the device. Therefore, it will produce a potential drip in the main or central scattering molecular region of the nanodevice. This scheme is acceptable in the researchers' forum for nanodevice modeling since the ionization method may cause structural distortion for the nanomaterials. In the case of conservative doping, there are numerous glitches may arise. Some of them are recorded below:

1. Feeble contact detained among dopant and mass molecules.
2. Compatibility problems arise.
3. Heat generation happens throughout the ion implantation technique.
4. Interatomic response may rise among the host and dopant.

These aforesaid matters arise when the conservative doping course is applied to style nanoscale devices by means of molecular thin films. The electrical doping process is an interesting phenomenon where no external or foreign molecules are introduced to attain the desired potential drop among the system. Whereas in this case a small amount of equal but opposite polarity molecules are necessary to introduce the potential drop, which acts as the driving force of the carrier movements. This is mostly important for thin film nanoscale devices. Due to the several advantages of this electrical doping process it is thus implemented in molecular-level device designing procedure. It is therefore used to modify and control the characteristics of thin film molecular interfaces. Even more, one can implement this process at room temperature. This process avoids any sort of contamination into the molecules. Thus it is also helpful for organic materials, which are sensitive to impurities and reactive to foreign molecules. This film technology mostly offers this electrical doping to get a more accurate and error-free evaluation of molecular devices. The successful implementation of this electrical doping phenomenon is important to provide more accurate results, as in this case, traditional p and n dopants are not required. To improve the device efficiency, both the attractive carrier injection and potential voltage drop are important features that act as the leading driving force of the molecular devices to exaggerate the carrier movements through the central molecular region. So eventually, electrical doping is the supplement of electron donation or electron acceptation to the tinny molecular films. Yamamori et al. well described the competence of the electrical doping technique, which is important for thin film molecular device designing. They demonstrated that n and p dopants of a hole transport poly-carbonate polymer with tris(4-bromophenyl)aminium hexachloroantimonate (TBAHA), which is allowed to spread the width of polymer up to 500 nm along with lowering the drive voltage to 5 V to 6 V. Nowadays organic LEDs (OLEDs) are garnering lots of attention from the scientists. The electroluminescence effect of OLEDs is effectively useful and demonstrated by the researchers (Gupta and Jaiswal 2015, Zahir et al. 2016). In III-V semiconductor compounds like GaN, AlGaN, etc., n-type doping development is important for the possible implementation of nanoscale semiconductor modeling. This is successfully done by the introduction of the electrical doping process. Introducing this doping phenomenon forward voltage, carrier injection, carrier recombination, and electrical resistance are conceivable without cracking in the interfacing layer (Ghosh et al. 2014). One of the most well-known areas where electrical doping is being introduced is in organic photovoltaic cells (OPVCs). Electrical doping is an important parameter when it acts on the conductivity of those cells as well as tunes the energy level arrangement in OPVC (Ghosh and Mahapatra 2013a). Carrier inoculation is improved with this electrical doping practice. This procedure helps to make a depletion area in metal and organic layers at which the quantum-ballistic tunneling process happens at room temperature. The carrier inclusion process is stimulated by the driving force, which is an impulse by the electrical doping phenomenon. A useful alternative for all these driving forces is initiated by the potential drop, which arises due to the electrical doping process in inorganic contacts (Lam and Liang 2009). Electrical doping widely changes the charge impartiality levels of the molecular interfaces without affecting the characteristics of the inorganic thin films. Generally, 0.1% to 1% molecular dopants can be assimilated using this type of doping procedure for molecular tinny and thin films. Hefty doping concentration normally helps the successful generation of degenerate semiconductors. This type of doping stops the formation of doping-induced bands, as well as delivering great doping concentrations (Bai et al. 2006).

When the molecular chain passes through the nanopore of a GaAs nanosheet then the hetero junction molecular chain is used to sense several gases. In that case, also electrical doping is induced at the two parts of this nanosheet. Due to the effective inductance this biomolecular chain demonstrates its aptitude for detecting the adsorbed foreign gas molecules (Khazaei et al. 2007). Vertex B-N mixture doping can be applied (Bai et al. 2007, Li et al. 2008). Other external dopants like Te, Sb, As, and Cr are applied during this conventional doping process (Bai et al. 2008). In the case of this conventional doping process B and N are truly important dopants as the limit of doping can be adjusted according to the need of experiment (Min et al. 2009, Liu et al. 2011). The case of nanoscale device designing also deals with the adsorption of the molecules. For instance, the adsorption of some volatile molecules into ZnO nanowire at 32°C temperature is investigated (El-Hendawy et al. 2011). A first principles paradigm which is based on DFT and NEGF calculations is used to design nano-FET and various structural modifications at the molecular level. Numerous characteristics of these nano-FETs are also proposed with their various activities for specimen, maximum obtainable current assessment, highest occupied molecular orbital (HOMO) and lowest unoccupied molecular orbital (LUMO) gaps measurement, scalability assessment and RF performance, linearity investigation, and many more (Dey et al. 2015, 2016, Dai et al. 2017, Dey, Roy, and De 2017a, 2017b, Harada et al. 2017, Wang et al. 2017, Dey and De 2018). Conjugated pairs of the cooligomers-based molecular diode are designed and proposed using DFT- and NEGF-based calculations. The cooligomers are therefore useful to be connected with two electrodes and form an arrangement of a molecular diode. The energy gap, current-voltage (I-V) characteristics, and spatial orientation are analyzed for this diode (Geng et al. 2009). Using the first principles approach, the geometrically optimized nanostructures of seven different junctions are derived from carbon nanotube (CNT) using different linkers (Chakraverty et al. 2013). Several types of diodes can be designed and their characteristics are analyzed using the first principles paradigm. For example, bipolar spin diode, diblock molecular diode, backward diode characteristics, Schottky diode, single molecular diode, spin current diode, and many more are therefore reported using this approach (Geng et al. 2008, Nadimi et al. 2010, Ghosh and Mahapatra 2013b, Chakraverty et al. 2016, Li et al. 2016, Dey, Roy, De, et al. 2017).

Nowadays, nanoscale device designing is a challenging facet for scientists. Transistor diode and logic gates are implemented at the molecular level. There is another possibility for researchers to contrivance nanobiosemiconductor devices at the molecular level. Some of these biomolecular devices have already been introduced in the arena of biomedicine. The Atomistix-Tool Kit and Virtual Nano Laboratory (ATK-VNL)–based QuantumWise software simulator is mainly used to design and simulate these nanoscale devices. This simulation results in the theoretical and analytical implications of the molecular and atomic structures (Rudaz and Serge L 1998, Zhong and Hane 2012, Mativetsky and Mativetsky 2015, Palla et al. 2016, Tabe et al. 2016, Hsu et al. 2017, Dey Debarati and Debashis De 2018). Even more, Quantum Cellular Automata (QCA) logic design can be theoretically proposed using DFT- and NEGF-based first principles paradigm (Purkayastha et al. 2016). Various types of logic gates and their characteristics at the nanoscale are defined, analyzed, and proposed at room temperature. It is possible to design universal logic gates using bioinspired molecules, and the analyzed results obtained from these theoretical investigations are validated with the results obtained from Multi-Sim or SPICE or other simulators (Dey Debarati and Debashis De 2018). The electrical doping process is one of the advantageous processes, and thus it steals the attention of many nanodimension researchers. In this process no external dopant or foreign molecules are required for the successive potential drop. Quantum-ballistic tunneling effect is somehow exaggerated or diminished due to the effect of the back-scattering effect. This phenomenon happens due to the channel length being smaller than the mean free path. Henceforth, electrical doping is an interesting and effective phenomenon by implementing which we can avoid the related problems with the conventional doping process. The dipole combination model is introduced at the metal-semiconductor interface molecular level for Schottky barrier tuning, which is also suggested by Har-Lavan et al. (2012). DFT approaches and the NEGF paradigm are also applicable and useful for magnetic tunnel junctions and for the investigation of their quantum electronic effects, which have been analyzed by Galanakis and Mavropoulos (2007). To calculate and to ensure the effect of leakage current through SiO_2 and SiO_xN_y-based MOSFET, researchers use the DFT- and NEGF-based first principles paradigm (Nadimi et al. 2010). The ab initio demonstration is an important approach that is applied to the modeling of Schottky barrier height tuning using the yttrium and nickel silicide atomic-scale interface (Gao and Qun 2014). Direct band-to-band tunneling effect in reverse biased direction for MOS_2 p-n junction nanoribbon can be demonstrated, investigated, and described using DFT and NEGF calculations (Azad et al. 2021). The effect of the amalgamation of opposite-polarity dopant atoms into the nanowire displays excellent electrical properties, which resemble Zener diode characteristics (Tutuc et al. 2006). It is important for any sensor device regarding the detection process of various gases using biomolecular heterojunction chain through multilayer GaAs nanopore, which is implemented by using DFT and NEGF analytical methods. The theoretical approach for the implementation of electrically doped biomolecular switch is designed, and their characteristics are analyzed at room temperature when single-wall carbon nanotubes (SWCNTs) act as electrodes. Analysis of graphene-based antidot resonant tunnel diode using NEGF formalisms helps to design and investigate its characteristics. Atomistic characteristics of two-dimensional silicon p-n junctions are demonstrated along with their electrical properties analysis using first principles formalisms (Padilha et al. 2017, Xu et al. 2019, Chabi and Kadel 2020, Dey et al. 2021). Diodes and transistors are the main functional building slabs of any electronic circuitry. Logic gates are also realized using diodes and transistors. Finally it is concluded that any logic can be implemented using first principles formalisms.

REFERENCES

A. Shah, K. and Rashid Dar, J., 2014. Investigation of Doping Effects on Electronic Properties of Two Probe Carbon Nanotube System: A Computational Comparative Study. *International Journal of Innovative Research in Science, Engineering and Technology*, 3 (11), 17395–17402.

An, Y., Wei, X., and Yang, Z., 2012. Improving electronic transport of zigzag graphene nanoribbons by ordered doping of B or N atoms. *Physical Chemistry Chemical Physics*, 14 (45) , 15802–15806.

Anasthasiya, A.N.A., Ramya, S., Balamurugan, D., Rai, P.K., and Jeyaprakash, B.G., 2018. Adsorption property of volatile molecules on ZnO nanowires: computational and experimental approach. *Bulletin of Materials Science*, 41 (1), 1–7.

Azad, Z.A., Shokri, A., and Khezrabad, M.S.A., 2021. First-principle study on the quantum transport in an armchair silicene nanoribbon-based p–n diode. *Materials Chemistry and Physics*, 257, 123483.

Bai, P., Chong, C.C., Li E.P., and Chen, Z., 2006. A molecular diode based on conjugated co-oligomers. International Journal of Nanoscience, 5 (04n05), 535–540.

Bai, P., Kai, T.L., Li, E., and Ken, K.F.C., 2007. A comprehensive atomic study of carbon nanotube Schottky diode using first principles approach. *In: Technical Digest – International Electron Devices Meeting*, 749–752.

Bai, P., Li, E., Lam, K.T., Kurniawan, O., and Koh, W.S., 2008. Carbon nanotube Schottky diode: An atomic perspective. *Nanotechnology*, 19 (11), 115203.

Biswal, S.M., Baral, B., De, D., and Sarkar, A., 2016. Study of effect of gate-length downscaling on the analog/RF performance and linearity investigation of InAs-based nanowire Tunnel FET. *Superlattices and Microstructures*, 91, 319–330.

Brahma, M., Kabiraj, A., Saha, D., and Mahapatra, S., 2018. Scalability assessment of Group-IV mono-chalcogenide based tunnel FET. *Scientific Reports*, 8 (1), 5993.

Calais, J.-L., 1993 Density-functional theory of atoms and molecules. *International Journal of Quantum Chemistry*, 47 (1), 101–101.

Callaway, J. and March, N. H., 1984. Density functional methods: theory and applications. *Solid State Physics*, *38*, 135–221.

Chabi, S. and Kadel, K., 2020. Two-dimensional silicon carbide: Emerging direct band gap semiconductor. *Nanomaterials*, 10 (11), 2226.

Chakraverty, M., Harisankar, P.S., Gupta, K., Ruparelia, V., and Rahman, H., 2016. Simulation of electrical characteristics of silicon and germanium nanowires progressively doped to zener diode configuration using first principle calculations. In *Microelectronics, Electromagnetics and Telecommunications: Proceedings of ICMEET 2015* (pp. 421–428). Springer India.

Chakraverty, M., Kittur, H.M., and Kumar, P.A., 2013. First Principle Simulations of Various Magnetic Tunnel Junctions \$\kern-2pt\$ for\$\kern-2pt\$ Applications in Magnetoresistive Random Access Memories. *IEEE Transactions on Nanotechnology*, 12 (6), 971–977.

Chen, L.N., Ma, S.S., Ouyang, F.P., Xiao, J., and Xu, H., 2011. First-principles study of metallic carbon nanotubes with boron/nitrogen co-doping. *Chinese Physics B*, 20 (1), 017103.

Dai, X., Zhang, L., Li, J., and Li, H., 2017. Metal-Semiconductor Transition of Single-Wall Armchair Boron Nanotubes Induced by Atomic Depression. *Journal of Physical Chemistry C*, 121 (46), 26096–26101.

Deng, X.Q., Zhang, Z.H., Tang, G.P., Fan, Z.Q., Qiu, M., and Guo, C., 2012. Rectifying behaviors induced by BN-doping in trigonal graphene with zigzag edges. *Applied Physics Letters*, 100 (6), 063107.

Dey, D. and De, D., 2018. A first principle approach toward circuit level modeling of electrically doped gated diode from single wall thymine nanotube-like structure. *Microsystem Technologies*, 24 (7), 3107–3121.

Dey, D. and De, D., Ahmadian, A., Ghaemi, F., and Senu, N., 2021. Electrically doped nanoscale devices using first-principle approach: a comprehensive survey. *Nanoscale Research Letters* 16, 1–16.

Dey, D., Roy, P., and De, D., 2016. Electronic characterisation of atomistic modelling based electrically doped nano bio p-i-n FET. *IET Computers and Digital Techniques*, 10 (5), 273–285.

Dey, D., Roy, P., and De, D., 2017a. Atomic scale modeling of electrically doped p-i-n FET from adenine based single wall nanotube. *Journal of Molecular Graphics and Modelling*, 76, 118–127.

Dey, D., Roy, P., and De, D., 2017b. Design and electronic characterization of bio-molecular QCA: A first principle approach. *Journal of Nano Research*, 49, 202–214.

Dey, D., Roy, P., De, D., and Ghosh, T., 2017. Detection of ammonia and phosphine gas using heterojunction biomolecular chain with multilayer GaAs nanopore electrode. *Journal of Nanostructures*, 7 (1), 21–31.

Dey, D., Roy, P., Purkayastha, T., and De, D., 2015. A First Principle Approach to Design Gated p-i-n Nanodiode. *Journal of Nano Research*, 36, 16–30.

Dey, D., and De, D., 2018. Electrically Doped Adenine Based Optical Bio Molecular pin Switch with Single Walled Carbon Nanotube Electrodes. *Journal of Active & Passive Electronic Devices*, 13 (2/3), 107–118.

El-Hendawy, M.M., El-Nahas, A.M., and Awad, M.K., 2011. The effect of constitutional and conformational isomerization on the electrical properties of diblock molecular diode. *Organic Electronics*, 12 (6), 1080–1092.

Galanakis, I. and Mavropoulos, P., 2007. Spin-polarization and electronic properties of half-metallic Heusler alloys calculated from first principles. *Journal of Physics Condensed Matter*, 19 (31), 315213.

Gao and Qun, 2014. *Contact resistance in semiconductor devices: Physics and modeling.* University of Florida.

Gao, W. and Kahn, A., 2003. Electrical doping: The impact on interfaces of π-conjugated molecular films. *Journal of Physics Condensed Matter*, 15 (38), 2757.

Geng, L., Magyari-Köpe, B., and Nishi, Y., 2009. Image charge and dipole combination model for the Schottky barrier tuning at the dopant segregated metal/semiconductor interface. *IEEE Electron Device Letters*, 30 (9), 963–965.

Geng, L., Magyari-Kope, B., Zhang, Z., and Nishi, Y., 2008. Ab initio modeling of Schottky-barrier height tuning by yttrium at nickel silicide/silicon interface. *IEEE Electron Device Letters*, 29 (7), 746–749.

Ghosh, R.K., Brahma, M., and Mahapatra, S., 2014. Germanane: A low effective mass and high bandgap 2-D channel material for future FETs. *IEEE Transactions on Electron Devices*, 61 (7), 2309–2315.

Ghosh, R.K. and Mahapatra, S., 2013a. Proposal for Graphene-boron nitride Heterobilayer-based tunnel FET. *IEEE Transactions on Nanotechnology*, 12 (5), 665–667.

Ghosh, R.K. and Mahapatra, S., 2013b. Direct band-to-band tunneling in reverse biased MoS2nanoribbon p-n junctions. *IEEE Transactions on Electron Devices*, 60 (1), 274–279.

Gupta, S.K. and Jaiswal, G.N., 2015. Study of Nitrogen terminated doped zigzag GNR FET exhibiting negative differential resistance. *Superlattices and Microstructures*, 86, 355–362.

Harada, N., Jippo, H., and Sato, S., 2017. Theoretical study on high-frequency graphene-nanoribbon heterojunction backward diode. *Applied Physics Express*, 10 (7), 074001.

Har-Lavan, R., Yaffe, O., Joshi, P., Kazaz, R., Cohen, H., and Cahen, D., 2012. Ambient organic molecular passivation of Si yields near-ideal, Schottky-Mott limited, junctions. *AIP Advances*, 2 (1), 012164.

Harrison, N. M. 2003. An introduction to density functional theory. *Computational Materials Science*, 187, 45.

Haruk, A. M. and Mativetsky, J. M., 2015. Supramolecular approaches to nanoscale morphological control in organic solar cells. *International Journal of Molecular Sciences* 16 (6), 13381–13406.

Hohenberg, P. and Kohn, W., 1964. Inhomogeneous electron gas. *Physical Review*, 136 (3B), B864.

Hsu, Y.C., Hung, Y.C., and Wang, C.Y., 2017. Controlling Growth High Uniformity Indium Selenide (In2Se3) Nanowires via the Rapid Thermal Annealing Process at Low Temperature. *Nanoscale Research Letters*, 12, 1–6.

Kahn, A., Koch, N., and Gao, W., 2003. Electronic structure and electrical properties of interfaces between metals and π-conjugated molecular films. *Journal of Polymer Science, Part B: Polymer Physics*, 41 (21), 2529–2548.

Kahn, A., Zhao, W., Gao, W., Vázquez, H., and Flores, F., 2006. Doping-induced realignment of molecular levels at organic-organic heterojunctions. *Chemical Physics*, 325 (1), 129–137.

Khazaei, M., Sang, U.L., Pichierri, F., and Kawazoe, Y., 2007. Computational design of a rectifying diode made by interconnecting carbon nanotubes with peptide linkages. *Journal of Physical Chemistry C*, 111 (33), 12175–12180.

Lam, K.T. and Liang, G., 2009. Computational study on the performance comparison of monolayer and bilayer zigzag graphene nanoribbon FETs. *In: Proceedings – 2009 13th International Workshop on Computational Electronics, 2009.*

Li, J., Gao, G., Min, Y., and Yao, K., 2016. Half-metallic YN $_2$ monolayer: dual spin filtering, dual spin diode and spin Seebeck effects. *Physical Chemistry Chemical Physics*, 18 (40), 28018–28023.

Li, Y., Yao, J., Liu, C., and Yang, C., 2008. Theoretical investigation on electron transport properties of a single molecular diode. *Journal of Molecular Structure: THEOCHEM*, 867 (1–3), 59–63.

Ling, Y.C., Ning, F., Zhou, Y.H., and Chen, K.Q., 2015. Rectifying behavior and negative differential resistance in triangular graphene p-n junctions induced by vertex B-N mixture doping. *Organic Electronics*, 19, 92–97.

Liu, H., Wang, N., Li, P., Yin, X., Yu, C., Gao, N., and Zhao, J., 2011. Theoretical investigation into molecular diodes integrated in series using the non-equilibrium Green's function method. *Physical Chemistry Chemical Physics*, 13 (4), 1301–1306.

March, N.H., 1999. *Electron Correlation in the Solid State*. Electron Correlation in the Solid State.

Michelini, M.C., Pis Diez, R., and Jubert, A.H., 1998. A density functional study of small nickel clusters. *International Journal of Quantum Chemistry*, 70 (4–5), 693–701.

Min, Y., Yao, K.L., Liu, Z.L., Cheng, H.G., Zhu, S.C., and Gao, G.Y., 2009. CrAs(0 0 1)/AlAs(0 0 1) heterogeneous junction as a spin current diode predicted by first-principles calculations. *Journal of Magnetism and Magnetic Materials*, 321 (4), 312–315.

Music, D., Geyer, R. W., and Schneider, J. M., 2016. Recent progress and new directions in density functional theory based design of hard coatings. Surface and Coatings Technology, 286, 178–190.

Nadimi, E., Planitz, P., Ottking, R., Wieczorek, K., and Radehaus, C., 2010. First principle calculation of the leakage current through SiO2 and SiOxNy gate dielectrics in MOSFETs. *IEEE Transactions on Electron Devices*, 57 (3), 690–695.

Padilha, J.E., Miwa, R.H., da Silva, A.J.R., and Fazzio, A., 2017. Two-dimensional van der Waals *p-n* junction of InSe/phosphorene. *Physical Review B*, 95 (19), 195143.

Palla, P., Ethiraj, A. S., and Raina, J. P. Resonant tunneling diode based on band gap engineered graphene antidot structures. AIP Conference Proceedings, vol. 1724, no. 1. AIP Publishing, 2016.

Purkayastha, T., De, D., Das, B., and Chattapadhyay, T. First principle study of molecular quantum dot cellular automata using mixed valence compounds. In *2016* 3rd International Conference on Devices, Circuits and Systems *(ICDCS)*, pp. 244–248. IEEE, 2016.

Rastegar, S.F., Hadipour, N.L., and Soleymanabadi, H., 2014. Theoretical investigation on the selective detection of SO2 molecule by AlN nanosheets. *Journal of Molecular Modeling*, 20 (9), 1–6.

Rastegar, S.F., Hadipour, N.L., Tabar, M.B., and Soleymanabadi, H., 2013. DFT studies of acrolein molecule adsorption on pristine and Al-doped graphenes. *Journal of Molecular Modeling*, 19 (9), 3733–3740.

Rudaz, S. L, inventor; Hewlett Packard Co, assignee. Maximizing electrical doping while reducing material cracking in III-V nitride semiconductor devices. United States patent US 5,729,029. 1998 Mar 17.

Segall, M.D., Lindan, P.J.D., Probert, M.J., Pickard, C.J., Hasnip, P.J., Clark, S.J., and Payne, M.C., 2002. First-principles simulation: Ideas, illustrations and the CASTEP code. *Journal of Physics Condensed Matter*, 14 (11), 2717.

Tabe, M., Tan, H.N., Mizuno, T., Muruganathan, M., Anh, L.T., Mizuta, H., Nuryadi, R., and Moraru, D., 2016. Atomistic nature in band-to-band tunneling in two-dimensional silicon pn tunnel diodes. *Applied Physics Letters*, 108 (9), 093502.

Tutuc, E., Appenzeller, J., Reuter, M.C., and Guha, S., 2006. Realization of a linear germanium nanowire p-n junction. *Nano Letters*, 6 (9), 2070–2074.

Wang, S., Wei, M.Z., Hu, G.C., Wang, C.K., and Zhang, G.P., 2017. Mechanisms of the odd-even effect and its reversal in rectifying performance of ferrocenyl-n-alkanethiolate molecular diodes. *Organic Electronics*, 49, 76–84.

Xu, Q., Li, W., Ding, L., Yang, W., Xiao, H., and Ong, W-J., 2019. Function-driven engineering of 1D carbon nanotubes and 0D carbon dots: mechanism, properties and applications. Nanoscale 11 (4), 1475–1504.

Yamamori, A., Adachi, C., Koyama, T., and Taniguchi, Y., 1998. Doped organic light emitting diodes having a 650-nm-thick hole transport layer. *Applied Physics Letters*, 72 (17), 2147–2149.

Yu, S., Frisch, J., Opitz, A., Cohen, E., Bendikov, M., Koch, N., and Salzmann, I., 2015. Effect of molecular electrical doping on polyfuran based photovoltaic cells. *Applied Physics Letters*, 106 (20), 203301.

Zahir, A., Pulimeno, A., Demarchi, D., Roch, M.R., Masera, G., Graziano, M., and Piccinini, G., 2016. EE-BESD: molecular FET modeling for efficient and effective nanocomputing design. *Journal of Computational Electronics*, 15 (2), 479–491.

Zhong, A. and Hane, K., 2012. Growth of GaN nanowall network on Si (111) substrate by molecular beam epitaxy. *Nanoscale Research Letters*, 7 (1), 1–7.

7 Nanoparticles in Biomedical Applications
MRI Contrast Agents

Panchanan Sahoo, Dipesh Choudhury,
Sudip Kundu, Snehasis Mishra,
Abhishek Mukherjee, and
Chandan Kumar Ghosh

7.1 INTRODUCTION

Nanotechnology is a rapidly emerging field of science that deals with the development of nanoparticles (NPs) of numerous chemical compositions, shapes, and sizes ranging between 10 and 1000 nm and their utility in human welfare (Kreuter 2007). NPs have gained their interest in different fields corresponding to electronic, optical, agriculture, biomaterials energy production, biomedical, etc., due to greater surface-to-volume ratio that provides them with some exceptional unique properties like electrical conductivity, thermal conductivity, melting point, scattering, wettability, light absorption, and catalytic activity (McNamara and Tofail 2015). However, it is commonly believed that the kernels for modern nanotechnology and nanomedicine were spread by Dr. Richard Feynman, who conveyed the famous lecture "There's Plenty of Room at the Bottom" during the annual meeting of the American Physical Society in 1959 (Junk and Riess 2006). He intellectualized the methods to influence specific atoms and molecules that would result in the alteration of their physical properties, flagging for molecular nanotechnology. In recent times, NPs promptly offer a wide range of biomedical applications such as biosensors, bioimaging, targeted drug delivery, cell labeling, gene delivery, and phototherapy due to some of their fascinating characteristics including nontoxicity, biocompatibility, chemical stability, structural property, defect induced property, near-infrared light absorption, and high saturation magnetization (Zheng et al. 2012, Jiang et al. 2014, Shah et al. 2015).

The progress in the field of nanobiotechnology is assisted by the development of dominant diagnostic techniques such as magnetic resonance imaging (MRI), computed tomography (CT), photoacoustic imaging (PAI), positron emission tomography (PET), and ultrasound imaging that can be used as biomarkers for disease identification, progress, and treatment response.

DOI: 10.1201/9781003323518-9

MRI is a very useful noninvasive diagnostic technique in radiology used for the visualization of the anatomy and physiological processes and internal structure of the body. However, it is superior to other imaging modalities including high-resolution imaging, nonionizing radiation, better soft tissue contrast, and multiplanar imaging capabilities (Griewing et al. 1992). This technique is mainly based on the basic principle of nuclear magnetic resonance (NMR) in the presence of magnetic field gradients used to encode signals in all three directions. NMR signal can be influenced by the water density and transverse (r_2) and longitudinal (r_1) proton relaxivities of protons of the tissue. Different types of contrast agents (CAs) can be used to control image contrast by locally changing relaxation rates. The contrasting efficiency of a contrast agent depends upon r_2 and r_1 value. For CAs with $r_2 / r_1 < 2$, treat as positive or T_1 CA provides white spot, where they localized in the body such as different paramagnetic Gd^{3+}-based complexes widely used as a positive CA. Commercially available ProHance is also positive CA used in MRI diagnosis. On the other hand if $r_2 / r_1 > 2$, then they provide a dark signal known as negative or T_2 CA such as superparamagnetic iron oxide nanoparticles (SPIONs) (Perrier et al. 2013).

However, there exist a few U.S. Food and Drug Administration (FDA)–approved MRI Cas, but out of them only eight Cas are commercially available. Most of these Cas contain mononuclear Gd^{3+} ion, coordinated with single water molecules (Bussi et al. 2020). Among them Gd-EOB-DTPA can be used as a liver imaging CA while Gd^{3+} complex MS-325 can be used for angiographic imaging. Besides these, Mn^{2+} complex, known as Mn-DPDP, got approval for liver imaging, while superparamagnetic iron oxide nanoparticles (SPIONs) can be used for liver imaging (ferumoxide) and gastrointestinal imaging (ferumoxsil). Generally, these CAs are limited for clinical use due to their reduced specificity, low relaxivities, and overall toxicity to the human body (Gale and Caravan 2018).

In this chapter we propose the theoretical model based on the inner sphere Solomon-Bloembergen-Morgan (SBM) theory and outer sphere diffusion theory to explain the contrast mechanism of MRI CAs and different strategies have been attributed to enhance contrast efficiency of a CA. Next we describe different factors which affect the r_1 and r_2 value of a CA. After that we review different types of CA along with the limitation of commercially available Gd(III)-based CA. Finally, we survey different approaches to develop new type of CAs, including direct targeting approach and stimuli responsive approach to develop smart CA. In this context, particular attention has been dedicated to the diverse multimodal imaging CAs encompassing both Prussian blue (PB) and Prussian blue analogous (PBA) NPs which serve as MR-PA imaging-based bimodal CAs.

7.2 THEORETICAL BACKGROUND

Clinically used MRIs employ a dominant magnet to yield a sturdy magnetic field and alignment of body protons with the magnetic field. In the presence of external radiofrequency, the body protons are both enthused and spin out from equilibrium against the magnetic field. When resonance frequency is turned off, protons are relaxed back to the initial state in two pathways: one is spin flip, which is called T_1 or spin-lattice relaxation, and another is from spin in-phase to diphase, which is called T_2 or spin-spin relaxation. MR scans of diverse tissues, each exhibiting distinct relaxivity values, can be generated even in the absence of an exotic CA which relies on a range of mechanisms, including variances in chemical shifts between bodily water and mobile lipids, fluctuations in water proton density across different tissues, certain physiochemical properties of water such as diffusion and velocity, and the exchange of magnetization between macromolecules and water. The relaxivity of a CA is the rate of change of relaxation time per unit concentration of the magnetic CA. A local magnetic field is produced in the presence of CA, which improves the contrast of an image by accelerating the proton relaxation rate. Theoretical statement for the calculation of relaxivity value for simple molecules is given below:

$$\frac{1}{T_j} = \frac{1}{T_d} + [C] r_j; j = 1, 2 \qquad (7.1)$$

where T_d is the relaxation time related to the diamagnetic solvent (water); T_1 and T_2 are longitudinal and transverse time related to solution, respectively; and C is the concentration of the CA (Li, Chen, et al. 2014). Here in this section we will briefly describe the interaction between CA and H_2O molecules that affect relaxivity based on some classical models.

7.2.1 SOLOMON-BLOEMBERGEN-MORGAN (SBM) THEORY

Bloembergen et al. (1948) pioneered the SBM theory, concerning dipole-dipole interactions, which are widely regarded as the most important mechanism behind the T_1 and T_2 relaxations of water protons; nevertheless, the initial formulation of dipolar interaction was confined to proton-proton interactions. Thereafter, Solomon, Bloembergen, and Morgan modified this theory by introducing proton-electron interaction for relaxivity (Bloembergen and Morgan 1961). Later it was observed that electron-proton interaction has a greater impact on both relaxivities due to the smaller mass and high spin magnetic moment of the electron. Dipole interaction is basically directed by the communication between spins such as relative motion, distance, angle, and type of spins (Chen et al. 2020). There are a lot of good literature reviews where dipole interaction between two protons has been explained with mathematical equations. Briefly, this interaction depends on the fourth power of proton gyromatic ratio, spectral density, Larmor frequency, correlation time, and inverse proportion to the sixth power of the distance between the magnetic metal and water protons. The motion of any molecules can be justified by translation, vibration, and rotation, where only rotational motion plays an important role in relaxivities as it happens in the range of Larmor frequency. In this context, it can be stated that both relaxations rates are associated with bulk H_2O molecules which are shorter than coordinated H_2O molecules and are responsible to create contrast in different organs like the liver,

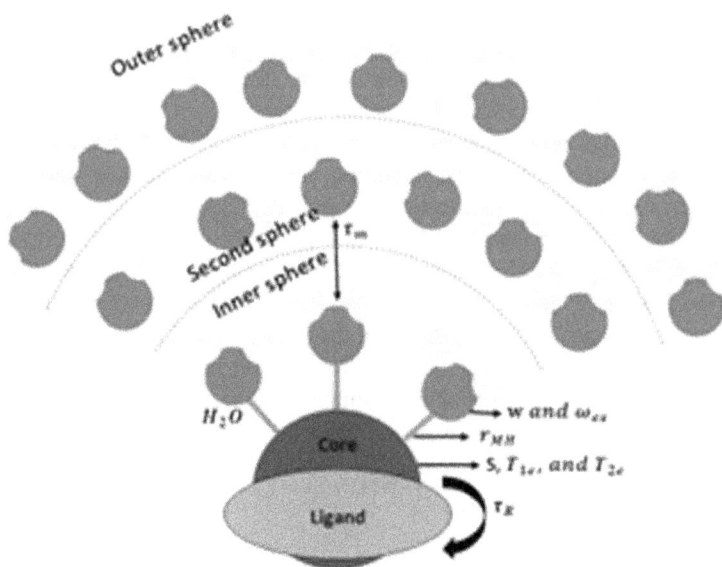

SCHEME 7.1   Water interaction in relaxivity mechanism.

brain, etc. However, electron-proton interaction between the paramagnetic metal ion and water proton is the core mechanism for T_1 relaxation.

From the chemical point of view, generally three contributions that accelerate relaxation can be taken under consideration (shown in Scheme 7.1). These are (1) inner-sphere relaxation, in which H_2O molecules directly coordinated to transition metal ion may be relaxed and this effect can be transmitted to the bulk water by exchangeable another H_2O molecules; (2) second-sphere relaxation, in which hydrogen-bonded zeolitic H_2O molecules with coordinated H_2O molecules or exchangeable hydrogen atoms can be relaxed and exchanged; and (3) outer-sphere relaxation, in which H_2O molecules may be diffused to the nearby paramagnetic center and can be relaxed. Therefore, it can be written as $r_1 = r_1^{IS} + r_1^{SS} + r_1^{OS}$ (Verwilst et al. 2015). Many factors are involved in such types of relaxation in the presence of CA such as (i) the number of paramagnetic metal ions existing in the CA; (ii) the effective electronic spin value for the operative paramagnetic site and its relaxation times, sometimes known as reversal frequency ($1/\tau_i$, $i = 1,2$); (iii) the number of coordinated H_2O molecules; (iv) the dynamics of the chemical exchange rate ($1/\tau_{ex}$, where τ_{ex} is the mean exchange correlation time) between H_2O molecules coordinated to metal ion and bulk water; (v) rotational dynamics, known as tumbling of the molecules (τ_R), which is also influenced by the size of the CA; (vi) the diffusion correlation time (τ_D) of H_2O molecules; (vii) metal-water hydrogen distance as an intrinsic factor; (viii) applied magnetic field strength and frequency; and (ix) temperature as an extrinsic factor. Because of the increment of Lamor frequency with the enhancement of the magnetic field, the relaxivity value decreases. In this connection Gd-DTPA, fast-tumbling small molecules, display a reasonable decrease in r_1 value with the field while MS-325, commercially available slow-tumbling molecules cramped to serum albumin, display an enhanced relaxivity value between 0.5 and 1 T as well as show a decrement of r_1 value with field (Rohrer et al. 2005). Briefly for most of the tissues, T_1 increases while T_2 remains relatively unchanged with increment of field strength. The shortest value of T_1 is present when molecular tumbling rate is nearly equal to the Lamor frequency, then its value further increases (Rohrer et al. 2005). Although Lamor frequency is directly proportional to the field strength, for highly mobile molecules such as free water, both T_1 and T_2 are less affected with magnetic field.

7.2.1.1 Inner Sphere Relaxivity

The inner sphere mechanism is mainly responsible for the enhancement of r_1 relaxivity in paramagnetic-type Cas, which is explained by some mathematical expressions, as follows:

$$r_1^{IS} = \frac{w \big/ [H_2O]}{\left(T_{1M} + \tau_m\right)} \tag{7.2}$$

$$r_2^{IS} = \left(w \big/ [H_2O] \right) * \left(\frac{1}{\tau_m} \right) * \left(\frac{T_{2M}^{-1}\left(\tau_m^{-1} + T_{2M}^{-1}\right) + \Delta\omega_{ex}^2}{(\tau_m^{-1} + T_{2M}^{-1})^2 + \Delta\omega_{ex}^2} \right) \tag{7.3}$$

$$\frac{1}{T_{1M}} = \left(\frac{2}{15}\right) * \left(\mu_0 \big/ 4\pi\right) * \left(\frac{\Upsilon_p^2 g^2 \mu_B^2 S(S+1)}{r_{MH}^6}\right) * \left[\frac{7\tau_{c2}}{\left(1 + \omega_e^2 \tau_{c2}^2\right)} + \frac{3\tau_{c1}}{\left(1 + \omega_p^2 \tau_{c1}^2\right)}\right] \tag{7.4}$$

$$\frac{1}{T_{2M}} = \left(\frac{1}{15}\right) * \left(\mu_0 \big/ 4\pi\right) * \left(\frac{\Upsilon_p^2 g^2 \mu_B^2 S(S+1)}{r_{MH}^6}\right) *$$
$$\left[4\tau_{c1} + \frac{13\tau_{c2}}{\left(1 + \omega_e^2 \tau_{c2}^2\right)} + \frac{3\tau_{c1}}{\left(1 + \omega_p^2 \tau_{c1}^2\right)}\right] + \frac{S(S+1)}{3} * \left(\frac{A}{h}\right)^2 * \tau_s \tag{7.5}$$

$$\frac{1}{\tau_{cj}} = \frac{1}{\tau_R} + \frac{1}{\tau_m} + \frac{1}{T_{je}}; \ j = 1,2 \tag{7.6}$$

$$\frac{1}{\tau_s} = \frac{1}{T_{1e}} + \frac{1}{\tau_m} \tag{7.7}$$

$$\frac{1}{T_{1M}} = \left(\frac{B}{r_{MH}^6}\right) * \left[\frac{3\tau_c}{(1+\omega_p^2 \tau_c^2)}\right] \tag{7.8}$$

$$r_1^{IS} = B * w * \tau_R \tag{7.9}$$

where the parameters are defined as:

T_{1M} = Longitudinal relaxation time

T_{2M} = Transverse relaxation time

τ_m = Mean exchange correlation time or residency time

w = no of H_2O molecules

ω_{ex} = Exchange correlation frequency

Υ_p = proton gyromagnetic ratio

g = Lande g factor

μ_B = Bohr magneton

S = electronic spin value

r_{MH} = metal – to – water hydrogen distance

τ_c = magnetic fluctuation

ω_e = Larmor frequency of electron

ω_p = Larmor frequency of proton

τ_s = corelation time for the interaction between transition metal ion and hydrogen

T_{je} = electronic relaxation time; $j = 1, 2$

B = physical constant

From SBM theory, the spin-lattice relaxation rate ($1/T_{1M}$) for bound nuclei depend on (i) scalar coupling and (ii) dipolar coupling between electron spin and nucleus. Scaler coupling between electron spin and nucleus has merely an effect with respect to dipolar coupling for Gd^{3+} metal ion but appreciable for Mn^{2+} and Fe^{3+} metal ion (shown in Equation 7.4). Moreover, this expression also indicates that T_{1M} mechanism directly depends upon electronic spin value and proton and electron gyromagnetic ratio. Since ω_e is very large compared to ω_p, Equation 7.4 reduces to Equation 7.8. Again T_{1M}, for simple metal CA containing coordinated H_2O molecules, is always high with respect to τ_m, so Equation 7.2 reduces to Equation 7.9. It is also observed that relaxivity decreases with the fast rotation of the CA. So relaxivity increases by enhancing the molecular weight of the CA. Again magnetic fluctuation (τ_c) can be caused by the coordinated H_2O molecules. So, in conclusion, the inner sphere relaxivity of a CA is greatly influenced by particle size, paramagnetic center, and the number of coordinated hydration states.

7.2.1.2 Electronic Relaxation

Besides this, SBM further includes that electronic relaxation rates are dominated by zero field splitting (ZFS) at a very low magnetic field (<0.1 T), which describes interelectronic interaction between unpaired electrons in the paramagnetic complex. The perturbed Hamiltonian for the lifting of the energy state can be written as follows:

$$\hat{H}_{ZFC} = \hat{S}.\hat{D}.\hat{S}D(\hat{S}_z^2 - \frac{1}{3}\hat{S}^2) + E(\hat{S}_+^2 + \hat{S}_-^2) \tag{7.10}$$

where E and D are the magnitude of D parallel and perpendicular to the z-axis. This perturbation involves lifting the energy state to 2S+1 degeneracy at zero field.

In case of high spin configuration, such as Mn^{2+}, Fe^{3+} complexes, ZFS energy can generally be disregarded in comparison to the Zeeman energy. However, some relaxometers like NMRD can be operated at a lower frequency, i.e., 10 kHz, then ZFS energy will appear, and then SBM theory explains the electronic relaxation effect. In this context it can be stated that transient ZFS will appear along with static ZFS due to solvent collision. The electronic relaxation rates are modeled by the following equations, where $-\tau_v$ is splitting correlation time, whereas Δ is known as ZFS energy. This effect can be modeled as follows:

$$T_{1e} = \frac{1}{25}\Delta^2\tau_v\left[4S(S+1)-3\right]\left[\frac{1}{1+\omega_e^2\tau_v^2} + \frac{4}{1+4\omega_e^2\tau_v^2}\right] \tag{7.11}$$

$$T_{2e} = \frac{1}{25}\Delta^2\tau_v\left[4S(S+1)-3\right]\left[\frac{5}{1+\omega_e^2\tau_v^2} + \frac{2}{1+4\omega_e^2\tau_v^2} + 3\right] \tag{7.12}$$

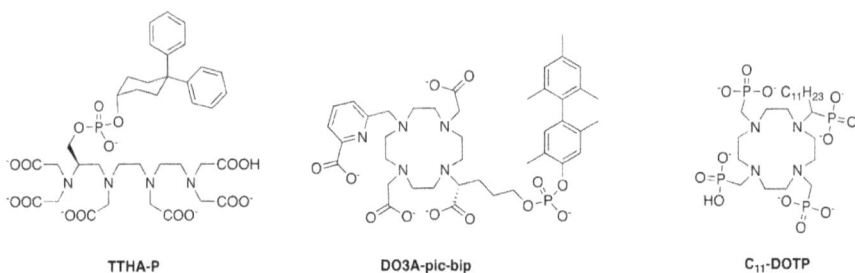

FIGURE 7.1 Albumin binding gadolinium-based complex with absence of coordinated water molecules.

Besides this, r_1^{SS} is associated with the influence of the second sphere H_2O molecules, which are hydrogen bonded with the coordinated H_2O molecules. Therefore, in some extreme cases, the contribution of r_1^{SS} becomes significant in the absence of any coordinated H_2O molecules or presence of an equal number of both types of H_2O molecules. The mechanistic behavior of zeolitic-type H_2O molecules in r_1 relaxivity is similar to the coordinated-type H_2O molecules. In addition, the existence of the zeolitic-type H_2O molecules or exchangeable protons follows the same behavior as the inner sphere mechanism, but they have larger residency time in contrast to diffusion time. Some selected examples in albumin binding groups with the molecular structure are shown in Figure 7.1, where the second sphere mechanism plays an important role in relaxivity measurement while the number of coordinated H_2O molecules is zero. However, DO3A-pic-bip exhibits r_1 ~3.1 mM^{-1} s^{-1} in buffer and 7.0 mM^{-1} s^{-1} in HSA, while protein-tagged triethylenetetr aaminehexaacetato (TTHA-P) shows r_1 ~2.1 mM^{-1} s^{-1} in buffer and 8.0 mM^{-1} s^{-1} in HSA. Similarly, another one complex is known as $[Gd(C_{11} - DOTP)]^{5-}$. Herein, it is claimed that due to the presence of negatively charged phosphonate surface, which has a strong second sphere effect, $[Gd(C_{11} - DOTP)]^{5-}$ shows high relaxivity compared to $[Gd(TTHA - P)]^{4-}$ (Bonnet and Tóth 2021). Finally, the authors claim that r_1^{SS} contributed 10–30% to the r_1 in the absence of any r_1^{IS} activity (Chen et al. 1998). Other than r_1^{IS} and r_1^{SS}, $r_1^{OS}H_2O$ molecules also participate in the variation of r_1 relaxivity, where water protons do not directly interact with the paramagnetic center. The outer sphere relaxation mechanism was first modeled in 1975 by Freed and Hwang and has been successively refined day by day. This theory mainly focused on relative translational diffusion and rotational motion. This effect may be neglected for paramagnetic molecules with fast molecular tumbling effects (Bonnet et al. 2010).

7.2.2 OUTER-SPHERE DIFFUSION-BASED RELAXIVITY MODEL

Outer-sphere relaxivity theory is generally applied in weakly magnetized para-magnetic metal complexes where the translational diffusive motion of bulk H_2O molecules, noncoordinated to the magnetic metal center, plays a major role. The superparamagnetic nanoparticles having high magnetic susceptibility are likely to produce a local magnetic field in the existence of an externally applied magnetic field. As a result, this local field is able to shorten T_2 relaxation time by perturbing phase coherence of bulk H_2O molecules. The diffusive theoretical model is also applied for superparamagnetic nanomaterials, which are small enough to satisfy the motional averaging regime (MAR) theory. So the outer-sphere diffusion-based r_2 relaxivity by MAR theory is given as follows:

$$\frac{1}{T_2} = \frac{\left(\dfrac{256\pi^2 \gamma_p^2}{405} \right) \kappa M_s^2 a^2}{D\left(1 + \dfrac{1}{a}\right)} \qquad (7.13)$$

where γ_p and M_s are the gyromagnetic constant of hydrogen and saturation magnet-ization constant, respectively; in $\kappa = \dfrac{V^*}{C}$, V^* is volume fraction and C is the total magnetic metal concentration; and D and "r" represent the diffusion coefficient of H_2O molecules and effective radius of the particles, which is the sum of the actual radius (a) of the core and thickness (l) of the impermeable surface coating, i.e., $r = a + l$, which forms an impermeable surface to limit diffusion of H_2O molecules. Herein, τ_D defines translational diffusion time required for diffusion of H_2O molecules at a distance of $\sqrt{2}a$; $\Delta\omega_r$ is the rms angular frequency shift at the nanoparticle surface. The MAR theory states the relation between τ_D and $\Delta\omega_r$, that is, $\tau_D \Delta\omega_r < 1$, where $\tau_D = \dfrac{r^2}{D}$. On the other hand, for paramagnetic complexes, the size of the particles plays a crucial role in proton relaxation in MAR when other parameters seem to be constant, as the motion of both the magnetic center and H_2O molecules is too fast to move through an effective dephasing process. Apparently, it is easier to replace relaxed H_2O molecules by bulk H_2O molecules in an altered magnetic field around smaller particles rather than larger particles. So, larger particles are more proficient in short T_2 relaxation time in MAR. As a result, r_2 is positively correlated with M_s and radius of the magnetic nanoparticles.

However, when the size of the magnetic nanoparticles exceeds its threshold limit or the applied magnetic field is strong enough, then they no longer gratify the MAR condition ($\tau_D \Delta\omega_r < 1$). This situation is governed as static dephasing regime (SDR), where H_2O molecules are entirely dephased before they enter an alternate magnetic field after diffusion at a critical distance. So, SDR theory is ignored in T_2 relaxation. Herein, it has been observed that the range of $\tau_D \Delta\omega_r$ for SDR is $1 < \tau_D \Delta\omega_r < 5$ (Gillis et al. 2002). Additionally, between $5 < \tau_D \Delta\omega_r < 20$ the relaxivity of a mag-netic particle is independent of particle size. This situation governed as echo-limited situation (ELR) (Brooks 2002).

In this context, Zhang and his group explicated the reduction of r_1 due to functionalization on the basis of relaxivity contribution from the outer sphere according to the following expression:

$$r_1^{OS} = \frac{128\pi^2\gamma_p^2 M_n}{405\rho}\left(\frac{1}{1+\frac{1}{a}}\right)^3 M_s^2 \tau_D \, J_A\left(\sqrt{2\omega_I\tau_D}\right) \tag{7.14}$$

where M_n, ρ, M_n, and J_A are the molarity (mole liter^{-1}), the density of the core, saturation magnetization, and Ayant's spectral density, respectively (Zhang et al. 2016).

7.3 PARAMETRIC OPTIMIZATION FOR ENHANCING RELAXIVITY

Depending upon the mode of action, MRI CA can be distinguished as T_1 and T_2 CA. As previously described, T_1 CAs are primarily characterized by a decrement of spin-lattice or longitudinal relaxation time at which the excited proton transfers its energy to the nearby medium while T_2 CAs are associated with spin-spin or transverse relaxation time at which protons become out of phase with each other. The efficiency of a CA is directly proportional to the concentration; a brightening effect appears for T_1 CA while darkening effect is observed for T_2 CA. Generally, Gd^{3+}, Mn^{2+}-based paramagnetic CAs are known as T_1 CAs. Currently, there are seven FDA-approved Gd^{3+}-based ECF CAs for clinical purposes with relaxivity values 3.6 to 6.3 mM^{-1} s^{-1} at a magnetic field of 1.5 and 3 T. Most commercially available gadolinium-based CAs possess relatively low relaxivity values compared to theoretical values. Besides this, some limitations are also observed, like the accumulation of Gd^{3+} in the central nervous system (CNS), which will cause a serious issue to human health. In this context it may be stated that in 2017 European Medicines Agency suspended the use of three CAs, namely Gd-DTPA, Gd-DTPA-BMA, and Gd-DTPA-BMEA, due to their accumulation in the brain; as well, they announced the restrictive use of Gd-DTPA-EOB and Gd-BOPTA in liver imaging. Furthermore, illimitable use of these CA would enhance nephrogenic systemic fibrosis risk, leading to chronic kidney disease.

So, the current research is focused on designing novel CAs with enhanced human safety. Researchers are dedicated to developing T_1 CAs, employing various approaches. One such approach involves creating targeted Gd^{3+}-based CAs, which are designed to possess a high relaxivity value at minimal concentration while also exhibiting enhanced stability against Gd^{3+} dissociation in biological environment. Moreover, they are focused on the development of new-generation CAs other than gadolinium-based CAs with high efficiency. In this context it may be stated that there exist some molecular parameters which should be tuned to improve the relaxivity value. These are a number of coordinated H_2O molecules (w), rotational correlation time (τ_R), and mean residency time (τ_m).

7.3.1 ENRICHMENT OF THE NUMBER OF COORDINATED WATER MOLECULES

From the previous discussion we are already aware that inner sphere relaxivity is directly proportional to the number of coordinated H_2O molecules. So, an increase

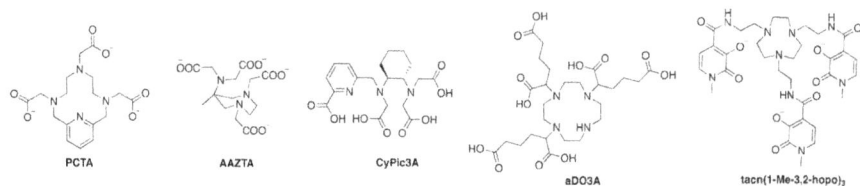

FIGURE 7.2 Thermodynamically stable Gd(III)-based ligand complexes. (Reproduced with permission from *American Chemical Society*, copyright from Ref. (Wahsner et al. 2019).

in the number of H_2O molecules will enhance the relaxivity value, leading to an improved brightening effect, which indeed depends upon the synthesis condition. In this context it may be stated that clinically used gadolinium-based CAs consist of a single H_2O molecule. However, an increment of H_2O molecules is often responsible for the decrement of thermodynamic stability or kinetic inertness and the enhancement of displacement of two cis-coordinated water ligands through coordinating anion, leading to a reduction in r_1 relaxivity (Polasek and Caravan 2013). Herein, some FDA-approved stable gadolinium-based ligand systems with augmented coordinated H_2O molecules are PCTA, AAZTA, CyPic3A, and aDO3A and $tacn(1-Me-3,2-hopo)_3$ and their structures are shown in Figure 7.2. Although Gd-PCTA ($r_1 = 6.9$ mM^{-1} s^{-1} at 20 MHz) is thermodynamically less stable compared to Gd-DOTA, it is highly kinetically inert toward the dissociation of Gd^{3+} (Dunbar and Heintz 1997, Reedijk and Poeppelmeier 2013, Wahsner et al. 2019). Moreover, ongoing recent research has focused on FDA-approved enriched H_2O molecules containing Prussian blue (PB) or Gd^{3+}, Mn^{2+}-doped Prussian blue analogous (PBA) nanostructures, elaxivity and thermodynamically stability attributed to high ligand field stabilization energy (LFSE) within the covalently bonded $[Fe(CN)_6]^{4-}$ matrix anion, rendering them highly inert (Dunbar and Heintz 1997). A brief description of the efficiency of PB and PBA mixed valence coordination framework as an MRI CA in both in vitro and in vivo studies is given in a later section. Recently, it has been demonstrated that P03277, a Gd-PCTA derivative, exhibits r_1 (12.8 mM^{-1} s^{-1} at 1.5 T and 11.6 mM^{-1} s^{-1} at 3 T) in human blood plasma, which is 2.5-fold higher than gadobutrol, a clinically approved paramagnetic Gd^{3+} chelate complex. Currently P03277 is approved for human trials (Wahsner et al. 2019).

7.3.2 OPTIMIZATION OF ROTATIONAL CORRELATION TIME

As discussed in the section on SBM theory for inner sphere relaxivity, in macromolecules, when the accessible field is greater than 0.1 T, τ_m (~10^{-9}–10^{-7} s) is much greater than τ_R (~10^{-12} s), i.e., $\tau_{cl} \approx \tau_R$; hence Equation 7.2 may be written

as $r_1^{IS} \propto \dfrac{3\tau_R}{1+\omega_L^2 \tau_R^2}$. As clinical MRI operates at 127.74 MHz frequency generally,

hence neglecting $\omega_L^2 \tau_R^2$ in the denominator, it may be simplified as $r_1^{IS} \propto \tau_R$. In a

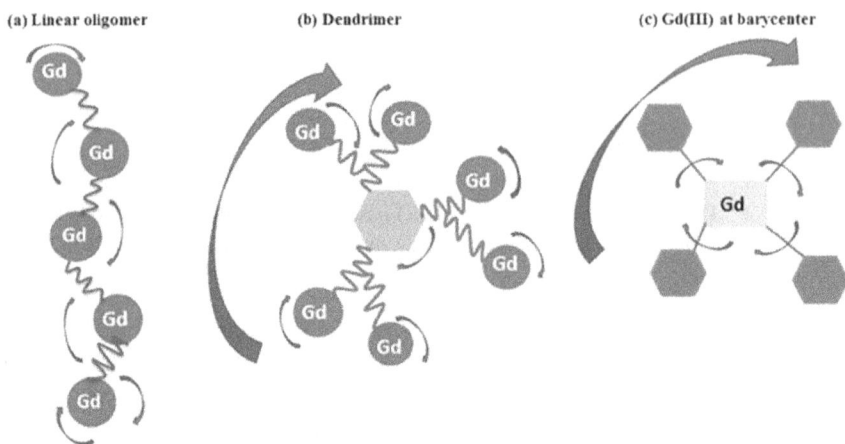

FIGURE 7.3 Different types of internal motion for Gd-DO3A derivatives.

similar way second sphere relaxivity is also directly related with τ_R. Thus, enrichment in the relaxivity of a complex can be achieved by slowing down its rotational motion. Furthermore, the Debye-Stokes relation, $\tau_R = \dfrac{4\pi\eta r^3}{3k_BT}$, where, η, k_B, and T represent the viscosity of the medium, Boltzmann constant, and absolute temperature, respectively, indicates that τ_R is directly related to particle size (Chen et al. 2013). So, higher relaxivity of a CA can be achieved by enlargement of particle size or covalent/noncovalent attachment to a macromolecule (Aaron et al. 2010, Liu et al. 2011). Helm et al. established the discrepancy of τ_R and r_1 of monohydrate Gd-DOTA derivatives with molecular weight in a linear fashion (Helm Lothar 2017).

Another way to maximize τ_R is to associate multiple MR complexes together; increment of magnetic centers per molecule leads to improved contrast. However, in this strategy, spherical-shaped coupling (e.g., dendrimers) is more effective than coupling in a linear fashion due to internal motion. The rotation of the dendrimers is isotropic, whereas for linear oligomers, the rotation is anisotropic. In anisotropic motion the rotation of the oligomers along the long axis is faster compared to whole molecules, leading to limitation in the relaxivity of the molecules. Herein, the different types of motion have been explained in Figure 7.3 for a Gd-DO3A derivative. When it is assembled in a linear fashion, the anisotropic motion limits the relaxivity to 40% compared to the isotropic motion of spherical dendrimers. Moreover, when the gadolinium complex exists at the barycenter of the molecule, then both will rotate at an equal rate, giving the highest relaxivity (Rohrer et al. 2005).

7.3.3 MINIMIZING INTERNAL MOTION

Lipari and Szabo developed a model based on the reduction of relaxivity due to add-itional internal motions of CAs, which are non-/covalently linked to a macromolecule (Lipari and Szabo 1982). They proposed two types of motion for an isotropic rotation, namely overall molecular rotational motion, known as τ_0, and rotational motion of the linker, known as local rotation, τ_1. Their contribution in T_1 relaxivity is given by the following expression:

$$\frac{1}{T_1} = \left(\frac{B}{r_{MH}^6} \right) \cdot \left(\frac{3F\tau_{co}}{1+\omega_p^2 \tau_{co}^2} + \frac{3(1-F)\tau_{cl}}{1+\omega_p^2 \tau_{cl}^2} \right) \tag{7.15}$$

$$\frac{1}{\tau_{co}} = \frac{1}{\tau_0} + \frac{1}{T_{1e}} + \frac{1}{\tau_m} \tag{7.16}$$

$$\frac{1}{\tau_{cl}} = \frac{1}{\tau_{co}} + \frac{1}{\tau_1} \tag{7.17}$$

where B and F denote the physical constant and degree of isotropic motion, respect-ively. F = 0 represents that linker is entirely decoupled from the macromolecule, while F = 1 is associated with immobilization from the macromolecule.

However, two strategies can be effectively taken to moderate the internal motion of a complex system. One way is to assemble the system in such a way that the rota-tion of the metal-water proton should be isotropic with the overall molecule. P792, a blood pool agent, shows high relaxivity followed by this (Port et al. 2001). Another way is to link a CA to the macromolecules with two points attachment. In this con-text, Zhang and his team revealed that tetrameric Gd-DTPA oligomers with two points binding to HAS protein show high relaxivity due to restricted local rotation (shown in Figure 7.4) (Zhang et al. 2005).

7.3.4 OPTIMIZATION OF WATER RESIDENCY TIME

Enhancement of r_1 relaxivity indeed depends on the tuning of water residency time. The coordinated H_2O proton relaxation effect is rapidly transmitted to the bulk H_2O molecules, leading to limitation in the relaxivity. In this context, it may be stated that for coordinated H_2O molecules, $\tau_m \approx 10^{-9} - 10^{-7}$ s, while T_{1M}^{IS} varies between 10^{-4} and 10^{-6} s; thus, for all practical purposes $T_{1M}^{IS} \gg \tau_m$; hence Equation 7.2 may be simplified into $r_1^{IS} \propto \dfrac{1}{T_{1M}^{IS}}$. So, it is concluded that for fast-tumbling CAs, τ_m has a very petite influence on relaxivity. However, τ_m is very vital when τ_R is high, which influences the relaxivity to be diminished.

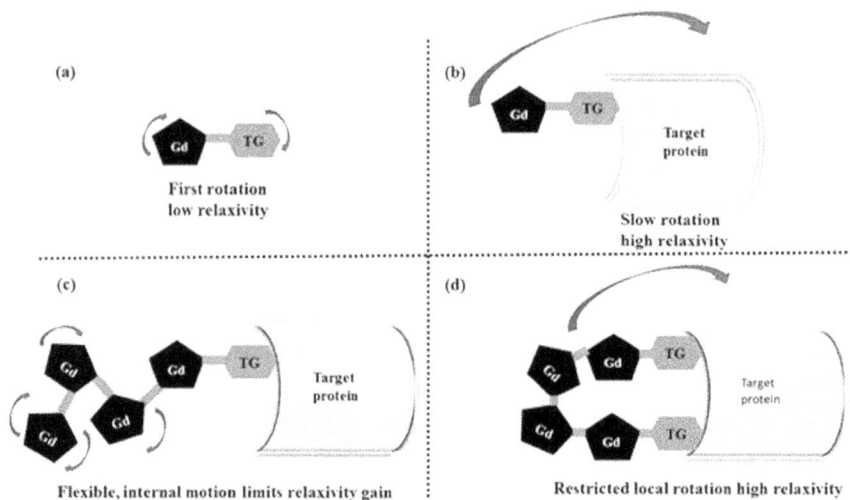

FIGURE 7.4   Strategies for enhancement of relaxivity by minimizing internal motion.

Herein, the relaxivity value of MS-325 MRI CA bound to HSA is higher compared to unbound MS-325 as τ_R enriched to 10.1 ns from 115 ps. But this relaxivity enhancement after HAS binding is not so much as predicted as τ_m plays an important role here, which limits the r_1 relaxivity. It has been well reported that $\tau_m \approx 69$ and 170 ns for unbound and bound MS-325, respectively (Caravan et al. 2002).

In this context, it can be stated that steric compression has a significant role in the variation of relaxivity. Merbach and his team first demonstrate that the enhancement of steric compression at the binding site of H_2O molecules will accelerate the water exchange rate, such as Gd-DTPA, Gd-EPTPA-type Gd-based chelates (Helm and Merbach 2005). Moreover, Dumas and coworkers have found that steric crowding can be accelerated by introducing additional groups. Such variation of τ_m can be written as follows when an acetate group of gadolinium-based chelate can be replaced by a donor group: pyridyl ~ acetamide > sulphonamides > alfa-substituted acetates > phosphonates ~ phenolates (Dumas et al. 2010).

7.4   FACTORS AFFECTING r_1 AND r_2 RELAXIVITY

Recently advanced nanotechnology has stimulated the development of CAs, which have two main features: dominant quantum confinement effect and greater surface area. Size-dependent optical, electronic, and magnetic properties originating from quantum confinement effects can be used in designing high-performance MRI CAs. Additionally, the large surface area of NPs offers enhanced chemical reactivity for surface modifications. The following factors of NPs affect the relaxivities of MRI contrast agents (shown in Scheme 7.2).

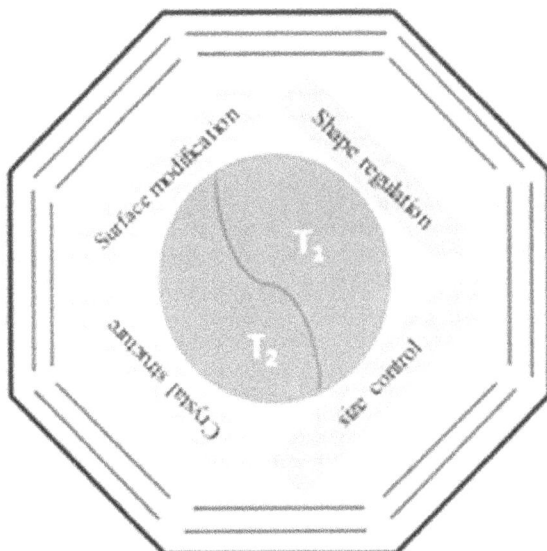

SCHEME 7.2   Factors affecting r_1 and r_2 relaxivity of MRI CA.

7.4.1   SIZE EFFECT

Before quantum confinement theory, "nano" was the parameter of size that came into consideration. The relation between the size of magnetic nanoparticles (MNPs) and magnetization is well established (Li et al. 2017). The involvement mechanism between sizes of MNPs and proton relaxation will surely lead to an advantage in designing novel T_1 and T_2 contrast agents for greater contrasting efficiency in MR imaging. Here in this section, we will discuss the size of the MNPs vs. the effect on r_1 and r_2 relaxivity separately.

7.4.1.1   Effect on r_2 Relaxivity

Magnetization of a certain MNP is defined as a divide between the arithmetic sum of magnetic moments and the sample volume. Magnetization is independent of grain size. The electronic configuration of an atom on the outer surface and the inner surface is different, which is known as the spin canting effect. The magnetic moment of MNPs is calculated by an integrated sum of the long-range order magnetic spin on the center and spin-canted spins at the boundary of the MNPs. Depending on crystallinity, the spin canting region of maghemite varies from 0.5 nm to 0.9 nm, which gives rise to size-varying magnetic moment as well as variation in T_2 relaxivity (Linderoth et al. 1994). As previously discussed, the outer sphere diffusive theory for the MAR identifies two factors regarding size effect that affects T_2 relaxivity, namely saturated magnetization (M_s) and diameter (r). The size-dependent r_2 relaxivity is universal for many kinds of MNPs including alloyed or doped MNPs such as europium-engineered IO, FeCo/graphitic shell, manganese ferrite, etc. Size-dependent T_2 relaxivity reaches

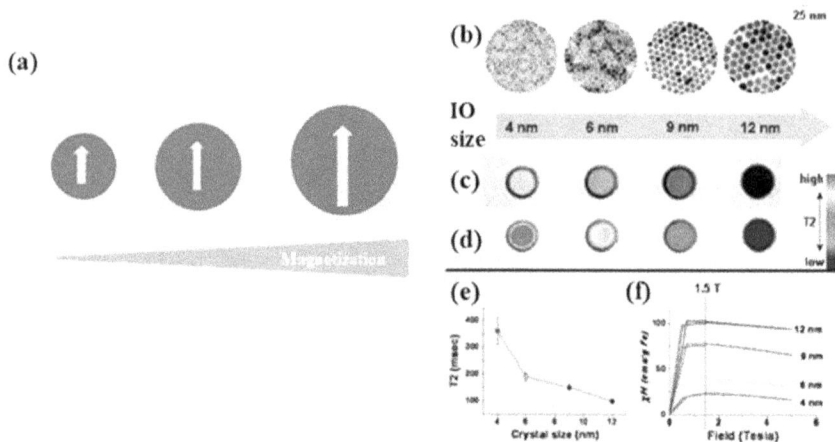

FIGURE 7.5 Size-dependent r_1 relaxivity: (a) The image presenting size affecting magnetization property of MNPs. (b) TEM images; (c, d) Phantom images (aqueous solution, 1.5 T); (e) T_2 relaxation time; (f) magnetizations of different sized IONP. (Reproduced with permission from *American Chemical Society*, copyright from Ref. (Jun et al. 2005).

an end when MNPs approach their critical size. SDR theory also plays an important role in MRI signal formation by gradient echo sequences. In echo-limited-regime (ELR), r_2 relaxivity reduces as the size of the MNP decreases. The threshold size of ELR depends on the magnetic property of MNPs, which is generally observed in multidomain MNP clusters. Herein, Cheon and coworkers developed different-sized IONPs which exhibit size-dependent magnetization along with r_2 relaxivity (shown in Figure 7.5) (Jun et al. 2005).

7.4.1.2 Effect on r_1 Relaxivity

From SBM theory it is well known that the enhancement of T_1 relaxation depends on models of paramagnetic molecules. In comparison with paramagnetic materials, MNPs also exhibit strong magnetization due to outer sphere contribution to T_1 relaxivity. According to the surface-to-volume ratio smaller MNPs provide a greater amount of surface-exposed magnetic atoms for chemical exchange; as well, proton coordination of water molecules leads to high r_1 relaxivity compared to larger MNPs. Herein it may be stated that Prussian blue nanocube (PBNC) with different particle sizes causes variation in crystallinity. Thus for smaller-size PBNP exhibits the highest r_1 relaxivity value (shown in Figure 7.6) (Feng et al. 2021). During r_1 relaxivity measurement, r_1 magnetization recovery is deeply connected with spin dephasing effect due to transition from longitudinal to transverse signal. Paramagnetism of MNPs is the key factor for the acceleration of r_1 relaxivity due to the transverse field fluctuation near Larmor frequency. It also explains why superparamagnetic NPs do not show T_1 relaxation at the high magnetic field. Besides this, the reduction of

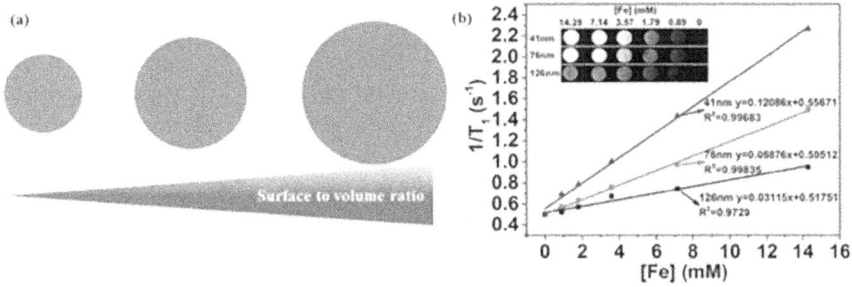

FIGURE 7.6 Size-dependent r_1 relaxivity: (a) Cartoon presenting size affecting surface-to-volume ratio. (b) Size-dependent r_1 relaxivity curve with phantom images of PBNC dispersed in water. (Reproduced with permission from *American Chemical Society*, copyright from Ref. (Feng et al. 2021).

particle size accelerated rotational motion leads to the decrement of the relaxivity of a CA. The effect of τ_R on relaxivity is sounder in longitudinal relaxivity compared to transverse relaxivity.

For many MNPs, like MnO, NaGdF$_4$, and iron oxide (IO), it has been reported that r_1 value depends on the size. T_1 relaxivity depends not only on size-dependent volume-to-surface ratios but also on colloidal stability in ambient temperature. Reducing the size of MNPs increases the surface energy and further decreases the colloidal stability. Thus, the interior spin-canting action and colloidal stability were both given the increased amount of the suitable T_1 relaxation. Dominating value of T_2 and T_1 relaxivity is irrespective of reducing the particle size of MNPs. As a fundamental rule of measuring MRI and signal processing, T_1 signal occurs when T_2 shortening effect is shrunk. So tuning the stability, variation of r_1 and r_2 relaxivity in MNPs is a new wing of modern research work. Depending on size, many MNPs like pure IO, metal-doped IO, and FeCO alloy show T_1, T_2, and T_1-T_2 dual mode relaxivity. For example, near-5-nm diameters of IO show T_1-T_2 dual mode, whereas a diameter near 3 nm shows T_1 relaxivity and a diameter near 10 nm dominates T_2 relaxivity (Jung et al. 2014). For superparamagnetic IONPs (ŞPIONs), large T_1 contrast magnification was observed by using an ultralow magnetic field (ULF) irrespective of size effect (Yin et al. 2018).

7.4.2 SHAPE

Designing of differently shaped colloidal nanoparticles has garnered great interest in various fields like plasmonics, magnetization, sensing, catalysis, and also biomedical applications. From classical electrodynamics, it is established that only ellipsoidal-shaped bodies exhibit homogenous magnetization, while certain shape distortion includes extra external energy for the stabilization of particle anisotropy (Boyer 1975). The shape of any MNP has a great impact on magnetic anisotropic character. So the discussion on various shapes of MNPs like cube, concave, tripod, plate, octopod, flower, etc., dealing with magnetic properties and potential impact on MRI relaxivity is very much necessary. In this context, it may be stated that the local magnetic field produced in the presence of a nonspherical MNP is generally directed in the parallel direction of particle magnetization while it is closely condensed to the MNP in the other direction in comparison with spherical ones (Tejedor et al. 1995).

7.4.2.1 Cubes

As an example, magnetic spins extend at the corner of the cubes as in a flower-like state. So water diffusion and relaxation near a cubic magnetic nanoparticle greatly provide the advantage of the emerging difficulty of the local induced magnetic field, which causes irreversible dephasing in routine Carr-Purcell-Meiboom-Gill sequences in MRI. This effect enhances the r_2 relaxivity with respect to spheres of equivalent size and magnetic moment. In this context it can be stated that coating with phospholipid polyethylene glycol ferromagnetic iron oxide nanocubes (FIONs) exhibits a highly sensitive MRI contrasting agent with a length edge of 22 nm, exhibiting a high r_2 value of 761 mM^{-1} s^{-1} under a 3 T clinical scanner (Lee et al. 2012). Core-shell-type cubic nanoparticles with antiferromagnetic core and ferromagnetic shell provide both MRI effect and hyperthermia effect (Walter et al. 2014). Many research groups reported the enhanced r_1 and $r_1 - r_2$ dual-modal relaxivity for a well-defined shape and surface of cubic-shaped nanoparticles (Yang et al. 2015).

7.4.2.2 Plates

Plate-like MNPs have also garnered great interest in MR imaging. Ferromagnetic exchange interaction has been noticed in the excessively thin layered structures of magnetic crystals, which leads to an interest in tailoring the spin exchange interactions in developing magnetic devices. It is feasible that the conformation change from sphere to plate-like structure affects the density of the surface, exposing metal ions, which is favorable for the coordination and chemical exchange of water protons, leading to variation in relaxivity.

7.4.2.3 Octopods

The octopod-shaped IONPs with 30 nm edge length exhibit an ultrahigh r_2 relaxivity of 679.3 ± 30 mM^{-1} s^{-1} at 7.0 T. This expressed MR contrast imaging with high sensitivity at low doses. By simulation, it was observed that the effective radius of octopod-like NPs is about 2.4 times higher than the spherical-like NPs with equivalent solid volume. This further expresses that nonspherical MNPs may exhibit a larger area of effective spin perturbation than spherical MNPs with an equivalent M_s value. Recently some researchers have shown that tripod shape NPs also exhibit enhanced T_2 relaxivity.

7.4.2.4 Other Shapes

Through enormous changes in precursors and reaction ambient conditions, researchers have succeeded in growing nanoparticles of various shapes. Researchers observed a broad variation in both relaxivities depending on nanoparticle shapes. For example, hollow-shaped manganese oxide nanoparticles (HMONs) showed markedly higher relaxivities for both T_1 and T_2 than water-soluble manganese oxide nanoparticles (WMONs) (Shin et al. 2009). It can be explained that the number of manganese ions exposed on the surface of HMONs was higher than the number of ions exposed on the surface of WMONs. For the case of IO nanoparticles, hollow IO nanoparticles were developed as T_2 contrasting agent, but they are limited because of their low r_2 and r_1 values.

7.4.3 CRYSTAL STRUCTURE

Specific arrangement of magnetic atoms and alignment of magnetic moments generate the magnetic property of NPs. It depends on several factors of a crystal structure such as crystal phase, domain boundary, crystallinity, superexchange effect, and crystallinity anisotropy. In the following section we will discuss their effects on T_1 and T_2 relaxivities.

7.4.3.1 Crystal Dopants

During synthesis, various dopant controls have been well established in the crystals. These dopings also change magnetic properties significantly. For example, Co ion–doped $(Fe_{1-x}Co_x)_3BO_5$ nanorods significantly changed the magnetic ordering from antiferromagnetic at low temperature to ferromagnetic at room temperature (Reynolds et al. 2005). Again in the spinel or inverse spinel structure of magnetite, Fe(II) and Fe(III) are distributed in either octahedral (O_h) or in tetrahedral (T_h). Magnetic spins in O_h and T_h are parallel and antiparallel to the external magnetic field, respectively. Therefore, net magnetization comes from the remaining spin magnetization after antiferromagnetic coupling. Some researchers showed that nonmagnetic metals like Zn-doped IO drastically enhance magnetic moment to 175 emu g^{-1} with respect to IONPs.

Sometimes due to some mismatching of coordination environment, doping metal ions, atomic size etc. doping occurs in the embedded form inside the NPs. These types of cases show partial paramagnetism compared to pure magnetic NPs of equivalent size (Zhou et al. 2012). For example, gadolinium-embedded iron oxide (GdIO) nanoparticles showed synergistic properties of T_1-T_2 dual-modal MRI as contrasting agents. Moreover, decreasing the size of GdIO significantly reduces r_2 relaxivity and enhances T_1 contrasting. A similar phenomenon was observed for europium-engineered iron oxide (EuIO). Based on the size and doping ratio of EuIO nanocubes, r_1 and r_2 values can be tuned. These particles also showed improved retention time and efficient tumor passive targeting and quick renal clearance (Zhou et al. 2013).

7.4.3.2 Crystal Phase

Due to the superexchange effect, spinel ($A^{2+}B_2^{3+}O_4^{2-}$; A may be magnesium, iron, zinc, manganese, or nickel; B may be aluminum, chromium, or iron; and O is oxygen) allows magnetic atoms to form long-range order spin states. Within crystal the degree of spin order directly contributes to the induced magnetic field and the magnetization effect.

Crystallinity means the degree of structural order in solid materials. It depends on the reaction environment during crystal formation. The correlation between crystallinity as well as crystal phase and T_2 relaxivities of MNPs in MRI follow a general rule that the higher order of magnetic spin within the crystal shows higher T_2 relaxivity.

7.4.4 SURFACE MODIFICATION

Generally, MNPs were synthesized through various processes which require several modifications for stability, biocompatibility, and targeting (Xie et al. 2011) criteria for biomedical applications. Nowadays, a huge amount of research attention has been paid to the surface engineering of MNPs. Here our focus is on several approaches to surface modification that will modify the MRI relaxivity of MNPs.

7.4.4.1 Anchoring Structure

Generally, the stability of MNPs in the solvent depends on surface ligands to avoid interparticle agglomeration, which is either connected to MNPs through chemical coordination or surface coating by physical forces like van der Waals force, electrostatic force, etc. Various kinds of chelating molecules are used for surface modifications of MNPs. For example, IONPs coated with polyethylene glycol (PEG) polymers via a catecholate-type anchoring moiety maintained magnetic moment, whereas other anchoring moieties like carboxylate, dopamine, and phosphonate decreased the magnetic moment of IONPs (Smolensky et al. 2013). The T_2 relaxivity is connected to the anchoring nature of surface molecules on IONPs because of the alteration of the spin canting effect through specific modifications. Recently some researchers showed that the binding affinity of anchoring ligands is highly correlated to the magnetic moments of IONPs. Particularly higher binding affinity of ligands shows lower magnetic moment of IONPs.

7.4.4.2 Organic Polymers

Polymeric coating on the surface of MNPs has an effect on the relaxivity in MRI. T_2 relaxivity hugely depends on the interaction between MNPs and bulk water molecules around them. The magnetic field induced by MNPs in gradually decreases with the distance away from the core, which states that a dense layer of surface polymer coating may exclude water molecules from undergoing diffusion around the magnetic field. On the other hand, winding structure and unique folding decreases the diffusion efficiency of water molecules on the polymers, which results in the enhanced T_2 relaxivity of MNPs. In a research work, Monte Carlo simulations suggested that surface coating thickness affects T_2 relaxivities by two competitive factors: the physical exclusion of water protons away from MNPs and the increased residence time of water molecules within the polymer layer (LaConte et al. 2007). Various forms of protein also have been explored as coating materials on the surface of MNPs as a modification. Casein is one of them which is made mainly of bovine milk, which

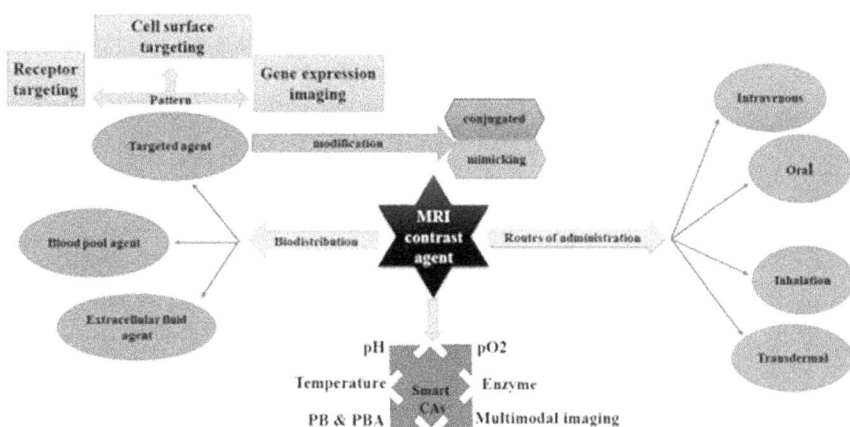

SCHEME 7.3 Schematic representation of different types of CAs.

belongs to the family of phosphoproteins. The hydrated functional group on casein enhances the exchange efficiency between hydrated water molecules and bulky ones. κ-Casein is the most suitable variant in the casein family members, accountable for the enhanced T_2 relaxivity (Zhu et al. 2017).

7.4.4.3 Inorganic Layer

Nanomagnetic inorganic materials like gold, silica, and metal oxides have been rigorously engineered for theranostic as well as for MRI relaxivity. A dense inorganic layer may reduce the magnetic field induced by MNPs and prohibit the interaction between the MNPs and their surroundings. So it can be stated that both T_1 and T_2 relaxivity could reduce in MNPs after dense inorganic layer coating. This phenomenon is called "magnetic dilution." Thicker coating shows reduced values for both r_1 and r_2 (Zhou et al. 2014).

7.5 TYPES OF CONTRAST AGENTS

Magnetic resonance imaging plays a vital role in diagnostic medicine as well as in monitoring disease progression and treatment. In recent years MRI CAs have evolved as a smart agent, adapting their response based on diverse factors like temperature, pH, enzyme, partial pressure of oxygen (pO_2), etc. In this chapter, we also describe the various pathways of administration of MRI CAs (shown in Scheme 7.3). The classification and fate of MRI CAs will also be described here, with their advantages and limitations *in vivo* and *in vitro*.

7.5.1 ROUTES OF ADMINISTRATION OF MRI CONTRAST AGENT

The first question about MRI CAs that comes to our mind is how they can reach our body or, more specifically, the target site. For instance, some contrast agents reach any specific organ to make their therapeutic as well as their diagnostic impact. To address

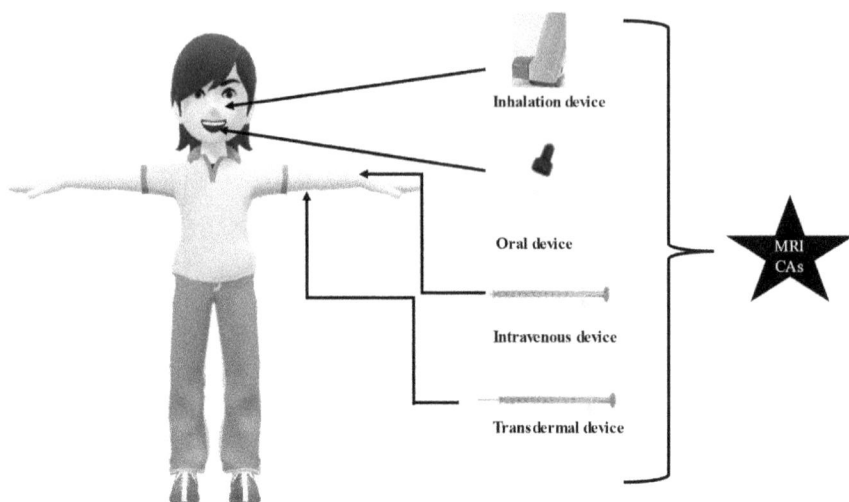

SCHEME 7.4 Routes of administration of MRI contrast agent.

these inquiries, it is necessary to comprehend the manner in which MRI CAs infiltrate physiological circumstances or the routes through which they are administered. So based on these criteria MRI CAs can be classified as either 1) transdermal, 2) oral, 3) via inhalation, or 4) intravenous. Scheme 7.4 shows the various pathways for the administration of MRI CAs.

7.5.1.1 Transdermal

Skin is the largest organ in the human body, covering a surface area of 1.8–2 m^2 (Gallo 2017). It is made up of three layers: epidermis, dermis, and hypodermis. The outermost layer of the epidermis is called the stratum corneum, which is basically made up of dead keratinocytes cells. The main goal of MRI CAs is to penetrate the stratum corneum layer. To overcome this problem there are so many techniques involved, like the use of microneedles, thermal ablation, microdermabrasion, high-pressure jets, laser, electroporation, ultrasound, etc. These techniques are very effective in delivering macromolecules and some vaccines. Illustratively, EPI-SPION (epirubicin–superparamagnetic iron oxide nanoparticle) serves as a transdermal agent, capable of surmounting obstacles and infiltrating subcutaneous tissue under the influence of external magnetic field. *In vivo* cytotoxicity study of these CAs also indicates good biocompatibility. So in the future this type of MRI CA can be utilized as a promising candidate in skin cancer therapy (Rao et al. 2015). In recent days many research groups have been working on the development of transdermal MRI CAs for cancer therapy as well as imaging purposes.

7.5.1.2 Oral Routes

The oral route is the most preferred route for drug delivery in the human body. Due to its pain avoidance, efficacy, greater convenience, and reduction of needle injuries it is the most common practice in the modern world. However, the oral route also suffers from the degradation of the drug because of the acidic environment and enzymatic secretion in the gastrointestinal tract (GIT). On the basis of contrast mechanism, gastrointestinal MRI CAs can also be positive and negative CAs. After administration, the MRI CAs will encounter the acidic environment of the GIT. Herein, paramagnetic agents including ferric ammonium citrate, ferric chloride, gadolinium-DTPA, etc., and oil emulsions or sucrose polyester with short T_1 relaxation time are used as positive GIT MRI CAs, while perfluorooctyl bromide (C_8BrF_{17}) is the only fluorine-based negative MRI CA that has been experimented with for oral use in humans (Mattrey 1989). Apart from this, magnetite albumin microspheres, magnetic nanoparticles, and superparamagnetic iron oxide are also be used as negative orally injected GIT MRI CAs. An example of oral MRI CA is AMI-227, an ultrasmall superparamagnetic iron oxide nanoparticle (Rogers et al. 1994).

There are some limitations to MRI CAs like dilution of positive contrast agents occurring in the upper GI tract if they are miscible with water because of gastrointestinal secretions. This allows for the use of a small dose but will cause a loss of signal intensity as the concentration decreases. Immiscible positive agents using oils, especially nonabsorbable ones, will not experience the loss of signal with dilution. They will probably require a larger volume to replace any residual bowel contents (Hoad et al. 2015). Another disadvantage of a positive oral contrast agent is the possibility of residual material in the bowel simulating a mass when surrounded by a bright signal and vice versa. A bright mass (such as a lipoma) might be obscured by the contrast agent. Lastly these MRI CAs are costly and have low availability. Some examples of T_1 and T_2-weighted oral MRI CAs with their relaxivity values are presented in Table 7.1.

7.5.1.3 Inhalation Routes

In the inhalation process the MRI CAs directly pass throughout the respiratory tract and reach the lung without any oral process or injection. After reaching the lung MRI CAs are distributed into systemic circulation. The distribution of those agents depends on the sedimentation rate in bronchi, bronchioles, and alveoli. Deposition of an agent in the respiratory tract is maintained by diffusional alteration due to the thermal motion of air molecules interacting with particles in the inhaled and exhaled air streams. There are three types of inhalation techniques: (i) pressurized metered-dose inhalers (pMDI); (ii) nebulizers; and (iii) dry powder inhalers (DPI), currently available commercially. Examples of MRI CAs administered through the inhalation technique along with their relaxivity values are revealed in Table 7.1.

7.5.1.4 Intravenous Delivery

Systemic delivery of drugs through intravenous injection is also a popular technique. In this process drugs are injected through intravenous injection and directly circulate through the bloodstream. The first intravenously administered nanoparticulate product, Abraxane® (a reformulation of paclitaxel), was approved by the FDA in 2006 (Wong et al. 2008). But the problem is that small nanoparticles (<5 nm) undergo renal clearance after intravenous administration and the uncoated nanoparticles suffer from opsonization and macrophage uptake. To overcome this problem, nanoparticle

TABLE 7.1
Examples of Some MRI Contrast Agents on the Basis of Their Route of Administration

Generic Name	Trade Name	Enhancement	Route	Reference
Ferric ammonium citrate	Ferriseltz	Positive	Oral GI imaging	(Xiao et al. 2016)
Ferristene	Abdoscan	Negative	Oral GI imaging	(Xiao et al. 2016)
Manganese chloride	LumenHance	Positive	Orally	(Xiao et al. 2016)
Ferumoxsil	Lumirem	Negative	Orally	(Xiao et al. 2016)
Gadoteridol injection solution	ProHance	Positive	intravenous	(Xiao et al. 2016)
Gadobutrol	Gadavist	Positive	intravenous	(Xiao et al. 2016)
Gadoterate meglumine	Clariscan	Positive	intravenous	(Xiao et al. 2016)
inert perfluorinated gases			Inhalation	(Pintaske et al. 2006)

or intravenously injected drug can be modified using various polymers and cholesterol. In the case of MRI CAs there are two types of intravenously administrated agents: ionic and nonionic intravenous CAs. Ionic intravenous CAs are hyperosmolar, while nonionic agents are relatively hypoosmolar (Manouchehr Saljoughian 2012). Omniscan, a nonionic intravenous CA, has two-fifths the osmolality of Gd-DTPA, which is the first approved intravenously used ionic agent in humans due to its stability and reliability. Most paramagnetic metal ions such as Gd^{3+}, Mn^{2+}, etc., which are suitable for MRI CAs, show a cytotoxic effect when injected intravenously, but chelation of these ions reduces the toxicity and has made them usable for the long term for imaging purposes. In this context, it may be stated that Gd-DTPA is distributed in the intravascular and extracellular fluid spaces while they do not cross the blood-brain barrier and are rapidly excreted through glomerular filtration (Kaminsky et al. 1991). Some examples of T_1 and T_2-weighted MRI CAs intravenously administered along with their relaxivity values are presented in Table 7.1.

7.5.2 BASIS OF BIODISTRIBUTION

Biodistribution is a tracking method of where the compounds of interest travel in an experimental subject (either experimental animal or in human species). The previous section of this chapter shows the administration pathway of MRI CAs. In this section, we explain the ultimate fate of these agents in physiological conditions. In a biological system MRI CAs may be classified as three broad spectra on the basis of their biological distribution: (A) extracellular fluid (ECF); (B) blood pool; and (C) target-specific agents.

7.5.2.1 Extracellular Fluid Agent

Extracellular fluid (ECF) agents are a type of compound mainly distributed between the intravascular and cellular phases. These compounds are mainly injected

TABLE 7.2
Some Examples of Positive Extracellular Fluid Agents at 1.5 T and 37°C

Generic name	Trade name	Types	r_1 (mM^{-1}s^{-1})	r_2 (mM^{-1}s^{-1})
Gadopentate dimeglumine	Magnevist	Gd-DTPA	4.1	4.6
Gadobutrol	Gadovist	Gd-DO3A-butrol	3.3	3.9
Gadoterate meglumine	Dotarem	Gd-DOTA	2.9	3.2
Gadodiamide injection	Omniscan	Gd-DTPA-BMA	3.3	3.6
Gadoversetamide	OptiMARK	Gd-DTPA-BMEA	3.8	4.2
Gadoteridol	ProHance	Gd-DO3A-HP	4.1	5.0

Source: Edelman et al. 1989.

intravenously. After injection into the body, these agents are first transported into the heart through systemic circulation and then through porous blood vessels, and finally, these CAs are leaked into the extravascular and extracellular phases. The primary aim of ECFs is to diagnose arterial abnormalities, disrupted blood vessels, defected endothelium tissue, etc. In the United States approximately 98% of commercially available contrast agents are ECF agents (Mosbah et al. 2008). The ECF agents are rapidly excreted from the body through the renal pathway. For example, gadolinium-derived CAs enter the nephron through the afferent artery and are then excreted through the kidney by renal excretion. Other gadolinium nanomaterials like Gd_2O, Gd-DOTA and Gd-DTPA are also distributed as extracellular fluid agents (Wahsner et al. 2019). All of these materials are most abundantly used for imaging purposes (Ahmad et al. 2015). Some examples of ECF agents along with their relaxivity values are given in Table 7.2.

7.5.2.2 Blood Pool Contrast Agents

Blood pool agents (also known as intravascular contrast agents) are differentiated from other contrast agents due to their high molecular weight and higher relaxivities (Geraldes and Laurent 2009). These agents are present in blood vessels and circulate in the body and stay for a prolonged time in our body than other types of contrast agents. In the case of magnetic MRI CAs, they mainly depend on their specific requirement for their plasma half-life and their final distribution in the body (Laurent et al. 2017). In the future these CAs may be useful for imaging of blood vessels or angiography of the heart. NC-100150, Gadomer-17, and Gd-BOPTA are some examples of these kinds of agents (Schalla et al. 2002). There are several advantages of this type of agent, i.e.: (a) information about capillary permeability; (b) being useful for permeability changes in the case of tumor or metastatic cells; and (c) long-term imaging. Several CAs like chromium-labeled red blood cells, SPION, and Gd-DTPA Dextran are mainly used for blood pool CAs. For instance, paramagnetic Cr(III)-labeled red bold cells (RBCs) have the potential to serve as a blood pool CA for MR imaging (Duguet et al. 2006). An experiment in dogs showed that significant enhancement of the liver and spleen is noted while minimal enhancement occurred in

TABLE 7.3
Examples of Some Blood Pool Agents at 37°C at 1.5 T

Generic name	Trade name	Enhancement	Reference
Ferucarbotran (USPIO)	Supravist	Positive	(Kan et al. 2022)
Gadobenate dimeglumine	MultiHance	Positive	(Schwickert et al. 1995)
Gadomer-17	Gadomer	Positive	(Schwickert et al. 1995)
Gadofosveset trisodium	Ablavar	Positive	(Xiao et al. 2016)
PEG-feron (USPIO)	Clariscan	Positive	(Xiao et al. 2016)
ferumoxtran-10 (USPIO)	Sinerem/Combidex	Both	(Pintaske et al. 2006)
ferumoxytol (USPIO)	Ferahem	Negative	(Xiao et al. 2016)

the kidneys (Creasman et al. 1966). Some examples of blood pool agents along with their relaxivity values are shown in Table 7.3.

7.5.2.3 Target-Specific Agents

These types of agents are based on their capabilities for targeting specific organs, tissues, or cells; for example, Gd-EOB-DTPA and Gd-DTPA-mesoporphyrin target the liver and myocardium, respectively (Akhtar et al. 2021). Specific organs like the liver, spleen, kidney, etc., are the main experimental parts for target-specific MRI CAs in the *in vivo* model. These agents are used in drug delivery, cell killing, apoptosis, internalizations, cellular uptake, etc. Some examples of targeted MRI CAs are shown in Table 7.4 (Wahsner et al. 2019). Targeted MRI contrast agents may be classified on the basis of their modification or on the basis of their targeting pattern as shown in Scheme 7.3.

7.5.2.3.1 Basis of Modification

On the basis of surface modification, the targeted contrast agent can be subdivided into the following:

Conjugated MRI CAs: The important thing that can be considered to design target-specific MRI CAs is the conjugation of the contrast-enhancing agent with a targeting moiety such as antibody/polymer/peptide that interacts with the target site (shown in Scheme 7.5).

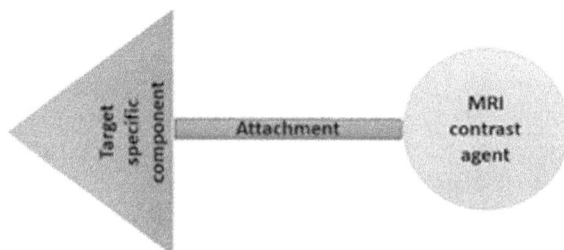

SCHEME 7.5 Strategy for development of conjugated MRI contrast agent.

TABLE 7.4

Some Examples of Targeting Agents with Their Targeting Sites along with Relaxivity Value

Targeting agent	Target site	r_1 (mM⁻¹s⁻¹)	r_2 (mM⁻¹s⁻¹)	Reference
Primovist	Liver	6.9 (1.5 T)	8.7 (1.5 T)	(Pintaske et al. 2006)
Eovist	Liver	6.9 (1.5 T)	8.7 (1.5 T)	(Vlahos et al. 1994)
Resovist	Liver	25.4 (1.5 T)	151 (1.5 T)	(Akhtar et al. 2021)
Teslascan	Liver	5.2 (1.5 T)	8.9 (1.5 T)	(Vlahos et al. 1994)
Multihance	Liver	6.7 (1.5 T)	8.9 (1.5 T)	(Vlahos et al. 1994)
Feridex	Liver	29.9	98.3	(Akhtar et al. 2021)
Dextran-coated SPIO AMI-25	Reticuloendothelial system (liver)	24 (0.4 T)	107 (0.4 T)	(Akhtar et al. 2021)
OMPg	Gastrointestinal (GI)			(Rijcken et al. 1994, Jung and Jacobs 1995)
Dextran-coated USPIO AMI-227	Lymph nodes	23 (0.4 T)	53 (0.4 T)	(Bulte et al. 1999)
PEGylated magnetoliposomes	Bone barrow	3 (1.5 T)	240 (1.5 T)	(Vogl et al. 1996)
Sinerem/Combidex	myocardium, brain perfusions	10	60	(Akhtar et al. 2021)

This conjugation generally affects the formation of biologically stable ethers, amides, thioesters, or triazoles. In biological systems fibrin and collagen are more prone to targeted imaging through MRI CAs because they are abundantly present in wound healing processes or angiogenesis or thrombosis. The cyclic decapeptide cCGLIIQKNEC (CLT1) conjugated to Gd(III) DPTA (r_1 ~4.22 mM^{-1} s^{-1}, 3 T) is an example of conjugated targeting agent. It is a fibrin-fibronectin labeling agent (Chowdhury 2017).

Various reports are available on monomeric and multimeric tumor-targeted CAs. Conjugation of steroids with CAs permits the targeting of overexpressed receptors on the surface of tumor cells. 21-Hydroxyprogesterone and 17β-estradiol are steroids that interact with the progesterone and estrogen receptors and are overexpressed by prostate, uterine, ovarian, and breast carcinomas. The conjugation of these two steroids with Gd(III)-based chelates produces different monomeric contrast agents with high relaxivity. In this context, it may be stated that although larger-sized multimeric CAs take a longer time to interact compared to monomeric CAs, they can be used as biomarkers to image any low expression site with a high larger relaxivity value due to the tumbling effect. Thus tetrameric fibrin targeted CA, 11-amino acid conjugated with four Gd(III)-DOTA show a longitudinal relaxivity value of ~10.1 mM^{-1} s^{-1} per gadolinium ion, while the value increases to 17.8 mM^{-1} s^{-1} per gadolinium ion upon binding with fibrin (1.41 T dispersed in tris-buffered saline) (Ye et al. 2008). Here some other conjugated MRI CAs can be represented such as EP-2104R, a fibrin-targeted Gd(III)-based nanoparticle, while EP-2104R detects various thrombus in the heart, carotid arteries, veins, etc. Apart from this some other proteins like serum albumin, collagen, elastin, and fibronectin are also targeting sites for MRI CA. In this context, it may be stated that MS-325 can be used for the detection of lung disease in MR imaging, while the use of gadofosveset enhances lung MRI to assess ongoing interstitial lung disease (ILD) (Montesi S et al. 2017). Gd-DTPA and Gd-DOTA chelate derivatives like EP-3533, 490 EP-3600,491, and CM-101 are used as collagen targeted CAs and are effective in the detection of fibrosis and myocardial infractions (Helm et al. 2008). In this context, Swanson and coworkers developed a tumor-specific nanocomposite conjugated with folic acid for folate receptor-based targeting and functionalized with polyamidoamine (PAMAM) dendrimer (Frias et al. 2004). The modified nanocomposite has been further functionalized by bifunctional DOTA-NCS (2-(4-isothiocyanatobenzyl)-1,4,7,10-tetraazacyclododecane-1,4,7,10-tetraacetic) chelator that makes a stable complex with gadolinium (Gd^{3+}). Due to their prolonged clearance time compared with other gadolinium-based CAs, this dendrimer nanocomposite plays a key role in targeted cancer imaging. Other than gadolinium-based CAs, IO-based nanoparticles can also be used as a targeted MRI CA (Xiao et al. 2016).

Mimicking MRI CAs: In the case of mimicking-type CAs, they mimic the structure of a ligand and bind with the target site (specific tissue or receptor molecules) without any conjugated molecules. But to make MRI CAs that mimic a biologically active molecule that acts as a ligand for some receptor or cell surface is challenging work. Enormous works of research are going on to create several MRI CAs which mimic HDL (high-density lipoprotein), porphyrin, and ferritin.

HDL NPs are one important class of mimicking targeted CAs that mimic natural molecular entities to gain selectivity. Synthesis procedures of HDL NPs are shown in Figure 7.7(a,b), which contains Gd^{3+} atoms on the outer phospholipid coating or encapsulation of core with IO. After teil-vein injection with a dose of 4.36 μmol kg^{-1} of NP shown in Figure 7.14(a), the *in vivo* imaging of hyperlipidemic (ApoE KO) mice reflects r_1 ~10.4 mM^{-1}s^{-1} per particle at 1.5 T, 25°C, and water at pH 7.4, while for the NP shown in Figure 7.7(b), r_2 ~94.2 mM^{-1} s^{-1} per particle at 1.41 T and 40°C with a dose of 0.54 mmol Fe kg^{-1} (Frias et al. 2004, Cormode et al. 2008, Zhang et al. 2009, Lee et al. 2010).

Porphyrins selectively gather at necrotic tissues and further interact with neurons. Accordingly, numerous porphyrin-based CAs are reported for tumor imaging. Mn^{3+}-containing porphyrin-dextran complex was used to visualize tumors in vivo. Another Mn^{3+}-containing porphyrin (shown in Figure 7.7(c)) has been stated as a brain tissue–specific CA, which is cell permeable and labels neuronal cell bodies in the hippocampus (Zhang et al. 2009). However, after intravenous application of this contrast agent to rat brains (0.017–0.02 μmol kg^{-1} dose), it is retained for a long time, which suggests interesting areas for future research. Moreover, two Gd^{3+}-based porphyrins (shown in Figure 7.7(d)) were described for *in vivo* melanoma imaging (r_1 ~16.3 and 31.7 mM^{-1} s^{-1} at 7 T, 23°C, and dispersed in aqueous medium) (Lee

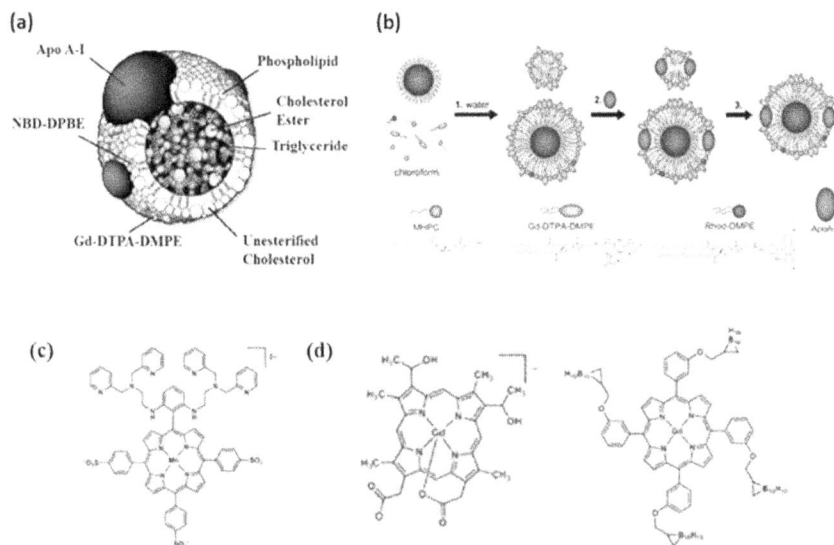

FIGURE 7.7 Development of targeted MRI CAs using mimicking strategy. (a) Recombinant HDL like CA: a special CA for MR imaging of atherosclerotic plaques. (Reproduced with permission from *American Chemical Society*, copyright from Ref. (Frias et al. 2004). (b) Synthesis of nanocrystal core high-density lipoproteins. (Reproduced with permission from *American Chemical Society*, copyright from Ref. (Cormode et al. 2008). (c) Mn^{3+} containing porphyrin-based CA. (d) Gd^{3+} containing porphyrin-based CA.

et al. 2010). After intravenous injection, *in vivo* imaging in the mice model (0.1 mmol kg^{-1}) revealed contrast enhancement in tumors due to definite accumulation. However, Gd^{3+}-based porphyrins are kinetically unstable in the biological environment due to the discrepancy between the porphyrin cavity and the metal ion (Shahbazi-Gahrouei 2006). So, nowadays expanded porphyrin systems, like texaphyrins, are widely used to make stable Gd^{3+}-based complexes.

Another example of porphyrin mimicking agent is copper phthalocyanine dye Luxol fast blue MBS (LFB MBS) (relaxivity ~0.15 mM^{-1} s^{-1}, 4.7 T, and ethanol) (Lee et al. 2010). Another example of a mimicking-type contrast agent is ferritin. It is a protein that stores iron. It is used to detect some ferritin-related diseases with the help of MRI (Shahbazi-Gahrouei et al. 2001). It has also been used in *in vitro* as a targeted CA for macrophage (Cerasa et al. 2012).

Targeting agents are also further subdivided on the basis of their targeting pattern: a) the cell surface; b) other target-specific receptor; and c) gene expression imaging.

7.5.2.3.2   Basis of Targeting Pattern

The categorization of target-specific CAs can be delineated into the subsequent division:

Cell surface targeting: Tumor cells in various organs like the brain, lung, or metastatic cells have a negatively charged surface compared to normal cells (Mehrishi 1969, Anghileri et al. 1976, Uchida et al. 2008). Most MRI CAs are also negatively charged, so the electrostatic bond between the cell surface and these agents is very poor. So numerous studies are ongoing regarding the modification of MRI CAs with positively charged amino acids (e.g., polyornithine) or polymers that can lead to the accumulation of them onto the specific tumor/cancer cell surface for therapeutic as well as imaging purposes.

A group of researchers developed a gastrin-releasing peptide (GRP) fused protein-based MRI contrast agent (ProCA1). It has been reported that a protein-based MRI contrast agent with tumor targeting modality can specifically target GRP receptor-positive prostate cancer (Yan et al. 2007). As per the report, it has been observed that MRI signal was increased after 48 h of treatment and the drug was retained in the prostate cancer cell line, so the CA displayed a prolonged retention time in the tumor (Yasui and Nakamura 2000).

Receptor targeting: In the case of receptor targeting MRI CA, the main goal is to modify the agents in such a way that they can attach to a specific receptor. Antibody-tagged MRI CAs are also used as receptor-targeting agents. As per present reports, manganese porphyrins, water-soluble meso-tetrasulfonatophenyl porphyrin (TPPS), and gadolinium texaphyrins are useful for antibody tagging (Sessler J. L. 2016). Manganese porphyrins are also used to target LDL (low-density lipoprotein) receptors for affecting tumor cells because tumor cells express a high number of LDL receptors (Nandi et al. 2017). Moreover, TPPS is used in Walker carcinosarcoma (Winkelman 1962). Additionally, Gd(III) and Mn(II) based TPPS chelates demonstrate potential as tumor-targeting contrast agents (Nandi et al. 2017). Furthermore, nanoparticles are modified with different external agents like folic acid, hyaluronic acid, and peptides to target the folate receptor, CD44 receptor, GRP receptor, etc.,

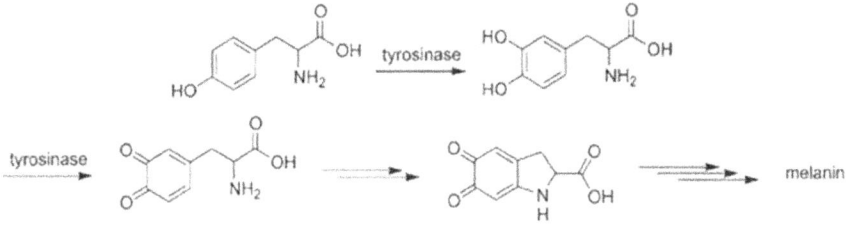

FIGURE 7.8 Synthesis of melanin from tyrosine.

which are generally overexpressed at the surface of specific cancer cells (Nandi et al. 2017). But the main disadvantage of targeting antibody-tagged MRI CAs is very slow accumulation at the target site. So modification of this type of agent may be further required for theranostics purposes.

Gene expression imaging: *In vivo* monitoring of gene expression and other genetic work is still a challenging factor in medical science. So if we developed some MRI CAs that can monitor gene expression it is very useful for gene therapy, for example, the synthesis of melanin from tyrosine, a pathway that may be monitored by MRI CA because melanin has a high affinity with iron. Figure 7.8 explains the pathway of melanin synthesis from tyrosine.

Scientists hypothesized that tyrosinase gene expression could be imaged by the product of this expression, i.e., melanin (Weissleder 1999, Pillaiyar et al. 2017). But this approach is quite tough because the tyrosinase gene is large and difficult to conjugate with a delivery vector. Secondly, the *in vivo* production of melanin is low for efficient MR detection of its metal complex (Pillaiyar et al. 2017).

7.5.3 SUPERPARAMAGNETIC IRON OXIDE NANOPARTICLES (SPIONS)

Iron oxide nanoparticles are a decent candidate to use as MRI CAs due to their magnetic property and biocompatibility. These particles can also be used as a conjugated target agent after coating with polysaccharide, chitosan, albumin, starch, etc. Furthermore, iron oxide nanoparticles enter the targeted cell through the process of phagocytosis, most of them used as a T_2 CA with efficient contrast efficiency. Those with diameters ranging from 5 to 300 nm are called SPION (Mahmoudi et al. 2011). SPIONs can also be classified into three groups on the basis of their size: (i) ultrasmall superparamagnetic iron oxide (USPIO) particles, d < 50 nm; (ii) small superparamagnetic iron oxide (SPIO) particles, d > 50 nm; and (iii) micron-sized particles of iron oxide (MPIO), diameter in excess of a micron (Laurent et al. 2008, Ma et al. 2020). Some examples of SPIONs are shown in Table 7.5. But in the biological system nanoparticles with less than 50 nm are more useful for imaging purposes since they have less cytotoxic effect (Xiao et al. 2016). SPIONs are also designed as conjugated CAs with the help of some antibody, peptide, protein, or polymer. This process increases the target specificity of SPION and sometimes makes them more biocompatible. Except for some target-specific MRI CAs that are clinically approved (e.g., MS-325) or some in the clinical trials phase (texaphyrins), most

TABLE 7.5

Some Examples of Superparamagnetic Iron Oxide Nanoparticles

Name	r_1 (mM^{-1}s^{-1}) (37°C, 1.5 T)	r_2 (mM^{-1}s^{-1})	Type
Ferumoxytol (carboxymethyldextran)	15	89	(USPIO)
VSOP-C184 (citrate)	14	33.4	(USPIO)
SHU-555C, Supravist® (carboxydextran)	10.7	38	(USPIO)
AMI-25, Endorem®, Feridex®, Ferumoxides (dextran)	10.1	158	(SPIO)
AMI-121, Ferumoxil, Lumirem®, Gastromark® (silicone)	2	47	(SPIO)
AMI-227, Sinerem®, Combidex® (dextran)	19.5	87.6	(USPIO)
Feruglose, Clariscan® (pegylated starch)	20	35 (0.47 T)	(USPIO)
SHU-555A, Ferucarbodextran, Resovist® (carboxydextran)	9.7	189	(SPIO)

Source: Morcos 2005.

of the targeted agents are in preclinical trial levels; however, *in vivo* imaging was reported with many preclinical studies (Lobbes et al. 2010). So it is an area of interest for researchers to create more biocompatible and targeted CAs.

7.5.3.1 Limitations – Side Effects of MRI Contrasting Nanoparticles

MRI is a useful tool for diagnostic purposes in various physical abnormalities. Various contrasting agents are used for MRI. Gadolinium-based contrasting agents (GBCAs) are one of the most useful candidates for this purpose (Do et al. 2020). In human use, GBCAs are the safest pharmaceutical products among other contrasting agents, but in 2006 it was found that nephrogenic systemic fibrosis (NSF) is a progressive multiorgan fibrosing condition that occurs in some individuals with reduced kidney function. NSF is characterized by thickening and hardening (fibrosis) of the skin, subcutaneous tissues, and, sometimes, underlying skeletal muscle. The patient who has been exposed to GBCA and has bad kidney function is more prone to NSF (Edwards et al. 2014). In 2018 the U.S. Food and Drug Administration (FDA) further added that gadolinium can remain in the body for months to years after receiving these CAs for MRI purposes (Fotenos 2018). Other minor effects of GBCA are nausea, itching, rash, headaches, and dizziness (Neeley et al. 2016). Besides these, most commercially available MRI CAs cause serious issues for healthy cells due to their lack of specificity. Therefore researchers are trying to develop a new type of targeted MRI contrasting agent that minimizes the risk of NSF and toxicity. Certain reports suggest that the intravenous administration of Gd-derived MRI CAs can result in a reduction of T_1 relaxation time within the dentate nuclei. Examples of some commercially available MRI CAs are given as follows:

Gadobutrol is an MRI contrast agent, administrated intravenously, and has shown good tolerability with no clinically important changes in blood or urine sample. It has also shown good feedback on cardiac conditions (tested through electrocardiogram) (Scott 2018). It has also been reported that 0.3 mmol kg^{-1} dose is safe for humans (Endrikat et al. 2016). Moreover, some rare side effects have also been reported, such as discoloration of the skin, tingling feeling, irregular heartbeat, itching, seizures, etc. (Neeley et al. 2016).

Eovist is a gadolinium-based liver-specific MRI CA that can also be used to iden-tify focal liver lesion in adults. It is a biphasic CA with dynamic phase imaging and functional hepatobiliary phase information. The administrative root of Eovist is intra-venous injection. The most frequent (≥0.5%) side effects associated with the use of Eovist are headache, feeling hot, dizziness, back pain, and nausea (Pitchaimani et al. 2016).

Magnevist, an MRI CA, is injected to visualize lesions and abnormalities in the human brain, spine, and tissues. The most common adverse effects of Magnevist are headache (4.8%), nausea (2.7%), and coldness (2.3%) (Zheng et al. 2015). But for practical purposes, it should not be administered to patients with chronic, severe kidney disease (GFR < 30 mL min^{-1} 1.73 m^{-2}) or acute kidney injury (Schieda et al. 2018).

Apart from this, the targeting approach also has some difficulties. Sometimes it requires direct injection on the specific delivery site, but it is not a suitable practice for medical application (Bloembergen et al. 1948). In the case of targeting agents some limitations arise such as the following: (a) for antibody targeting agents, high immunogenicity, limited tumor penetration, affinity varying with conjugation approach, and being large in size; (b) for aptamers targeting agents, rapid degradation while unmodified, cross-reactivity, and binding affinity susceptible to the environ-ment; (c) for peptides targeting agents, variation in toxicology and biodistribution; and (d) for small molecules targeting agents, lack of specificity and nonsystemic approaches for development (Valcourt et al. 2018). Moreover, most Gd(III)-based clinically approved MRI CAs consist of single coordinated water molecules to enhance thermodynamical stability leading to low relaxivity. So, now researchers are trying to develop new types of CAs that are biocompatible and have better targeting, high relaxivity, and superior therapeutic value. Thus the development of smart con-trast agents is required to fulfill the above criteria.

7.5.4 Smart Contrast Agents

Smart CAs, often referred to as intelligent agents, are primarily composed of Gd(III)-based nanocomposites which display fluctuations in their in their relaxivity value, influenced by different stimuli such as temperature, pH, ion flux, chemical potential, enzymatic activity, etc. These dynamic changes in relaxivity hold the potential to be modulated during various disease processes. The most important thing that can be remembered here is the relative change in relaxivity due to the smart CAs. Due to the presence of such biological stimuli, they reveal biochemical or physiological anomalies that produce inimitable insights into the human body for precise and early detection of cancer. As previously discussed, coordinated hydration state (w), rotational correlation time (τ_R), and mean residency time (τ_m) are the central parameters responsible for the tuning of r_1 relaxivity. Herein, it can be stated that biochemical stimulus lead to changes in these molecular parameters which exhibit variation in relaxivity. Several examples are given below.

7.5.4.1 Temperature

Recently photothermal therapy (PTT), enhancement of temperature at the tumor site, has become an important therapeutic approach in cancer treatment (Han and Choi 2021). Thus, it is necessary to construct a CA for prolonged monitoring of the entire therapy in such a situation. Moreover, MRI has already been proposed for noninvasive temperature monitoring. In this context, Zheng and his team already developed a thermosensitive microgel-type smart CA using manganese porphyrin and poly(N-isopropylacrylamide) (shown in Figure 7.9). During enrichment of lower critical temperature (LCT), the volume of this microgel rapidly swells or shrinks, leading to variation in rotation correlation time, which influences inner-sphere r_1 relaxivity. Consequently, r_1 relaxivity was accelerated to 73% due to slight enhancement in temperature (Fossheim et al. 2000). Herein, it has also been stated that the chemical exchange rate highly depends on temperature. Thus paramagnetic chemical exchange saturation transfer (CEST) CAs show temperature-responsive relaxivity.

Besides this, another group has also synthesized a temperature-responsive liposomal MRI CA, which shows a fascinating phase-changing character from gel to liquid at a particular temperature (de Smet et al. 2010). This phase transition is mainly associated with the variation in membrane permeability for water molecules. If the temperature is lower than the critical value, it shows low relaxivity due to slow H_2O molecule exchange rate, while in the case of high temperature, high relaxivity

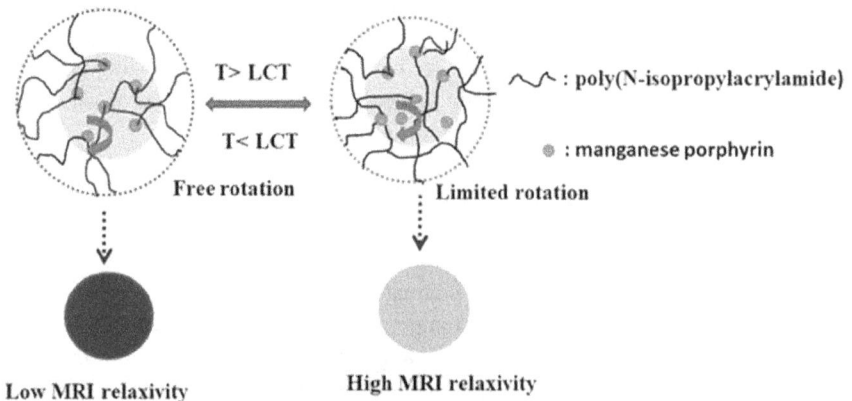

FIGURE 7.9 Conceptual scheme for a thermosensitive smart contrast agent.

has been observed. So, by adjusting the composition of the liposomal lipid membrane, the temperature can be finely tuned to get useful temperature-sensitive smart CAs. Moreover, Fe^{2+}, Gd^{3+}, and Co^{2+}-based paramagnetic CAs behave as inspiring smart agents in thermosensitive MR imaging (Zhang et al. 1999, Tsitovich et al. 2016, 2018).

7.5.4.2   pH

In the case of pH-sensitive smart CAs, the changes in pH at the cellular or extracellular level play a major role in the generation of these types of CAs. Generally, the extracellular part or the environment of the normal cells is relatively basic, but in the cancer cell environment this is more acidic compared to normal cells. The intracellular pH is more or less the same in both cells because of their homeostatic mechanism. In this context, Zhang and coworkers produced a gadolinium-based complex of a DOTA tetramide derivative and they observed an interesting behavior of its relaxivity versus pH sensitiveness. The relaxivity value increases from pH 4 to pH 6, then decreases until reaching 8.5. However, between pH 8.5 and 10, it increases once again (Aime et al. 2000). With the help of NMR spectroscopy, it has been concluded that the unusual behavior of pH and relaxivity is due to presence on the ligand of noncoordinated phosphonate group. The adjustment and the modification of Gd(III)-based CA may be very useful for future use.

7.5.4.3   Partial Pressure of Oxygen (pO$_2$)

Oxygen is an important parameter in living systems. Variations in the partial pressure of oxygen have been linked to several biological reactions. The partial pressure of oxygen (pO$_2$) plays a key role in the respiratory system. Lower or higher pO$_2$ has an impact on developing smart MRI CAs. An example that can show the role of pO$_2$ in MRI CAs is Gd-DOTP chelate derivatives, which allosterically modify hemoglobin, and the partial pressure of oxygen is also a dependent factor for the saturation level of oxygen in hemoglobin. These agents can bind more efficiently with T (tensed) state of hemoglobin rather than R (Relaxed) state of hemoglobin. This binding efficiency enhances the relaxivity of these agents (Louie et al. 2000). Some other paramagnetic agents like Eu(II)/Eu(III)-based systems have the same electronic configuration as Gd(III), and in the future it would be a replacement for Gd-based MRI CAs.

7.5.4.4   Enzyme

The biological role of an enzyme is like a catalyst; in most biological reactions a specific enzyme modulates the rate of reaction. So researchers are focused on developing MRI CAs which can modulate their action in the presence of particular enzymes. The interaction between enzymes and MRI CAs is strong due to the increase in relaxivity and the rotational correlation time. In *Xenopus laevis*, Gd-based MRI agents are used for monitoring the gene expression in vivo (Randall B et al. 1997). β-Galactoside, an enzyme encoded by lacZ, increases the relaxivity of Gd(III)-based contrast agent by approximately 20% (Louie et al. 2000). So in the presence of this enzyme, the smart contrast agent can monitor several types of cellular mechanisms, i.e., translation and

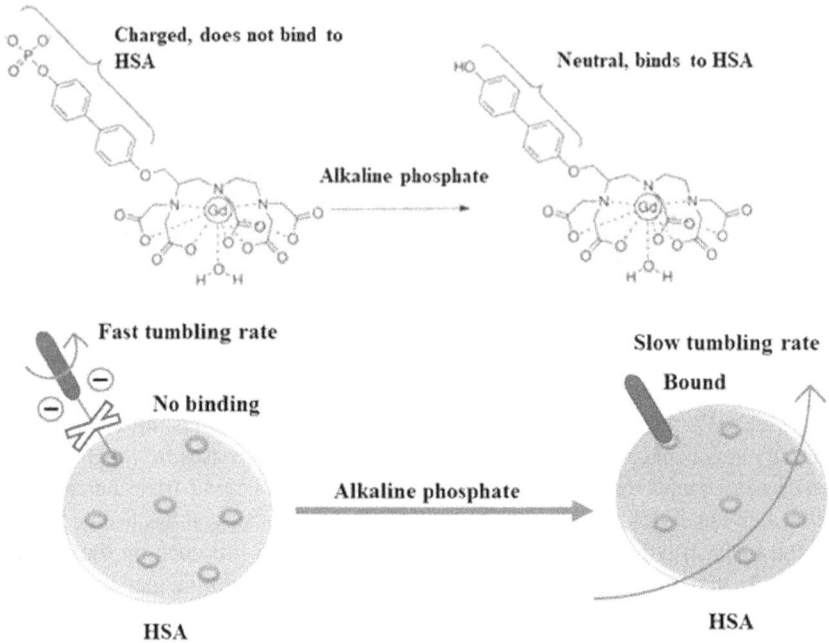

FIGURE 7.10   Activation of Gd-DTPA by alkaline phosphatase.

gene expression, and it also helps in tissue imaging. Another example is Gd-DTPA with a terminated hydrophobic sidearm by phosphate group. The effect of this modification is revealed in Figure 7.10, which shows the activation of Gd-DTPA by alkaline phosphate. Lauffer and coworkers observed that the modification of the complex between HSA and the hydrophobic MRI contrast agent increases its relaxivity by about 70% (Shokouhimehr, Soehnlen, Khitrin, et al. 2010).

7.5.4.5   Other Types of Smart Contrast Agent

There are some other types of CA, given as follows:

PB and PBA nanoparticles: Nowadays researchers have been focusing on developing a new kind of CA with enhanced proton relaxivities for MR imaging at a lower dose with less toxicity and high stability to suppress the limitation of clinically available CA. In this respect, the next generation of T_1 CAs having enhanced efficiency on locally contrast MR imaging and high proton relaxivities have been developed, known as PB or lanthanides (Gd^{3+} , Mn^{2+} , etc.) doped PBA nanostructure. These NPs carry a special structure that contains more coordinated water molecules to accelerate T_1 relaxation. These type of CAs show their potential in different diagnosis processes with minimal side effects and enhanced efficiency including magnetic resonance imaging (MRI), photoacoustic imaging (PAI), or scintigraphy (Siddique and Chow 2020).

Pioneering articles reveal that citric acid stabilized and 5(aminoacetamido) fluorescein dye conjugated PB nanocubes (PBNCs) with a size of about 13 nm show

$r_1 = 0.079$ mM^{-1} s^{-1} and $r_2 = 0.488$ mM^{-1} s^{-1} with $\dfrac{r_2}{r_1} = 6.1$ at 1.5 T, whereas citric acid stabilized antibody conjugated PBNC having size 22 nm shows $r_1 = 0.14$ mM^{-1} s^{-1} at 9.4 T, which can be used as both T_1 and T_2 CA (Shokouhimehr, Soehnlen, Hao, et al. 2010a, Li, Zeng, et al. 2014, Wheeler et al. 2019). However, these types of cyanobridged polymers are very much stable under various biological environments with a wide pH range and less toxicity compared to clinically used mononuclear Gd^{3+}-based CAs. In order to get enhanced longitudinal and transverse relaxivity values researchers have focused their investigation on PBNP and high spin transition metal ion or lanthanide-doped PBA nanostructures. In this connection, it has been observed that amino-terminated glycol (PEG-NH$_2$) functionalized Mn^{2+} (S = 5/2) doped PBA nanostructure (PB:Mn 5%) exhibit higher r_1 relaxivity of ~7.08 mM^{-1} s^{-1} at 3 T, which can vary depending on the doping concentration of Mn^{2+} in the lattice site of PB (Table 7.6) (Zhu et al. 2015). Similarly Gd^{3+}-doped PBA nanostructures display improved r_1 relaxivity value in the order of 2 or 3 compared to clinically available Gd^{3+} chelate complexes (Table 7.6). In this respect some articles claim that chitosan-coated Gd^{3+}/[Fe(CN)$_6$]$^{3-}$ nanostructure exhibits six times higher r_1 value compared to clinically FDA-approved Gd^{3+} complexes (Guari et al. 2008, Chelebaeva et al. 2011, Cai et al. 2016).

In vitro cellular phantom imaging (shown in Figure 7.11(a)) has been performed on $\left(K_{x-3}Gd\right)\left(Gd_nFe^{3+}_{y-1}\right)[Fe(CN)_6]_z$ nanoparticle, which reveals that with increasing Gd^{3+} ion concentration in the lattice site of PB in place of Fe^{3+} ion, the r_1 contrasting efficiency increased (Cai et al. 2016). *In vivo* study (shown in Figure 7.11(b)) has also been investigated on Mn^{2+}-doped PBNP functionalized with amino-terminated glycol (PEG-NH$_2$), which confirmed that these nanoparticles act as efficient T_1 CA, with a high longitudinal relaxivity value, allowing a remarkable brightening effect at the tumor site (Zhu et al. 2015, Qian et al. 2011).

In vitro toxicity test, cell uptake assay, and stability test in different biological environments for these type of nanoparticles have been studied, which revealed that (i) stabilized or bulk PB and PBA show less toxicity with high concentration for the normal cell under both dark and light conditions; (ii) these type of nanoparticles aggregate in the presence of NaCl, but after surface functionalization with various stabilizing agents the stabilized nanoparticles show high stability efficiency in different biological environments without leaching of any metal or CN$^-$ ion for at least 2 weeks (Verma et al. 2008); and (iii) these types of nanoparticles show good cell internalization via endocytosis, confirmed through confocal imaging (Perrier et al. 2013).

In summary, PB and PBA nanoparticles have remarkable potential as CAs in MR imaging. In most of the events they exhibit (i) an excellent cell uptake with proper localization after suitable functionalization; (ii) less toxicity to the cell under dark conditions both *in vitro* and *in vivo* and stable in biological environments; (iii) an enhanced performance in T_1 and T_2 weighted MR imaging when their structure is doped with high spin paramagnetic metal ions, leads to heightened blood circulation time and enhanced contrasting efficiency; and (iv) thermodynamic stability as well as kinetic inertness.

TABLE 7.6
PB and PBA Nanoparticles with Their Relaxivity Value

Composition	Functionalization	r_1 (mM^{-1}s^{-1})	r_2 (mM^{-1}s^{-1})	Reference
$Fe_4[Fe(CN)_6]_3$	PEG-NH$_2$	5.73 (3 T)	–	(Zhu et al. 2015)
PB:Mn (5%)	PEG-NH$_2$	5.85 (3 T)	–	(Zhu et al. 2015)
PB:Mn (15%)	PEG-NH$_2$	7.08 (3 T)	–	(Zhu et al. 2015)
PB:Mn (25%)	PEG-NH$_2$	7.64 (3 T)	–	(Zhu et al. 2015)
$Gd_3[Fe(CN)_6]_3$	PEG400	12.6 (4.7 T, 200 MHz); 12.1 (7 T)	16.0 (4.7 T, 300 MHz); 10.7 (7 T)	(Perrier et al. 2013)
$Gd[Fe(CN)_6]$	PEG1000	13.3 (4.7 T)	20.1 (4.7 T)	(Perrier et al. 2013)
$Gd[Fe(CN)_6]$	PEG-NH$_2$	11.2 (4.7 T)	12.2 (4.7 T)	(Perrier et al. 2013)
$Gd[Fe(CN)_6]$	*NADG	9.9 (4.7 T)	13.1 (4.7 T)	(Perrier et al. 2013)
$Gd[Co(CN)_6]$	PEG400	4.8 (4.7 T)	5.5 (4.7 T)	(Perrier et al. 2013)
$Tb[Fe(CN)_6]$	PEG400	0.11 (4.7 T)	2.7 (4.7 T)	(Perrier et al. 2013)
$Y[Fe(CN)_6]$	PEG400	0.01 (4.7 T)	0.05 (4.7 T)	(Perrier et al. 2013)
$Ni_3[Fe(CN)_6]_2$	PEG1000	0.08 (4.7 T)	1.32 (4.7 T)	(Perrier et al. 2013)
$Cu_3[Fe(CN)_6]_2$	PEG1000	0.26 (4.7 T)	1.09 (4.7 T)	(Perrier et al. 2013)
$Fe_4[Fe(CN)_6]_3$ @AuNP	SH-PEG-COOH	7.48 (3 T)	8.02 (3 T)	(Dou et al. 2017)
$Fe_4[Fe(CN)_6]_3$ @ Fe_3O_4			58.9 (3 T)	(Fu et al. 2014)
Gd^{3+}-doped PB	PVP	30.4 (7 T)		(Cai et al. 2016)
Hollow Gd^{3+}-doped PB	PVP	32.8 (7 T)		(Cai et al. 2016)
Hollow mesoporous PB	PVP	0.14 (7 T)		(Cai et al. 2016)
$Gd_{0.1}Fe_{2.9}[Fe(CN)_6]_3$	Citrate	7.6 (7 T)		(Shokouhimehr et al. 2010)
$Gd(H_2O)_4[Fe^{III}(CN)_6]$	Chitosan	6–7 (4.7 T)		(Guari et al. 2008)
$KGd(H_2O)_2[Fe^{III}(CN)_6].H_2O$	PVP	16.8 (7 T); 35.2 (1.4 T)	23.9 (7 T); 38.4 (1.4 T)	(Perera et al. 2014)

Note: *NADG= N-acetyl-D-glucosamine

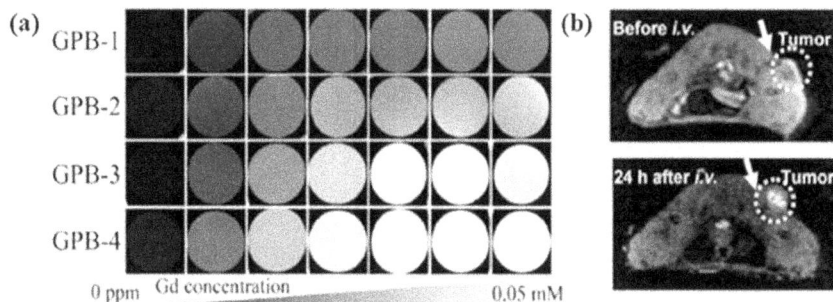

FIGURE 7.11 PB and PBA as efficient MRI CA. (a) *In vitro* T_1-weighted phantom images of Gd(III)-doped PBNP. (Reproduced with permission from *American Chemical Society*, copyright from Ref. (Cai et al. 2016). (b) *In vivo* T_1-weighted MR images of Mn(II)-doped PBNP. (Reproduced with permission from *American Chemical Society*, copyright from Ref. (Zhu et al. 2015).

Multimodal imaging smart contrast agent: In the recent past, nanotechnology has been widely used in many areas to protect from health hazards from various chronic diseases, caused by environmental pollution and unhygienic foods. Cancer is one of them, which exhibits different growth mechanisms because of its abnormal hyperplasia and neovascularization. The main problem that researchers are facing in this regard is the reliable early-stage detection of cancer followed by different side effects of commercially available particular CA for a particular diagnostic technique. However, these drawbacks can be overcome using nanotechnology in recent times as this newly developed technology offers much reliable diagnosis techniques such as magnetic resonance imaging (MRI), computed tomography (CT), photoacoustic imaging (PAI), positron emission tomography (PET), ultrasound imaging, fluorescence imaging (FI), etc. Furthermore, multimodal imaging permits examining more than a single molecule at a time so that cellular events may be studied concurrently or the progression of these events can be monitored in real time. To achieve a synergistic diagnostic effect, recent work has focused on assembling more nanomaterials on a single particle like a core-shell structure. In this context, PB@Au core-shell nanostructure has been developed, which exhibits an effective tumor localization after intravenous injection (shown in Figure 7.12) by inert targeted accumulation. These phenomena are confirmed by both T_1- and T_2-weighted MR and CT imaging, so PB@Au may be used as a dual-modal CA (Dou et al. 2017). Furthermore, PB@MnO$_2$ and PB@HA-based hybrid nanostructures have been synthesized, which can be used as both MRI (T_1- and T_2-weighted) and PAI CA in mice model (Peng et al. 2017, Zhou et al. 2018). Recently, *in vivo* triple-modal PAI, MRI, and fluorescence imaging has been observed using graphene–iron oxide nanoparticles (Feng et al. 2013, Qian et al. 2019). In this connection, some other nanocomposites which exhibit combinational diagnosis techniques are shown in Table 7.7.

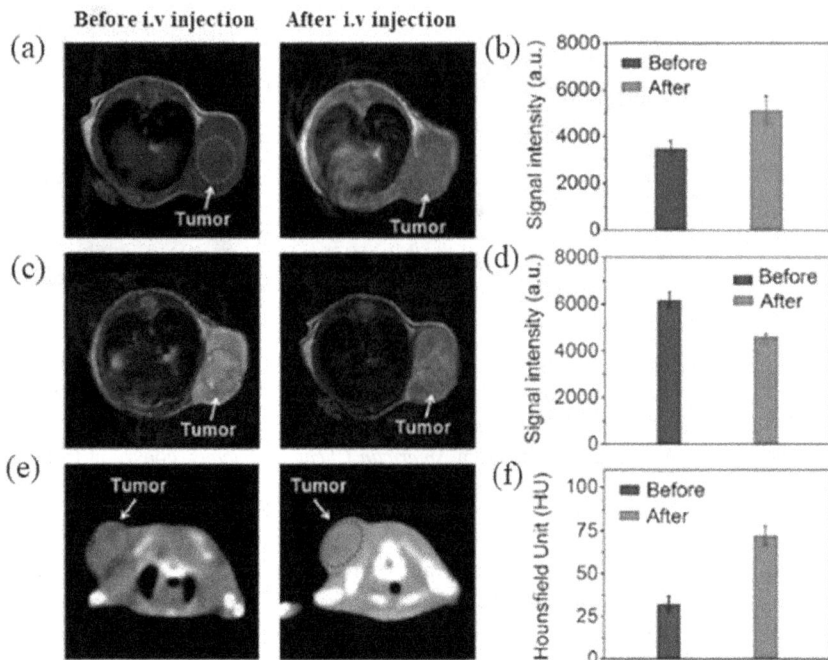

FIGURE 7.12 (a,b) *In vivo* T_1 -weighted MRI images and corresponding signal intensity plot. (c,d) *In vivo* T_2 -weighted MRI images and corresponding signal intensity plot. (e,f) *In vivo* CT images and corresponding signal intensity plot of Au@PB core-shell structure. (Reproduced with permission from *American Chemical Society*, copyright from Ref. (Dou et al. 2017).

TABLE 7.7
Examples of Some Multimodal Smart Contrast Agents

Composite	Synergistic diagnosis technique	Target specific	Reference
PB@Au	MRI and CT	4T1 tumor-bearing mice model	(Dou et al. 2017)
PB@HA	MRI and PAI	4T1 tumor-bearing mice model	(Zhou et al. 2018)
rGO@MnFe$_2$O$_4$	MRI and SPECT	4T1 tumor-bearing mice model	(Qian et al. 2019))
Mn^{2+}-doped PB	MRI and PAI	MCF-7 tumor-bearing mice model	(Zhu et al. 2015)
Gadolinium oxide nanocrystal	MRI and NIRF	4T1 tumor-bearing mice model	(Wang et al. 2015)
Gd(III)-doped Au@SiO$_2$	MRI, PAI, CT	MDA-MB-231 and MCF-7 tumor-bearing mice model	(Wang et al. 2016)
Co-loaded IO	MRI, PAI, FI	C6 tumor-bearing mice model	(Zhou et al. 2014)
PB@MnO$_2$	MRI and PAI	MCF-7 tumor-bearing mice model	(Peng et al. 2017)

7.6 CONCLUSION AND FUTURE PROSPECTS

Every day, within the realm of nanomedicine, numerous novel NPs continuously being developed to address definite challenges inherent within the human body. While the unique characteristics of the developed NPs hold promise for favorable biomedical applications, it is crucial to subject them to scrutiny for potential adverse effects. This entails conducting comprehensive *in vitro* and *in vivo* toxicology studies to assess their possible negative impacts.

In diagnostics, the factual task is to synthesize pithy and reasonable point-of-care systems that can be used for the early detection of human abnormalities with high specificity and sensitivity. In recent years, MRI has become a precious diagnostic technique that can create a better contrast between fat, muscle, water, and other soft tissue based on the abundance of water protons. The presence of any external agent will produce a local magnetic field which will further enhance the contrasting efficiency. The clinically used FDA-approved CAs, either gadolinium- and manganese-based chelates or iron oxide nanoparticles, have encouraged researchers to develop their understanding of the contrasting mechanism which depends on water interaction, composition, structure, and motion of molecules in the presence of an externally applied magnetic field. In the first section of this chapter we broadly discuss the theoretical model based on inner-sphere SBM theory and outer-sphere diffusion theory to explain the contrast mechanism of MRI CAs and different strategies for enhancing the contrast efficiency of a CA, and it was concluded that rotational correlation time, number of hydration states and their interaction, number of paramagnetic metal ions and their inherent properties like spin, magnetization value, etc., play an important role in the enhancement of relaxivity value. Most clinically used MRI CAs show low relaxivity value and they are limited due to human health issues. So, with modern technology researchers are focused on developing new types of CAs to overcome these drawbacks. Here, in the last section of this chapter we broadly discuss different types of CAs. Recent research on nanotechnology offers some paramagnetic nanoparticles including FDA-approved PB and PBA having high contrasting efficiency that may be used as finest MRI CAs as well as therapeutic agents for future use. Moreover, multimodal imaging techniques offer the ability to examine more than a single molecule at a time so that cellular events may be studied concurrently or the progression of these events can be monitored in real time.

In summary, nanotechnology possesses pronounced potential to provide multimodal diagnostic techniques for the early detection of abnormalities in the scenario of human health. Efforts are presently underway to transmit these scientific improvements into the clinic and should bring new archetypes in biomedicine.

REFERENCES

Aaron, A.J., Bumb, A., and Brechbiel, M.W., 2010. Macromolecules, dendrimers, and nanomaterials in magnetic resonance imaging: The interplay between size, function, and pharmacokinetics. *Chemical Reviews*, 110 (5), 2921–2959.

Ahmad, M.W., Xu, W., Kim, S.J., Baeck, J.S., Chang, Y., Bae, J.E., Chae, K.S., Park, J.A., Kim, T.J., and Lee, G.H., 2015. Potential dual imaging nanoparticle: Gd2O3 nanoparticle. *Scientific Reports*, 5, 8549.

Aime, S., Botta, M., Gianolio, E., and Terreno, E., 2000. A p(O2)-Responsive MRI contrast agent based on the redox switch of manganese(II/III)-porphyrin complexes. *Angewandte Chemie – International Edition*, 39 (4), 747–750.

Akhtar, K., Javed, Y., Akhtar, M.I., and Shad, N.A., 2021. Applications of iron oxide nanoparticles in the magnetic resonance imaging for the cancer diagnosis. *Nanopharmaceuticals: Principles and Application*, 1, 115–158.

Anghileri, L.J., Heidbreder, M., and Mathes, R., 1976. Accumulation of 57Co poly L lysine by tumors: an effect of the tumor electrical charge. *Journal of Nuclear Biology and Medicine*, 20 (2), 79–83.

Bloembergen, N. and Morgan, L.O., 1961. Proton relaxation times in paramagnetic solutions. Effects of electron spin relaxation. *The Journal of Chemical Physics*, 34 (3), 842–850.

Bloembergen, N., Purcell, E.M., and Pound, R. v., 1948. Relaxation effects in nuclear magnetic resonance absorption. *Physical Review*, 73 (7), 679.

Bonnet, C.S. and Tóth, É., 2021. Metal-based environment-sensitive MRI contrast agents. *Current Opinion in Chemical Biology*, 61, 154–169.

Bonnet, C.S., Fries, P.H., Crouzy, S., and Delangle, P., 2010. Outer-sphere investigation of MRI relaxation contrast agents. Example of a cyclodecapeptide gadolinium complex with second-sphere water. *Journal of Physical Chemistry B*, 114 (26), 8770–8781.

Boyer, T.H., 1975. Random electrodynamics: The theory of classical electrodynamics with classical electromagnetic zero-point radiation. *Physical Review D*, 11 (4), 790.

Brooks, R.A., 2002. T2-shortening by strongly magnetized spheres: A chemical exchange model. *Magnetic Resonance in Medicine*, 47 (2), 388–391.

Brooks, R.A., Moiny, F., and Gillis, P., 2002. On T2-shortening by strongly magnetized spheres: A partial refocusing model. *Magnetic Resonance in Medicine*, 47 (2), 257–263.

Bulte, J.W.M., Cuyper, M. de, Despres, D., and Frank, J.A., 1999. Preparation, relaxometry, and biokinetics of PEGylated magnetoliposomes as MR contrast agent. *Journal of Magnetism and Magnetic Materials*, 194 (1), 204–209.

Bussi, S., Coppo, A., Celeste, R., Fanizzi, A., Fringuello Mingo, A., Ferraris, A., Botteron, C., Kirchin, M.A., Tedoldi, F., and Maisano, F., 2020. Macrocyclic MR contrast agents: evaluation of multiple-organ gadolinium retention in healthy rats. *Insights into Imaging*, 11 (1), 1–10.

Cai, X., Gao, W., Zhang, L., Ma, M., Liu, T., Du, W., Zheng, Y., Chen, H., and Shi, J., 2016. Enabling Prussian Blue with Tunable Localized Surface Plasmon Resonances: Simultaneously Enhanced Dual-Mode Imaging and Tumor Photothermal Therapy. *ACS Nano*, 10 (12), 11115–11126.

Caravan, P., Cloutier, N.J., Greenfield, M.T., McDermid, S.A., Dunham, S.U., Bulte, J.W.M., Amedio, J.C., Looby, R.J., Supkowski, R.M., Horrocks, W.D.W., McMurry, T.J., and Lauffer, R.B., 2002. The interaction of MS-325 with human serum albumin and its effect on proton relaxation rates. *Journal of the American Chemical Society*, 124 (12), 3152–3162.

Cerasa, A., Cherubini, A., and Peran, P., 2012. Multimodal MRI in neurodegenerative disorders. *Neurology Research International*, 1–2.

Chelebaeva, E., Larionova, J., Guari, Y., Ferreira, R.A.S., Carlos, L.D., Trifonov, A.A., Kalaivani, T., Lascialfari, A., Guérin, C., Molvinger, K., Datas, L., Maynadier, M., Gary-Bobo, M., and Garcia, M., 2011. Nanoscale coordination polymers exhibiting luminescence properties and NMR relaxivity. *Nanoscale*, 3 (3), 1200–1210.

Chen, J.H., Haghmoradi, A., and Althaus, S.M., 2020. NMR Intermolecular Dipolar Cross-Relaxation in Nanoconfined Fluids. *Journal of Physical Chemistry B*, 124 (45), 10237–10244.

Chen, J.W., Belford, R.L., and Clarkson, R.B., 1998. Second-sphere and outer-sphere proton relaxation of paramagnetic complexes: From EPR to NMRD. *Journal of Physical Chemistry A*, 102 (12), 2117–2130.

Chen, Y., Yang, H., Tang, W., Cui, X., Wang, W., Chen, X., Yuan, Y., and Hu, A., 2013. Attaching double chain cationic Gd(iii)-containing surfactants on nanosized colloids for highly efficient MRI contrast agents. *Journal of Materials Chemistry B*, 1 (40), 5443–5449.

Chowdhury, M.A., 2017. Metal-organic-frameworks for biomedical applications in drug delivery, and as MRI contrast agents. *Journal of Biomedical Materials Research Part A*, 105 (4), 1184–1194.

Cormode, D.P., Skajaa, T., van Schooneveld, M.M., Koole, R., Jarzyna, P., Lobatto, M.E., Calcagno, C., Barazza, A., Gordon, R.E., Zanzonico, P., Fisher, E.A., Fayad, Z.A., and Mulder, W.J.M., 2008. Nanocrystal core high-density lipoproteins: A multimodality contrast agent platform. *Nano Letters*, 8 (11), 3715–3723.

Creasman, W.T., Duggan, E.R., and Lund, C.J., 1966. Absorption of transfused chromium-labeled erythrocytes from the fetal peritoneal cavity in hydrops fetalis. *American Journal of Obstetrics and Gynecology*, 94 (4), 586–588.

de Smet, M., Langereis, S., den Bosch, S. van, and Grüll, H., 2010. Temperature-sensitive liposomes for doxorubicin delivery under MRI guidance. *Journal of Controlled Release*, 143 (1), 120–127.

Do, C., DeAguero, J., Brearley, A., Trejo, X., Howard, T., Escobar, G.P., and Wagner, B., 2020. Gadolinium-Based Contrast Agent Use, Their Safety, and Practice Evolution. *Kidney360*, 1 (6), 561.

Dou, Y., Li, X., Yang, W., Guo, Y., Wu, M., Liu, Y., Li, X., Zhang, X., and Chang, J., 2017. PB@Au core−satellite multifunctional nanotheranostics for magnetic resonance and computed tomography imaging in vivo and synergetic photothermal and radiosensitive therapy. *ACS Applied Materials and Interfaces*, 9 (2), 1263–1272.

Duguet, E., Vasseur, S., Mornet, S., and Devoisselle, J-M., 2006. Magnetic nanoparticles and their applications in medicine. *Nanomedicine*, 1 (2), 157–168.

Dumas, S., Jacques, V., Sun, W.-C., Troughton, J.S., Welch, J.T., Chasse, J.M., Schmitt-Willich, H., and Caravan, P., 2010. High Relaxivity Magnetic Resonance Imaging Contrast Agents Part 1. *Investigative Radiology*, 45 (10), 600–612.

Dunbar, K.R. and Heintz, R.A., 1997. Chemistry of transition metal cyanide compounds: Modern perspectives. *Progress in Inorganic Chemistry*, 45, 283–292.

Edelman, R.R., Siegel, J.B., Singer, A., Dupuis, K., and Longmaid, H.E., 1989. Dynamic MR imaging of the liver with Gd-DTPA: Initial clinical results. *American Journal of Roentgenology*, 153 (6), 1213–1219.

Edwards, B.J., Laumann, A.E., Nardone, B., Miller, F.H., Restaino, J., Raisch, D.W., McKoy, J.M., Hammel, J.A., Bhatt, K., Bauer, K., Samaras, A.T., Fisher, M.J., Bull, C., Saddleton, E., Belknap, S.M., Thomsen, H.S., Kanal, E., Cowper, S.E., Abu Alfa, A.K., and West, D.P., 2014. Advancing pharmacovigilance through academic-legal collaboration: The case of gadolinium-based contrast agents and nephrogenic systemic fibrosis – A research on adverse drug events and reports (RADAR) report. *British Journal of Radiology*, 87 (1042), 20140307.

Endrikat, J., Vogtlaender, K., Dohanish, S., Balzer, T., and Breuer, J., 2016. Safety of gadobutrol: Results from 42 clinical phase II to IV studies and postmarketing surveillance after 29 million applications. *Investigative Radiology*, 51 (9), 537.

Feng, K., Zhang, J., Dong, H., Li, Z., Gu, N., Ma, M., and Zhang, Y., 2021. Prussian Blue Nanoparticles Having Various Sizes and Crystallinities for Multienzyme Catalysis and Magnetic Resonance Imaging. *ACS Applied Nano Materials*, 4 (5), 5176–5186.

Feng, L., Wu, L., and Qu, X., 2013. New horizons for diagnostics and therapeutic applications of graphene and graphene oxide. *Advanced Materials*, 25, 168–186.

Fossheim, S.L., Il'yasov, K.A., Hennig, J., and Bjornerud, A., 2000. Thermosensitive paramagnetic liposomes for temperature control during MR imaging-guided hyperthermia: In vitro feasibility studies. *Academic Radiology*, 7 (12), 1107–1115.

Fotenos, A., 2018. Update on FDA approach to safety issue of gadolinium retention after administration of gadolinium-based contrast agents 2018. *US Food and Drug Administration*.

Frias, J.C., Williams, K.J., Fisher, E.A., and Fayad, Z.A., 2004. Recombinant HDL-like nanoparticles: A specific contrast agent for MRI of atherosclerotic plaques. *Journal of the American Chemical Society*, 126 (50), 16316–16317.

Fu, G., Liu, W., Li, Y., Jin, Y., Jiang, L., Liang, X., Feng, S., and Dai, Z., 2014. Magnetic Prussian Blue Nanoparticles for Targeted Photothermal Therapy under Magnetic Resonance Imaging Guidance. *Bioconjugate Chemistry*, 25 (9), 1655–1663.

Gale, E.M. and Caravan, P., 2018. Gadolinium-Free Contrast Agents for Magnetic Resonance Imaging of the Central Nervous System. *ACS Chemical Neuroscience*, 9 (3), 395–397.

Gallo, R.L., 2017. Human skin is the largest epithelial surface for interaction with microbes. *Journal of Investigative Dermatology*, 137 (6):1213–1214.

Geraldes, C.F.G.C. and Laurent, S., 2009. Classification and basic properties of contrast agents for magnetic resonance imaging. *Contrast Media and Molecular Imaging*, 4 (1), 1–23.

Griewing, B., Hielscher, H., and Lutcke, A., 1992. The importance of MRI (magnetic resonance imaging) for the diagnosis of brainstem infarction. *Bildgebung/Imaging*, 59 (2), 94–97.

Guari, Y., Larionova, J., Corti, M., Lascialfari, A., Marinone, M., Poletti, G., Molvinger, K., and Guérin, C., 2008. Cyano-bridged coordination polymer nanoparticles with high nuclear relaxivity: Toward new contrast agents for MRI. *Dalton Transactions*, 28, 3658–3660.

Han, H.S. and Choi, K.Y., 2021. Advances in nanomaterial-mediated photothermal cancer therapies: Toward clinical applications. *Biomedicines*, 9 (3), 305.

Helm, L. and Merbach, A.E., 2005. Inorganic and bioinorganic solvent exchange mechanisms. *Chemical Reviews*, 105 (6), 1923–1960.

Helm, L., 2017. *Gadolinium-based contrast agents*. RSC Contrast agents for MRI: *Experimental Methods*, 13, 121.

Helm, P.A., Caravan, P., French, B.A., Jacques, V., Shen, L., Xu, Y., Beyers, R.J., Roy, R.J., Kramer, C.M., and Epstein, F.H., 2008. Postinfarction myocardial scarring in mice: Molecular MR imaging with use of a collagen-targeting contrast agent. *Radiology*, 247 (3), 788–796.

Hoad, C.L., Parker, H., Hudders, N., Costigan, C., Cox, E.F., Perkins, A.C., Blackshaw, P.E., Marciani, L., Spiller, R.C., Fox, M.R., and Gowland, P.A., 2015. Measurement of gastric meal and secretion volumes using magnetic resonance imaging. *Physics in Medicine and Biology*, 60 (3), 1367.

Jiang, Q.L., Zheng, S.W., Hong, R.Y., Deng, S.M., Guo, L., Hu, R.L., Gao, B., Huang, M., Cheng, L.F., Liu, G.H., and Wang, Y.Q., 2014. Folic acid-conjugated Fe3O4 magnetic nanoparticles for hyperthermia and MRI in vitro and in vivo. *Applied Surface Science*, 307, 224–233.

Jun, Y.W., Huh, Y.M., Choi, J.S., Lee, J.H., Song, H.T., Kim, S., Yoon, S., Kim, K.S., Shin, J.S., Suh, J.S., and Cheon, J., 2005. Nanoscale Size Effect of Magnetic Nanocrystals and Their Utilization for Cancer Diagnosis via Magnetic Resonance Imaging. *Journal of the American Chemical Society*, 127 (16), 5732–5733.

Jung, C.W. and Jacobs, P., 1995. Physical and chemical properties of superparamagnetic iron oxide MR contrast agents: Ferumoxides, ferumoxtran, ferumoxsil. *Magnetic Resonance Imaging*, 13 (5), 661–674.

Jung, H., Park, B., Lee, C., Cho, J., Suh, J., Park, J.Y., Kim, Y.R., Kim, J., Cho, G., and Cho, H.J., 2014. Dual MRI T1 and T2(*) contrast with size-controlled iron oxide nanoparticles. *Nanomedicine: Nanotechnology, Biology, and Medicine*, 10 (8), 1679–1689.

Junk, A. and Riess, F., 2006. From an idea to a vision: There's plenty of room at the bottom. *American Journal of Physics*, 74 (9), 825–830.

Kaminsky, S., Laniado, M., Gogoll, M., Kornmesser, W., Clauss, W., Langer, M., Claussen, C., and Felix, R., 1991. Gadopentetate dimeglumine as a bowel contrast agent: Safety and efficacy. *Radiology*, 178 (2), 503–508.

Kan, J., Milne, M., Tyrrell, D., and Mansfield, C., 2022. Lean body weight-adjusted intravenous iodinated contrast dose for abdominal CT in dogs reduces interpatient enhancement variability while providing diagnostic quality organ enhancement. *Veterinary Radiology & Ultrasound*, 63 (6), 719–728.

Kreuter, J., 2007. Nanoparticles – a historical perspective. *International Journal of Pharmaceutics*, 331 (1), 1–10.

LaConte, L.E.W., Nitin, N., Zurkiya, O., Caruntu, D., O'Connor, C.J., Hu, X., and Bao, G., 2007. Coating thickness of magnetic iron oxide nanoparticles affects R 2 relaxivity. *Journal of Magnetic Resonance Imaging*, 26 (6), 1634–1641.

Laurent, S., Forge, D., Port, M., Roch, A., Robic, C., van der Elst, L., and Muller, R.N., 2008. Magnetic iron oxide nanoparticles: Synthesis, stabilization, vectorization, physicochemical characterizations and biological applications. *Chemical Reviews*, 108 (6), 2064–2110.

Laurent, S., Henoumont, C., Stanicki, D., Boutry, S., Lipani, E., Belaid, S., Muller, R.N., and van der Elst, L., 2017. *Magnetic Properties. MRI Contrast Agents: From Molecules to Particles*. Springer, pp. 5–11.

Lee, N., Choi, Y., Lee, Y., Park, M., Moon, W.K., Choi, S.H., and Hyeon, T., 2012. Water- dispersible ferrimagnetic iron oxide nanocubes with extremely high r_2 relaxivity for highly sensitive in vivo MRI of tumors. *Nano Letters*, 12 (6), 3127–3131.

Lee, T., Zhang, X. an, Dhar, S., Faas, H., Lippard, S.J., and Jasanoff, A., 2010. In Vivo Imaging with a Cell-Permeable Porphyrin-Based MRI Contrast Agent. *Chemistry and Biology*, 17 (6), 665–673.

Li, Q., Kartikowati, C.W., Horie, S., Ogi, T., Iwaki, T., and Okuyama, K., 2017. Correlation between particle size/domain structure and magnetic properties of highly crystalline Fe_3O_4 nanoparticles. *Scientific Reports*, 7 (1), 9894.

Li, Y., Chen, T., Tan, W., and Talham, D.R., 2014. Size-dependent MRI relaxivity and dual imaging with Eu0.2Gd 0.8PO$_4$·H$_2$O Nanoparticles. *Langmuir*, 30 (20), 5873–5879.

Li, Z., Zeng, Y., Zhang, D., Wu, M., Wu, L., Huang, A., Yang, H., Liu, X., and Liu, J., 2014. Glypican-3 antibody functionalized Prussian blue nanoparticles for targeted MR imaging and photothermal therapy of hepatocellular carcinoma. *Journal of Materials Chemistry B*, 2 (23), 3686–3696.

Linderoth, S., Hendriksen, P. v., Bødker, F., Wells, S., Davies, K., Charles, S.W., and Mørup, S., 1994. On spin-canting in maghemite particles. *Journal of Applied Physics*, 75 (10), 6583–6585.

Lipari, G. and Szabo, A., 1982. Model-Free Approach to the Interpretation of Nuclear Magnetic Resonance Relaxation in Macromolecules. 1. Theory and Range of Validity. *Journal of the American Chemical Society*, 104 (17), 4546–4559.

Liu, G., Tse, N.M.K., Hill, M.R., Kennedy, D.F., and Drummond, C.J., 2011. Disordered mesoporous gadolinosilicate nanoparticles prepared using gadolinium based ionic liquid emulsions: Potential as magnetic resonance imaging contrast agents. *Australian Journal of Chemistry*, 64 (5), 617–624.

Lobbes, M.B.I., Heeneman, S., Passos, V.L., Welten, R., Kwee, R.M., van der Geest, R.J., Wiethoff, A.J., Caravan, P., Misselwitz, B., Daemen, M.J.A.P., van Engelshoven, J.M.A., Leiner, T., and Kooi, M.E., 2010. Gadofosveset-enhanced magnetic resonance imaging of human carotid atherosclerotic plaques: A proof-of-concept study. *Investigative Radiology*, 45 (5), 275–281.

Louie, A.Y., Hüber, M.M., Ahrens, E.T., Rothbächer, U., Moats, R., Jacobs, R.E., Fraser, S.E., and Meade, T.J., 2000. In vivo visualization of gene expression using magnetic resonance imaging. *Nature Biotechnology*, 18 (3), 321–325.

Ma, D., Shi, M., Li, X., Zhang, J., Fan, Y., Sun, K., Jiang, T., Peng, C., and Shi, X., 2020. Redox-Sensitive Clustered Ultrasmall Iron Oxide Nanoparticles for Switchable T2/T1-Weighted Magnetic Resonance Imaging Applications. *Bioconjugate Chemistry*, 31 (2), 352–359.

Mahmoudi, M., Sant, S., Wang, B., Laurent, S., and Sen, T., 2011. Superparamagnetic iron oxide nanoparticles (SPIONs): Development, surface modification and applications in chemotherapy. *Advanced Drug Delivery Reviews*, 63 (1–2), 24–46.

Mattrey, R.F., 1989. Perfluorooctylbromide: A new contrast agent for CT, sonography, and MR imaging. *American Journal of Roentgenology*, 152 (2), 247–252.

McNamara, K. and Tofail, S.A.M., 2015. Nanosystems: The use of nanoalloys, metallic, bimetallic, and magnetic nanoparticles in biomedical applications. *Physical Chemistry Chemical Physics*, 17 (42), 27981–27995.

Mehrishi, J.N., 1969. Effect of lysine polypeptides on the surface charge of normal and cancer cells. *European Journal of Cancer (1965)*, 5 (5), 427–435.

Montesi, S., Rao, R., Liang, L., Caravan, P., and Sharma, A., 2017. C78 Fibrosis: Mediators and modulators: Use of gadofosveset-enhanced lung MRI to assess ongoing lung injury in fibrotic interstitial lung disease. *American Journal of Respiratory and Critical Care Medicine*, 195, 1–2.

Morcos, S.K., 2005. Radiological contrast agents. *In*: Side Effects of Drugs Annual.

Mosbah, K., Ruiz-Cabello, J., Berthezène, Y., and Crémillieux, Y., 2008. Aerosols and gaseous contrast agents for magnetic resonance imaging of the lung. *Contrast Media and Molecular Imaging*, 3 (5), 173–190.

Nandi, R., Mishra, S., Maji, T.K., Manna, K., Kar, P., Banerjee, S., Dutta, S., Sharma, S.K., Lemmens, P., Saha, K. das, and Pal, S.K., 2017. A novel nanohybrid for cancer theranostics: folate sensitized Fe_2O_3 nanoparticles for colorectal cancer diagnosis and photodynamic therapy. *Journal of Materials Chemistry B*, 5 (21), 3927–3939.

Neeley, C., Moritz, M., Brown, J.J., and Zhou, Y., 2016. Acute side effects of three commonly used gadolinium contrast agents in the paediatric population. *British Journal of Radiology*, 89 (1063), 20160027.

Peng, J., Dong, M., Ran, B., Li, W., Hao, Y., Yang, Q., Tan, L., Shi, K., and Qian, Z., 2017. "One-for-all"-type, biodegradable Prussian blue/manganese dioxide hybrid nanocrystal for trimodal imaging-guided photothermal therapy and oxygen regulation of breast cancer. *ACS Applied Materials and Interfaces*, 9 (16), 13875–13886.

Perera, V.S., Yang, L.D., Hao, J., Chen, G., Erokwu, B.O., Flask, C.A., Zavalij, P.Y., Basilion, J.P., and Huang, S.D., 2014. Biocompatible nanoparticles of $KGd(H_2O)_2[Fe(CN)_6]\cdot H_2O$ with extremely high T1-weighted relaxivity owing to two water molecules directly bound to the Gd(III) center. *Langmuir*, 30 (40), 12018–12026.

Perrier, M., Kenouche, S., Long, J., Thangavel, K., Larionova, J., Goze-Bac, C., Lascialfari, A., Mariani, M., Baril, N., Guérin, C., Donnadieu, B., Trifonov, A., and Guari, Y., 2013. Investigation on NMR relaxivity of nano-sized cyano-bridged coordination polymers. *Inorganic Chemistry*, 52 (23), 13402–13414.

Pillaiyar, T., Manickam, M., and Namasivayam, V., 2017. Skin whitening agents: Medicinal chemistry perspective of tyrosinase inhibitors. *Journal of Enzyme Inhibition and Medicinal Chemistry*, 32 (1), 403–425.

Pintaske, J., Martirosian, P., Graf, H., Erb, G., Lodemann, K.P., Claussen, C.D., and Schick, F., 2006. Relaxivity of gadopentetate dimeglumine (Magnevist), gadobutrol (Gadovist), and gadobenate dimeglumine (MultiHance) in human blood plasma at 0.2, 1.5, and 3 Tesla. *Investigative Radiology*, 41 (3), 213–221.

Pitchaimani, A., Thanh Nguyen, T.D., Wang, H., Bossmann, S.H., and Aryal, S., 2016. Design and characterization of gadolinium infused theranostic liposomes. *RSC Advances*, 6 (43), 36898–36905.

Polasek, M. and Caravan, P., 2013. Is macrocycle a synonym for kinetic inertness in Gd(III) complexes? Effect of coordinating and noncoordinating substituents on inertness and relaxivity of Gd(III) chelates with DO3A-like ligands. *Inorganic Chemistry*, 52 (7), 4084–4096.

Port, M., Corot, C., Rousseaux, O., Raynal, I., Devoldere, L., Idée, J.M., Dencausse, A., le Greneur, S., Simonot, C., and Meyer, D., 2001. P792: A rapid clearance blood pool agent for magnetic resonance imaging: Preliminary results. *In: Magnetic Resonance Materials in Physics, Biology and Medicine*, 12, 121–127.

Qian, R., Maiti, D., Zhong, J., Xiong, S., Zhou, H., Zhu, R., Wan, J., and Yang, K., 2019. Multifunctional nano-graphene based nanocomposites for multimodal imaging guided combined radioisotope therapy and chemotherapy. *Carbon*, 149, 55–62.

Qian, W., Murakami, M., Ichikawa, Y., and Che, Y., 2011. Highly efficient and controllable PEGylation of gold nanoparticles prepared by femtosecond laser ablation in water. *Journal of Physical Chemistry C*, 115 (47), 23293–23298.

Randall, B., Lauffer Thomas, J., Mcmurry Stephen, O., Dunham Daniel, M., Scott David, J., and Stéphane, D., 1997. Bioactivated diagnostic imaging contrast agents, U.S. Patent Application No. 11/504,851.

Rao, Y.F., Chen, W., Liang, X.G., Huang, Y.Z., Miao, J., Liu, L., Lou, Y., Zhang, X.G., Wang, B., Tang, R.K., Chen, Z., and Lu, X.Y., 2015. Epirubicin-loaded superparamagnetic iron-oxide nanoparticles for transdermal delivery: Cancer therapy by circumventing the skin barrier. *Small*, 11 (2), 239–247.

Reedijk, J. and Poeppelmeier, K., 2013. *Comprehensive Inorganic Chemistry II (Second Edition): From Elements to Applications*. Elsevier.

Reynolds, F., O'Loughlin, T., Weissleder, R., and Josephson, L., 2005. Method of determining nanoparticle core weight. *Analytical Chemistry*, 77 (3), 814–817.

Rijcken, T.H.P., Davis, M.A., and Ros, P.R., 1994. Intraluminal contrast agents for MR imaging of the abdomen and pelvis. *Journal of Magnetic Resonance Imaging*, 4 (3), 291–300.

Rogers, J., Lewis, J., and Josephson, L., 1994. Use of AMI-227 as an oral MR contrast agent. *Magnetic Resonance Imaging*, 12 (4), 631–639.

Rohrer, M., Bauer, H., Mintorovitch, J., Requardt, M., and Weinmann, H.J., 2005. Comparison of magnetic properties of MRI contrast media solutions at different magnetic field strengths. *Investigative Radiology*, 40 (11), 715–724.

Saljoughian, M., 2012. Intravenous Radiocontrast Media: A Review of Allergic Reactions. *US Pharm*, 37 (5), 14–16.

Saljoughian, M., 2012. Intravenous Radiocontrast Media: A Review of Allergic Reactions. *US Pharm*, 37 (5), 14–16.

Schalla, S., Higgins, C.B., and Saeed, M., 2002. Contrast agents for cardiovascular magnetic resonance imaging: Current status and future directions. *Drugs in R and D*, 3, 285–302.

Schieda, N., Blaichman, J.I., Costa, A.F., Glikstein, R., Hurrell, C., James, M., Jabehdar Maralani, P., Shabana, W., Tang, A., Tsampalieros, A., van der Pol, C.B., and Hiremath, S., 2018. Gadolinium-based contrast agents in kidney disease: A comprehensive review and clinical practice guideline issued by the Canadian Association of Radiologists. *Canadian Journal of Kidney Health and Disease*, 5, 2054358118778573.

Schwickert, H.C., Stiskal, M., van Dijke, C.F., Roberts, T.P.L., Mann, J.S., Demsar, F., and Brasch, R.C., 1995. Tumor angiography using high-resolution, three-dimensional magnetic resonance imaging: Comparison of gadopentetate dimeglumine and a macromolecular blood-pool contrast agent. *Academic Radiology*, 2 (10), 851–858.

Scott, L.J., 2018. Gadobutrol: A Review in Contrast-Enhanced MRI and MRA. *Clinical Drug Investigation*, 38 (8), 773–784.

Sessler J. L., 2016. Texaphyrins: Life, Death, and Attempts at Resurrection. *John Wiley & Sons: West Sussex, UK*, 309–324.

Shah, R.R., Davis, T.P., Glover, A.L., Nikles, D.E., and Brazel, C.S., 2015. Impact of magnetic field parameters and iron oxide nanoparticle properties on heat generation for use in magnetic hyperthermia. *Journal of Magnetism and Magnetic Materials*, 387, 96–106.

Shahbazi-Gahrouei, D., 2006. Gadolinium-porphyrins: New potential magnetic resonance imaging contrast agents for melanoma detection. *Journal of Research in Medical Sciences*, 11 (4)., 217–223

Shahbazi-Gahrouei, D., Williams, M., Rizvi, S., and Allen, B.J., 2001. In vivo studies of Gd-DTPA-monoclonal antibody and Gd-porphyrins: Potential magnetic resonance imaging contrast agents for melanoma. *Journal of Magnetic Resonance Imaging*, 14 (2), 169–174.

Shin, J., Anisur, R.M., Ko, M.K., Im, G.H., Lee, J.H., and Lee, I.S., 2009. Hollow manganese oxide nanoparticles as multifunctional agents for magnetic resonance imaging and drug delivery. *Angewandte Chemie – International Edition*, 48 (2), 321–324.

Shokouhimehr, M., Soehnlen, E.S., Hao, J., Griswold, M., Flask, C., Fan, X., Basilion, J.P., Basu, S., and Huang, S.D., 2010a. Dual purpose Prussian blue nanoparticles for cellular imaging and drug delivery: a new generation of T1-weighted MRI contrast and small molecule delivery agents. *Journal of Materials Chemistry*, 20 (25), 5251.

Shokouhimehr, M., Soehnlen, E.S., Khitrin, A., Basu, S., and Huang, S.D., 2010. Biocompatible Prussian blue nanoparticles: Preparation, stability, cytotoxicity, and potential use as an MRI contrast agent. *Inorganic Chemistry Communications*, 13 (1), 58–61.

Siddique, S. and Chow, J.C.L., 2020. Application of nanomaterials in biomedical imaging and cancer therapy. *Nanomaterials*, 10 (9), 1700.

Smolensky, E.D., Park, H.Y.E., Zhou, Y., Rolla, G.A., Marjańska, M., Botta, M., and Pierre, V.C., 2013. Scaling laws at the nanosize: The effect of particle size and shape on the magnetism and relaxivity of iron oxide nanoparticle contrast agents. *Journal of Materials Chemistry B*, 1 (22), 2818–2828.

Tejedor, M., Rubio, H., Elbaile, L., and Iglesias, R., 1995. External Fields Created by Uniformly Magnetized Ellipsoids and Spheroids. *IEEE Transactions on Magnetics*, 31 (1), 830–836.

Tsitovich, P.B., Cox, J.M., Benedict, J.B., and Morrow, J.R., 2016. Six-coordinate Iron(II) and Cobalt(II) paraSHIFT Agents for Measuring Temperature by Magnetic Resonance Spectroscopy. *Inorganic Chemistry*, 55 (2), 700–716.

Tsitovich, P.B., Tittiris, T.Y., Cox, J.M., Benedict, J.B., and Morrow, J.R., 2018. Fe(II) and Co(II): N -methylated CYCLEN complexes as paraSHIFT agents with large temperature dependent shifts. *Dalton Transactions*, 47 (3), 916–924.

Uchida, M., Terashima, M., Cunningham, C.H., Suzuki, Y., Willits, D.A., Willis, A.F., Yang, P.C., Tsao, P.S., McConnell, M. v., Young, M.J., and Douglas, T., 2008. A human ferritin iron oxide nano-composite magnetic resonance contrast agent. *Magnetic Resonance in Medicine*, 60 (5), 1073–1081.

Valcourt, D.M., Harris, J., Riley, R.S., Dang, M., Wang, J., and Day, E.S., 2018. Advances in targeted nanotherapeutics: From bioconjugation to biomimicry. *Nano Research*, 11, 4999–5016.

Verma, A., Uzun, O., Hu, Y., Hu, Y., Han, H.S., Watson, N., Chen, S., Irvine, D.J., and Stellacci, F., 2008. Surface-structure-regulated cell-membrane penetration by monolayer-protected nanoparticles. *Nature Materials*, 7 (7), 588–595.

Verwilst, P., Park, S., Yoon, B., and Kim, J.S., 2015. Recent advances in Gd-chelate based bimodal optical/MRI contrast agents. *Chemical Society Reviews*, 44 (7), 1791–1806.

Vlahos, L., Gouliamos, A., Athanasopoulou, A., Kotoulas, G., Claus, W., Hatziioannou, A., Kalovidouris, A., and Papavasiliou, C., 1994. A comparative study between Gd-DTPA and oral magnetic particles (OMP) as gastrointestinal (GI) contrast agents for MRI of the abdomen. *Magnetic Resonance Imaging*, 12 (5), 719–726.

Vogl, T.J., Hammerstingl, R., Schwarz, W., Mack, M.G., Müller, P.K., Pegios, W., Keck, H., Eibl-Eibesfeldt, A., Hoelzl, J., Woessmer, B., Bergman, C., and Felix, R., 1996. Superparamagnetic iron oxide-enhanced versus gadolinium-enhanced MR imaging for differential diagnosis of focal liver lesions. *Radiology*, 198 (3), 881–887.

Wahsner, J., Gale, E.M., Rodríguez-Rodríguez, A., and Caravan, P., 2019. Chemistry of MRI contrast agents: Current challenges and new frontiers. *Chemical Reviews*, 119 (2), 957–1057.

Walter, A., Billotey, C., Garofalo, A., Ulhaq-Bouillet, C., Lefèvre, C., Taleb, J., Laurent, S., Elst, L. van der, Muller, R.N., Lartigue, L., Gazeau, F., Felder-Flesch, D., and Begin-Colin, S., 2014. Mastering the shape and composition of dendronized iron oxide nanoparticles to tailor magnetic resonance imaging and hyperthermia. *Chemistry of Materials*, 26 (18), 5252–5264.

Wang, J., Liu, J., Liu, Y., Wang, L., Cao, M., Ji, Y., Wu, X., Xu, Y., Bai, B., Miao, Q., Chen, C., and Zhao, Y., 2016. Gd-Hybridized Plasmonic Au-Nanocomposites Enhanced Tumor-Interior Drug Permeability in Multimodal Imaging-Guided Therapy. *Advanced Materials*, 28 (40), 8950–8958.

Wang, Y., Yang, T., Ke, H., Zhu, A., Wang, Y., Wang, J., Shen, J., Liu, G., Chen, C., Zhao, Y., and Chen, H., 2015. Smart Albumin-Biomineralized Nanocomposites for Multimodal Imaging and Photothermal Tumor Ablation. *Advanced Materials*, 27 (26), 3874–3882.

Weissleder, R., 1999. Molecular imaging: Exploring the next frontier. *Radiology*, 212 (3), 609–614.

Wheeler, S., Capone, I., Day, S., Tang, C., and Pasta, M., 2019. Low-Potential Prussian Blue Analogues for Sodium-Ion Batteries: Manganese Hexacyanochromate. *Chemistry of Materials*, 31 (7), 2619–2626.

Winkelman, J., 1962. The Distribution of Tetraphenylporphinesulfonate in the Tumor-bearing Rat. *Cancer Research*, 22, 589–596.

Wong, J., Brugger, A., Khare, A., Chaubal, M., Papadopoulos, P., Rabinow, B., Kipp, J., and Ning, J., 2008. Suspensions for intravenous (IV) injection: A review of development, preclinical and clinical aspects. *Advanced Drug Delivery Reviews*, 60 (8), 939–954.

Xiao, Y.D., Paudel, R., Liu, J., Ma, C., Zhang, Z.S., and Zhou, S.K., 2016. MRI contrast agents: Classification and application (Review). *International Journal of Molecular Medicine*, 38 (5), 1319–1326.

Xie, J., Liu, G., Eden, H.S., Ai, H., and Chen, X., 2011. Surface-engineered magnetic nanoparticle platforms for cancer imaging and therapy. *Accounts of Chemical Research*, 44 (10), 883–892.

Yan, G.P., Robinson, L., and Hogg, P., 2007. Magnetic resonance imaging contrast agents: Overview and perspectives. *Radiography*, 13, 5–19.

Yang, L., Zhou, Z., Liu, H., Wu, C., Zhang, H., Huang, G., Ai, H., and Gao, J., 2015. Europium-engineered iron oxide nanocubes with high T_1 and T_2 contrast abilities for MRI in living subjects. *Nanoscale*, 7 (15), 6843–6850.

Yasui, K. and Nakamura, Y., 2000. Positively charged liposomes containing tumor necrosis factor in solid tumors. *Biological and Pharmaceutical Bulletin*, 23 (3), 318–322.

Ye, F., Jeong, E.K., Jia, Z., Yang, T., Parker, D., and Lu, Z.R., 2008. A peptide targeted contrast agent specific to fibrin-fibronectin complexes for cancer molecular imaging with MRI. *Bioconjugate Chemistry*, 19 (12), 2300–2303.

Yin, X., Russek, S.E., Zabow, G., Sun, F., Mohapatra, J., Keenan, K.E., Boss, M.A., Zeng, H., Liu, J.P., Viert, A., Liou, S.H., and Moreland, J., 2018. Large T1 contrast enhancement using superparamagnetic nanoparticles in ultra-low field MRI. *Scientific Reports*, 8 (1), 11863.

Zhang, S., Wu, K., and Dean Sherry, A., 1999. A novel pH-sensitive MRI contrast agent. *Angewandte Chemie – International Edition*, 38 (21), 3192–3194.

Zhang, X., Blasiak, B., Marenco, A.J., Trudel, S., Tomanek, B., and van Veggel, F.C.J.M., 2016. Design and Regulation of NaHoF4 and NaDyF4 Nanoparticles for High-Field Magnetic Resonance Imaging. *Chemistry of Materials*, 28 (9), 3060–3072.

Zhang, Z., Greenfield, M.T., Spiller, M., McMurry, T.J., Lauffer, R.B., and Caravan, P., 2005. Multilocus binding increases the relaxivity of protein-bound MRI contrast agents. *Angewandte Chemie – International Edition*, 44 (41), 6766–6769.

Zhang, Z., He, R., Yan, K., Guo, Q. ni, Lu, Y. Guo, Wang, X. Xia, Lei, H., and Li, Z. Ying, 2009. Synthesis and in vitro and in vivo evaluation of manganese(III) porphyrin-dextran as a novel MRI contrast agent. *Bioorganic and Medicinal Chemistry Letters*, 19 (23), 6675–6678.

Zheng, S.W., Liu, G., Hong, R.Y., Li, H.Z., Li, Y.G., and Wei, D.G., 2012. Preparation and characterization of magnetic gene vectors for targeting gene delivery. *Applied Surface Science*, 259, 201–207.

Zheng, X., Qian, J., Tang, F., Wang, Z., Cao, C., and Zhong, K., 2015. Microgel-based thermosensitive MRI contrast agent. *ACS Macro Letters*, 4 (4), 431–435.

Zhou, B., Jiang, B.P., Sun, W., Wei, F.M., He, Y., Liang, H., and Shen, X.C., 2018. Water-Dispersible Prussian Blue Hyaluronic Acid Nanocubes with Near-Infrared Photoinduced Singlet Oxygen Production and Photothermal Activities for Cancer Theranostics. *ACS Applied Materials and Interfaces*, 10 (21), 18036–18049.

Zhou, Z., Huang, D., Bao, J., Chen, Q., Liu, G., Chen, Z., Chen, X., and Gao, J., 2012. A synergistically enhanced T1-T2 dual-modal contrast agent. *Advanced Materials*, 24 (46), 6223–6228.

Zhou, Z., Wang, L., Chi, X., Bao, J., Yang, L., Zhao, W., Chen, Z., Wang, X., Chen, X., and Gao, J., 2013. Engineered iron-oxide-based nanoparticles as enhanced T_1 contrast agents for efficient tumor imaging. *ACS Nano*, 7 (4), 3287–3296.

Zhou, Z., Zhao, Z., Zhang, H., Wang, Z., Chen, X., Wang, R., Chen, Z., and Gao, J., 2014. Interplay between longitudinal and transverse contrasts in Fe 3O4 nanoplates with (111) exposed surfaces. *ACS Nano*, 8 (8), 7976–7985.

Zhu, W., Liu, K., Sun, X., Wang, X., Li, Y., Cheng, L., and Liu, Z., 2015. Mn2+-doped Prussian blue Nanocubes for bimodal imaging and Photothermal therapy with enhanced performance. *ACS Applied Materials and Interfaces*, 7 (21), 11575–11582.

Zhu, X., Lin, H., Wang, L., Tang, X., Ma, L., Chen, Z., and Gao, J., 2017. Activatable T1 Relaxivity Recovery Nanoconjugates for Kinetic and Sensitive Analysis of Matrix Metalloprotease 2. *ACS Applied Materials and Interfaces*, 9 (26), 21688–21696.

8 Scheelite Materials in Cell Imaging and Bioanalysis

Nibedita Haldar, Dipesh Choudhury,
Tanmoy Mondal, and Chandan Kumar Ghosh

8.1 INTRODUCTION

Among various tungstates and molybdates, the scheelite-structured rare earth double tungstates or molybdates with generic formula $ARE(MO_4)_2$ (A: Li, Na, K; RE: La-Lu, M: Mo or W) have gained importance as they resemble very well chemical and physical reliability, assigned to their crystal structure, and distinct luminescence properties that originate from tungstate or molybdate (Wang et al. 2016). Due to tunable optical properties that mostly come from the 4f-4f and 4f-5d transitions, they are being widely used in the field of solid lasers, optoelectronics, quantum electronics, scintillators, solid-state light emitting diodes (LEDs), and many more (Durairajan et al. 2013, Gai et al. 2014, Sun 2015). The compounds also serve as excellent host lattices for upconversion luminescence, ascribed to M-O charge transfer and suitable energy transfer from the host to the doped rare-earth activator. Among various scopes, scheelites have gained attention to fulfill the growing energy demand for modern consumer electronics, hybrid electric automobiles, and stationary power grids; rechargeable batteries have reawakened interest in the last few decades. Because of their unmatched combination of high power, long cycle life, and lightweight induced volumetric/gravimetric energy densities, lithium-ion batteries are considered the most advanced rechargeable batteries among all battery chemistries. These properties are largely determined by the crystal structure stability of electrode materials, so there has been a lot of research into polyanionic materials with stable $(WO_4)_2$ or $(MoO_4)_2$ frameworks, resulting in new chemistry and higher capacity for Li-ion and Na-ion batteries (Macalik et al. 2000, Okubo et al. 2007, Guo et al. 2008, Mikhailova et al. 2010, Park et al. 2010, Lu et al. 2013, Deng 2015, Meng et al. 2017, Ozdogru and Capraz 2021). Because of high luminous efficiency, long lifetime, low power consumption, and eco-friendly nature, scheelite-based solid state phosphors as white light emitting diode have emerged as a promising and efficient platform for lighting and display systems.

Within this series of materials, the combination of Ce^{3+} and WO_4^{2-} in $LiCe(WO_4)_2$ has been reported to have enormous potential for use as a fluorescent material (Postema et al. 2011, Shivakumara et al. 2015). These scheelites are also found in

DOI: 10.1201/9781003323518-10

189

chip for light-emitting diode. Previously, blue-emitting GaN was combined with yellow YAG:Ce^{3+} phosphor to prepare while light emitting diode (WLED) chip; however, due to the absence of red light, WLED exhibited a low color rendering index (R$_a$80) and high correlated color temperature (>7000 K), hence their usage was limited. Later, a different strategy on the basis of codoping a sensitizer and an activator into the same host to control the emission color through energy transfer processes has been noticed to be most effective as this provides an efficient way to produce white light from a single-phase phosphor material. Sr$_3$La(PO$_4$)$_3$:Eu^{2+}, Mn^{2+} ; Ba$_3$GdK(PO$_4$)$_3$F:Tb^{3+}, Eu^{3+}; and Ca$_{20}$Al$_{26}$Mg$_3$Si$_3$O$_{68}$:Ce^{3+}, Dy^{3+} are examples for single-phase materials for WLED applications those had been developed so far (Cao et al. 2016, Han et al. 2016, Su et al. 2016, Zhai et al. 2016, Huang et al. 2017, Zhong et al. 2018). However, the majority of them were prepared using a traditional solid-state reaction method that required a high temperature and a reducing atmosphere, which resulted in high energy costs. Furthermore, the product obtained using this method had nonuniform particle size and morphology. The blue and yellow emission bands of Dy^{3+} ions are well known and they are attributed to $^4F_{9/2} \leftrightarrow {}^6H_{15/2}$ and $^4F_{9/2} \leftrightarrow {}^H_{13/2}$ transitions, respectively. Because the blue emission is insufficient in many hosts, Dy^{3+} ions emit yellow light in general. To adjust the yellow-to-blue emission intensity ratio, some rare-earth ions that emit blue light should be codoped in the host. The Tm^{3+} ions, according to previous reports, could produce blue light due to the $^1D_2 \leftrightarrow {}^3F_4$ transition (Tian et al. 2016). As a result, introducing Tm^{3+} into a Dy^{3+}-activated system could result in white light emission. Surprisingly, single-phase white light emitting phosphors based on the Tm^{3+} to Dy^{3+} energy transfer mechanism have been realized in scheelite hosts LiGd(WO$_4$)$_2$ and a few others like BaLaGa$_3$O$_7$ and Sr$_3$Y(PO$_4$)$_3$ (Wu et al. 2011, Li et al. 2015, Zeng et al. 2015, Wang, Lou, et al. 2017). Because of tunable structure, high UV absorption coefficient, and promising luminescence properties, scheelite structure double tungstate/molybdate phosphors have attracted a lot of attention in recent years and a wide range of micro-/nanostructured crystals with proper crystal phase, well-defined shapes, and controllable designs have been developed in this context. Using the hydrothermal method, NaLa(WO$_4$)$_2$:Eu^{3+} /Tb^{3+}/Tm^{3+} have been prepared in the presence of ethylenediaminetetraacetic acid (EDTA-2Na) surfactant (Liu, Hou, et al. 2014). By adjusting the amount of PVP ligand, the initial solution pH, and the reaction time, different morphologies (spindle-like microrods, microcuboids, and microflowers) of NaLa(WO$_4$)$_2$ were modulated (Huang, Wang, et al. 2012). 3D flower-like microarchitectures of NaY(WO$_4$)$_2$:Eu^{3+} have been prepared in the microwave-assisted hydrothermal method in a trisodium citrate–mediated process followed by heat treatment (Tian, Chen, et al. 2012). The Pechini method was used to make novel Ho^{3+}- and Tm^{3+}-doped gadolinium molybdate nanocrystals (Pan et al. 2007).

In generalized form, it can be stated that depending on the ionic radii of the cations A and RE, M of the types A^{2+}MO$_4$ and A$^+$RE^{3+}(MO$_4$)$_2$ crystallizes in either a tetragonal or triclinic structure (Munirathnappa et al. 2019). In the case of tetragonal crystal structure the cations A and RE are octahedrally coordinated to oxide ions with site occupancy of 0.5, while the M ion is tetrahedrally coordinated oxide ions. However, for triclinic structures, the M, RE, and A atoms are connected via oxygen atoms forming a cage-like structure where RE is coordinated with 8 oxygen atoms

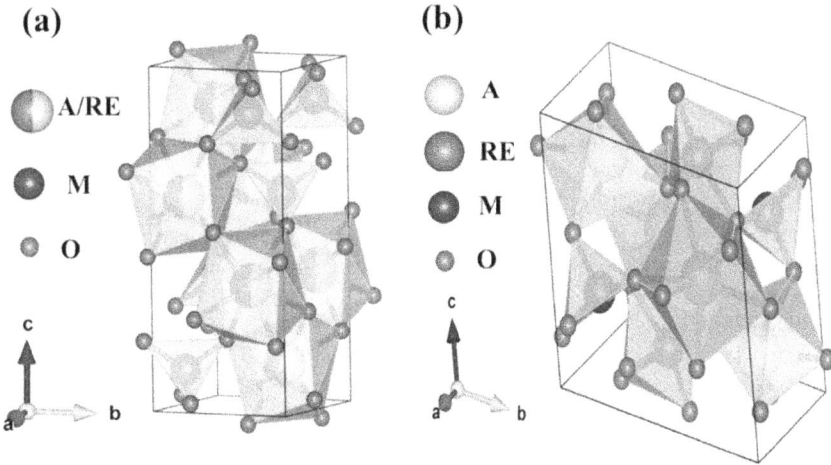

FIGURE 8.1 Crystal structures of scheelite nanomaterials: (a) tetragonal and (b) triclinic.

to form regular REO_8 polyhedra and A atoms form isolated AO_4 tetrahedra. In a unit cell, each AO_4 shares one edge with individual MO_6 octahedra and connects to REO_8 polyhedra via oxygen atoms. The two M atoms are in inversion symmetry with each other. The two M atoms form edge-sharing octahedral resulting in the formation of M_4O_{16} unit as a tetramer via M_2O_2 bridging oxygen atoms and the octahedral is found to be highly distorted in the crystal structure (represented in Figure 8.1). Mostly, the scheelites unit cell possesses inversion symmetry, but recent studies illustrate that they can also exist without inversion symmetry that makes scheelites optically tunable. As an example, $LiGd(WO_4)_2$ having $CaWO_4$ isostructure where W^{6+} is coordinated by four oxygen atoms at a tetrahedral site forming stable $(WO_4)_2$ tetrahedron, and the Li^+/Gd^{3+} ions are randomly distributed over the Ca^{2+} site and are coordinated by eight oxygen atoms from near four $(WO_4)_2$ of S_4 symmetry without inversion center. Herein, the inhomogeneous broadening of optical spectra is caused by the doping of Eu^{3+} ions occupying the randomly distributed Gd^{3+} ion sites. Because $Gd^{3+}(0.938)$ and $Eu^{3+}(0.947)$ ions have the same valence and almost identical ionic radius, Eu^{3+} ions can easily be substituted into the Gd^{3+} site of the host matrix (Kumar et al. 2013, Li et al. 2013, 2015, Balaji et al. 2014, Liu, Wang, et al. 2014, Gao et al. 2017).

Scheelite nanomaterials show various applications in clinical and biological fields. It was shown that $AgGd(MoO_4)_2$:Er^{3+}/Yb^{3+}@mSi (AGM) core-shell upconversion nanoplates serve as potential candidates for optical thermometry and biomedical applications (Pavitra et al. 2022). The researchers investigated the sensing properties of scheelite materials and they found that Yb_2WO_6 shows excellent electrochemical sensors (electrocatalyst) for the detection of drugs especially the antipsychotic drug chlorpromazine (CPZ) (Sundaresan et al. 2019). Again, $NaTb(WO_4)_2$, $BiVO_4$ along with Ag nanoparticles and $PbMoO_4$ show antibacterial activity due to greater colloidal and photostability and excellent biocompatibility

(Booshehri et al. 2014, Moura et al. 2018, Munirathnappa et al. 2020). Scientists developed a new type of radiosensitizers, namely polyethylene glycol-block-D, L-polylactic acid (PEG–PLA) coated and uncoated $CaWO_4$ radioluminescent particles (RLPs) for cancer treatment (Jo et al. 2018). Moreover, Parchur et al. have developed Eu^{3+}-doped $CaMoO_4$ nanomaterials, a potential candidate for bioimaging (optical imaging), providing luminescence in the red region and showing cytotoxicity after being conjugated with Fe_3O_4 (Parchur et al. 2014). Other efficient $NaEu(WO_4)_2$ nanomaterials were developed by Munirathnappa et al. for theranostic purposes due to their intense luminescence properties and nontoxicity (Munirathnappa et al. 2018). $BaWO_4$/GCN nanocomposite is one of the promising candidates for detecting various toxic materials present in food (Karuppusamy et al. 2021).

Furthermore, the properties of the as-elaborated double tungstates are strongly influenced by their morphologies and size distribution, as discussed in all of the above previous reports. As a result, developing simple and reliable synthetic methods for hierarchical architectures with designed chemical components and controlled morphologies, which have a significant impact on the properties of nano- or microstructures, remains a major challenge. Herein, we'll thoroughly discuss different methods to prepare scheelite-type nano-/microstructures and their scopes in biological fields.

8.2   SYNTHESIS STRATEGY

The photoluminescence properties of scheelite nanomaterials are closely connected to their preparative techniques. The nanoparticle size, morphology, and corresponding microstructure of luminescent materials depend on their different preparation method, which can affect their application in diversified directions. Among them, hydrothermal, solvothermal methods, conventional solid-state reaction, and sol-gel synthesis are the most often used methods. Herein, different synthesis protocols along with precursors and reaction conditions are summarized in Table 8.1.

8.3   PARTICLE GROWTH MECHANISM

Nanomaterials have been explored as promising tools for drug and gene delivery, biosensors and bioimaging, etc. Many properties of nanomaterials such as shape, size, surface structure, surface charge, aggregation, and colloidal solubility can greatly influence the interactions with cells and biomolecules. Nanomaterials with tuned size can affect the luminescent property. Therefore, the main target of controlling particle growth for the development of medicinal and bioapplication. Various factors can influence the growth of nanoparticles like organic ligands, reaction time, reaction temperature, pH of the reaction medium, rare earth sources, etc., discussed as follows:

8.3.1   INFLUENCE OF ORGANIC LIGANDS

It has been stated earlier that hydrothermal as well as microwave-assisted synthesis is most often used for the preparation of rare earth tungstate/molybdate nano-/

TABLE 8.1
Synthesis Procedures of Scheelite Nanomaterials

Method	Precursors materials	Reaction condition	Synthesized materials	References
Hydrothermal and solvothermal	A rare earth nitrate/chloride salt and a sodium/ ammonium tungstate or molybdate salt.	120–200°C for 1–72 hours, calcination at 800°C for 2–5 hours.	$La_2(MoO_4)_3$, $NaCe(MoO_4)_2$, $NaCe(WO_4)_2$, $NaY(MoO_4)_2$, $NaLa(WO_4)_2$, Tm^{3+} and Yb^{3+} codoped $NaGd(WO_4)_2$.	(Yi et al. 2002, Liu et al. 2012, Xu et al. 2012, Liu, Hou, et al. 2014, Wang et al. 2015, Dirany et al. 2016)
Microwave-assisted hydrothermal reaction	A rare earth nitrate salt and a sodium tungstate or molybdate salt.	Temperature 200°C with 90 W power of a microwave irradiation.	$NaY(WO_4)_2$, $NaLa(MoO_4)_2$.	(Zhang et al. 2011, Tian, Hua, et al. 2012)
Sol–gel synthesis, combined sol–gel synthesis, and electrospinning	A rare earth nitrate salt and an ammonium tungstate or molybdate salt.	The gel can be dried under critical conditions or by evaporation. Then calcined at high temperature.	$Gd_2(MoO_4)_3$, Gd_2MoO_6, $Tb_2(WO_4)_3$.	(Hou et al. 2011, Balaji et al. 2014)
Conventional solid-state reaction	A rare earth oxide and tungstate or molybdate oxide and alkali carbonates	1073 K for 12 hours at a rate of 250 K per hour.	$LiCe(WO_4)_2$, $NaCe(WO_4)_2$, $KCe(WO_4)_2$, $Gd_2(MoO_4)_3$, $NaEu(WO_4)_2$.	(He et al. 2010, Shimemura et al. 2017)
Molten salt synthesis	Rare earth oxides and tungstate or molybdate oxides, NaCl and KCl as reaction salt as well as solvent.	950°C for 6 hours.	Gd_2WO_6, Gd_2MoO_6	(Lei et al. 2008, 2009, Huang et al. 2011)[58–60]

micromaterials. In this synthetic procedure, an organic ligand is added to the reaction and forms a complex with the rare earth metal ions. Due to this complex formation, the nucleation and growth of the crystals slow down. Additionally, their functional groups bind on the surface of the materials, which affects the growth rate of certain crystal facets (Rasu et al. 2017). A large number of ligands have been employed for controlling the reaction for the preparation of tungstate/molybdate materials. Among all organic ligands, the most commonly used are cetyltrimethyl ammonium bromide, polyvinylpyrrolidone, ethylenediaminetetraacetic acid, trisodium citrate, ammonium oxalate disodium tartrate, and others. $NaCe(WO_4)_2$ rods and hierarchical spindles were synthesized by Dirany et al. via EDTA-assisted hydrothermal reaction at 200°C for 24 hours and calcination at 800°C for 5 hours (Xu et al. 2012). Preparation of $NaLa(MoO_4)_2$ in a hydrothermal synthesis method was carried out in the presence of mixed surfactants, i.e., oleic acid (OA) and oleylamine (OL) at temperature 140°C for 6 hours (de La Presa et al. 2006, Bu et al. 2009). A combination of OA and OL is a good organic solvent that is used in the synthesis of inorganic nanocrystals. In the presence of an eggshell (as template), the formed $BaWO_4$ is in the form of polyhedron morphology. But using organic additives (n-dodecanethiol, a-cyclodextrin, ethylenediamine, L-ascorbic acid, and polyformaldehyde) instead of eggshell the formed nanostructures were found to be like flower-like morphology, double-taper-like morphology, anchor-like morphology, sphere-like morphology, and fasciculus-like morphology respectively (Liu et al. 2005). When the reaction was carried out in the presence of trisodium citrate, 3D urchin-like microarchitectures were found. In the presence of polyvinylpyrrolidone, irregular microcrystals were obtained. When ammonium oxalate was used in the reaction the irregular microplates were found, whereas using disodium tartrate, nanoparticles were obtained. When ethylenediamine tetraacetic acid was employed, irregular nanoplates were observed (Figure 8.2). Again the formation of $Y_2(WO_4)_3$ nano-/microcrystal was obtained in the hydrothermal reaction using sodium dodecyl benzenesulfonate (SDBS) as organic solvent at 200°C for 20 hours, followed by calcination at 800°C for 2 hours (Huang, Zhang, et al. 2012). The researchers have found that different amounts of SDBS can control the shape of crystals. When 0.5 mmol SDBS was used in the reaction mixture, microstructures with uniform bowknot shaped, 8 mm in length, were found. Increasing the amount of SDBS to 0.75 mmol, bowknot-like structures were formed. It can be seen also that when the amount of SDBS is increased to 1.25 mmol, the nanosheet building blocks get formed with an arrangement into 3D microflower structures. On the basis of the above observation, it can be concluded that the amount of SDBS plays a crucial role in the evolution of the morphology of the $Y_2(WO_4)_3$ precursors. The possible mechanism for the formation of $Y_2(WO_4)_3$ nano-/microcrystal in the hydrothermal reaction process was presented by Xu et al. in the presence and absence of ethylenediaminetetraacetic acid (EDTA) (Xu et al. 2009). In the presence of EDTA, 3D flower-like architectures were formed. In the presence of different organic additives, different morphologies can be generated. In the presence of AO (as the organic additive) the formed $NaY(MoO_4)_2$ nanocrystals were irregular microplates; while adding PVP, no regular morphology was found (Xu et al. 2010). Using EDTA or Na_2tar as organic additives, mainly irregular nanoplate or nanoparticle shape morphologies were formed, respectively (shown in Figure 8.3).

FIGURE 8.2   SEM morphologies of the products: (A) no additive reagents; (B) adding 0.05 g of 0.01 mol/L n-dodecanethiol; (C) adding 0.04 g of 0.01 mol/L cyclodextrin; (D) adding 0.015 g of 0.01 mol/L ethylenediamine; (E) adding 0.044 g of 0.01 mol/L L-ascorbic acid; and (F) adding 0.0075 g of polyformaldehyde.

Source: Liu et al. 2005.

The above results prove that the organic ligands play an important role in the evolution of the size, shape, and morphology of the reaction product.

8.3.2   THE INFLUENCE OF pH

It was reported in the literature that the pH of the reaction mixture has a remarkable effect on the size of the final reaction product. The formation mechanism $NaY(MoO_4)_2$ microcrystals in the hydrothermal synthesis process was carried out by modifying the pH of the initial reaction solution; different morphologies could be obtained (Xu et al. 2010). When the molar ratio of the reactants $Y(NO_3)_3$ to Na_2MoO_4 was 1:7 and the pH value was 4, rectangular plate-like microcrystals were found. Maintaining the pH value between 5 and 6, rhombic morphologies were fabricated. Further increasing the pH to 7, sheet-like microcrystals were formed. It was logically concluded that the pH of the initial reaction solution has a significant effect on the size, shape, and morphology of the final reaction product. The $BiVO_4$ powders were achieved in microwave hydrothermal reaction at different pH conditions (Li et al.

FIGURE 8.3 SEM images of the as-prepared hydroxyl sodium yttrium molybdate samples formed in the presence of (A) AO; (B) PVP; (C) EDTA; and (D) Na$_2$tar.

Source: Xu et al. 2010.

2013). By changing the pH of the reaction medium various morphologies can be generated (shown in Figure 8.4). The NaCe(WO$_4$)$_2$ nanomaterial with various morphologies could be obtained including microspindles, microspheres, or microflowers of self-assembled nanoparticles at different pH (Tan et al. 2013). In the presence of 0.3 g of EDTA in a neutral environment at pH 7, spindle-like structures were obtained. While increasing the pH of the solution to 8, homogenous and self-assembled 3D hierarchical microspheres were found, provided all the reaction conditions remained unchanged.

8.3.3 THE INFLUENCE OF REACTION TEMPERATURE

A whole lot of literature works have established that the formation of rare earth tungstate/molybdate materials is often dependent on reaction temperature. The formation of NaEu(MoO$_4$)$_2$ nano-/microcrystals was obtained in a hydrothermal reaction for 24 hours by varying reaction temperature in the presence of EDTA. Different morphologies could be obtained by changing reaction temperature (Dirany et al. 2016). At a temperature of 120°C regular rhombic nanosheets 200 to 500 nm in size were found. By increasing the temperature to 140°C, irregular nanospindles with an average length of 3 mm and width of 500 nm were obtained. Further increasing the temperature to 180°C microrugbies with 0.4–1.3 mm in length were formed. In the formation of NaLa(MoO$_4$)$_2$ nano-/microcrystals, the synthesis was carried out

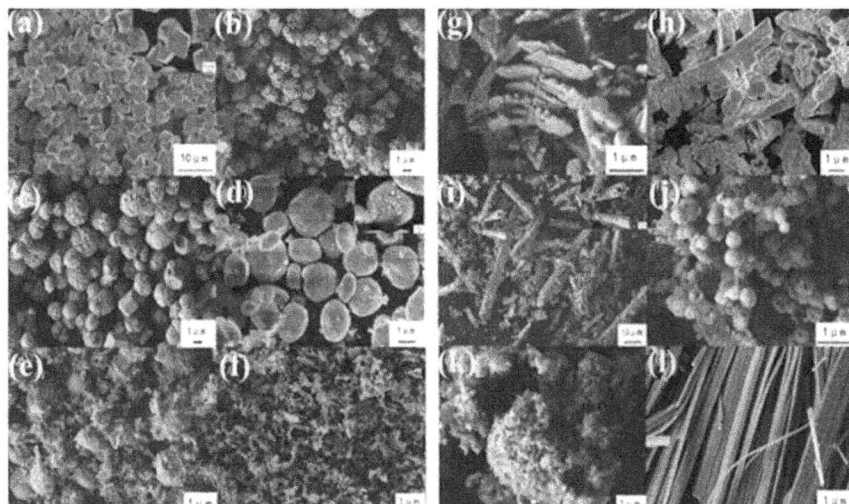

FIGURE 8.4   FESEM images of BiVO$_4$ powders prepared at different pH values: (a) 0.59; (b) 0.70; (c) 1.21; (d) 2.55; (e) 3.65; (f) 4.26; (g) 7.81; (g) 9.50; (h) 9.76; (i) 10.44; (j) 10.55; (k) 12.59; and (l) 12.93.

Source: Li et al. 2013.

in a microwave-assisted synthesis process in the absence of any organic ligands, at different temperatures for 10 minutes under magnetic stirring (Xu et al. 2011). Various types of morphologies were observed using different reaction temperatures. When the temperature of the reaction was 100°C or 120°C, octahedral bipyramidal-shaped particles were obtained. By increasing the temperature to 140°C or 160°C microcrystals were obtained with dendrite shape. Again, different shapes, sizes, and morphologies of Gd$_2$(WO$_4$)$_3$ microstructures were investigated by Zeng et al. (Zhang et al. 2011). The reaction was carried out in a hydrothermal reaction in the absence of an organic ligand for 36 hours. When the reaction temperature was 120°C, belt-like structures with 400–700 nm long and with an average thickness of 80 nm were obtained. Further increasing the reaction temperature to 160°C, star-like structures about 50 nm in thickness and around 3 mm in length were found. These structures were built from several leaf-like sheets with a common center.

8.3.4   THE INFLUENCE OF REACTION TIME

Reaction time is an important factor in the preparation of nanomaterial with different morphology. By varying reaction time in a microwave-assisted hydrothermal process, different morphologies of nano-/microstructures of NaY(MoO$_4$)$_2$ were obtained [65]. After 1 hour of completion of the reaction, the desired product was composed of large irregularly shaped and twisted pieces. Prolonging time to 3 hours, many rugged nanostrips were formed by assembling, radiating, and adhering with each other. After increasing the reaction time to 6-hour nanoflakes, more aggregates were detected.

FIGURE 8.5   SEM images of the as-prepared hydroxyl sodium yttrium molybdate samples at 180°C for different reaction times: (A) 0.5 h; (B) 3 h; (C) 6 h; and (D) 12 h.

Source: Xu et al. 2009.

Further increasing time to 12 h, the products transformed into uniform urchin-like morphology (shown in Figure 8.5). Having outstanding application in the field of ceramic and luminescence materials, etc., $BaWO_4$ crystals have been synthesized using novel supramolecule templates (eggshell and different organic additives) at mild conditions. Controlling the reaction time different morphologies can be tuned with different shapes and sizes of the crystals. In the presence of polyformaldehyde, with different time intervals various types of morphologies were found (represented in Figure 8.6) (Liu et al. 2005).

8.3.5   THE INFLUENCE OF THE RARE EARTH SOURCE

From most of the reported literature works, for the synthesis of scheelite types tungstates/molybdates materials, nitrate salts of rare earth elements are used in the reaction. There are only a few reports about the formation of such materials by varying the source of rare earth. The source of the rare earth–dependent experiment was investigated by Kaczmarek et al. (Kaczmarek et al. 2013). Using nitrates and acetate salts of the rare earth elements, the formation of $Y(WO_3)_2(OH)_3$ and $La_2(WO_4)_3$ nano-/microstructures were carried out in a DSS-assisted hydrothermal process. In the case of preparation of $Y(WO_3)_2(OH)_3$, when $Y(NO_3)_3$ was used as the source of yttrium at first irregular nanoparticles were formed. After 3 h completion of the reaction, nanorods with 300–500 nm long were obtained. These nanorods formed bundle-like

FIGURE 8.6 SEM morphologies of the products at different reaction times: (A) 1 h; (B) 4 h; (C) 6 h; (D) 10 h.

Source: Liu et al. 2005.

aggregates. With increasing reaction time the bundles persisted to develop, forming spherical microstructures. Further increasing the reaction time to 24 h, spherical microstructures in the range of 4–5 mm could be obtained. Whereas using $Y(OAc)_3$ as the source of yttrium, at first amorphous particles were found, which were parent materials for the evolution of shrubby-like structures built from nanorods illuminating from the 'trunk'. With increasing reaction time these shrubby-like structures aggregated to grow microstructures of a comparable size and shape as the materials obtained from $Y(NO_3)_3$. It can come to an end that the source of the rare earth has a significant effect on the size and shape of the nano building blocks. Microstructures obtained from the acetate salts showed tight packing of the nano building blocks as compared to nitrate salt, which was proved by almost no nitrogen absorption. Liu et al. reported the synthesis of $NaY(MoO_4)_2$ nano-/microcrystals using $Y(OH)_3$ nanorods as the yttrium source. They showed that the high crystallinity of the product was obtained without calcination. Various shapes, sizes, and morphologies were obtained, including sheaf-like structures, nanoflakes, cubes, truncated rhombic polyhedrons, tetragonal bipyramids, and perfect bipyramids (Liu et al. 2012).

8.3.6 THE INFLUENCE OF TUNGSTATE/MOLYBDATE AMOUNT

The size, shape, and morphology of materials depend on the amount of tungstate/molybdate source used in the reaction also. In the literature, the researchers reported the synthesis of $NaY(MoO_4)_2$ materials using an excess amount of Na_2MoO_4 in the hydrothermal process. The MoO_4^{2-} ions preferentially adsorbed on a particular plane (001 plane) of $NaY(MoO_4)_2$ crystal nuclei resulting in the change in the thickness of the $NaY(MoO_4)_2$ crystals from micro- to nanosized particles (Liu et al. 2011). In their literature they used different amounts of Na_2MoO_4. By varying the quantity of Na_2MoO_4 to 4, 6, 8, 10, and 14 mmol, the resulting products transitioned from uniform octahedra enclosed by (101) facets to octahedra truncated along (001) planes, subsequently transforming into quasicubes, tetragonal prisms, and finally thin plates with a thickness of 150 nm, respectively. Thus the amount of tungstate/molybdate used in the synthesis of $NaY(MoO_4)_2$ materials has a remarkable influence on the sizes, shapes, and morphologies of final products.

8.4 APPLICATION OF SCHEELITES IN BIOANALYSIS

For almost a few decades, imaging techniques for diagnosis and medical purposes such as MRI imaging, X-ray computed tomography, positron emission tomography, and fluorescence imaging have attracted huge attention in the diagnosis of cancer. Among various techniques, optical imaging based on fluorescent nanomaterials has been considered to be extraordinarily efficient over other techniques due to their distinct optical properties. In most of the biomedical applications, including gene therapy, drug delivery, biosensing, etc., the imaging technique using fluorescent nanoparticles has been readily accepted. Moreover, to improve the biological and clinical application the fluorescent nanomaterials are in focus owing to their luminescent efficiency, nontoxicity, colloidal stability, etc. Additionally, these materials have shown several techniques for surface modification with functional groups to increase the efficiency in bioimaging. So, RE activated nanoparticles have gained extreme interest as promising fluorescent probes, due to their high resistance to larger Stokes and anti-Stokes shifts, optical bleaching, blinking, and high quantum yield. Furthermore, these rare earth elements (activator) could efficiently absorb energy in the UV Visible–NIR range due to disordered (charge transfer band) and transmit energy to the available rare earth present in the host, resulting in a significant increase in emission intensities. Mostly for biomedical applications, particle size, shape, and morphology play an important role in the optical properties even in biocirculation and bioimaging. Therefore, $ARE(MoO_4)_2$ materials have broad applications in the biological and clinical field by controlling the preparative methods, shape, and size. The unique luminescence property due to intra 4f-4f and 4f-5d electronic transitions makes them useful for different biological applications like detecting some biological species (e.g. bacteria, protozoa, cancer cells, etc.), killing some pathogenic molecules, cancer therapy, and purification and processing in the food industry as schematically shown in Scheme 8.1 and they are briefly discussed in the following sections.

8.4.1 SCHEELITES AS A SENSOR FOR BIOLOGICAL SYSTEM

In biological systems, the measurement of various biological parameters like enzymes, proteins, or other chemical compounds is a very important part of research purposes. In modern days, many techniques like chromatography, photoelectrolysis, blotting, and

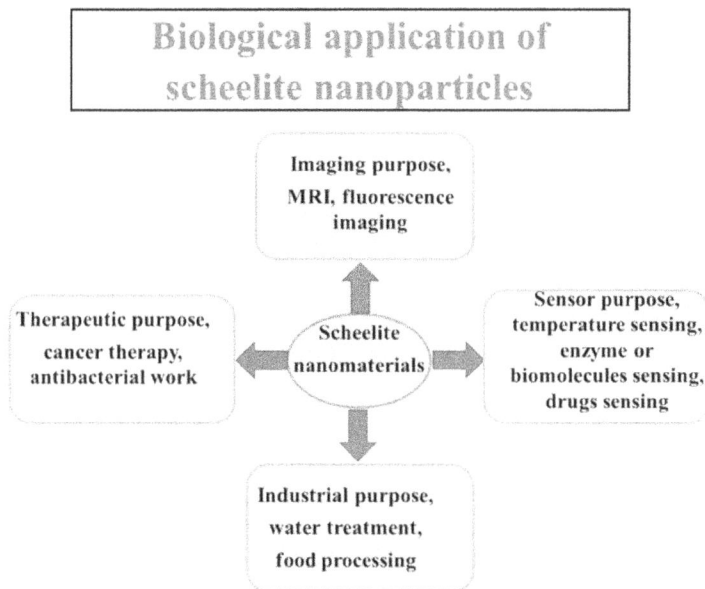

SCHEME 8.1 Various bioapplications of scheelite nanomaterials.

many more assays are available for detecting those compounds. But these techniques are time consuming, lack sensitivity, and are costly. So some groups of researchers are trying to develop new techniques which are more comfortable and fast. Scheelite nanoparticles are good candidates for those detection purposes. $CuWO_4$, a scheelite material that can mimic the peroxidase-like behavior, is used as a detector for nicotinamide adenine dinucleotide (NADH) (Aneesh et al. 2017). NADH is an essential molecule for the biological system; enhancement or reduction of NADH levels affects our brain and body. Many diseases like Parkinson, Alzheimer, fatigue, insomnia, etc., are directly regulated by the NADH level. So monitoring or sensing the level of NADH is an important diagnostic tool and scheelite nanoparticle $CuWO_4$ makes a good impact for detecting the level of NADH. On the other hand, bismuth vanadate ($BiVO_4$) nanomaterials have been successfully applied as a glucose detector (Wang et al. 2019). $BiVO_4$ was used as an electrode with fluorine-doped tin oxide by electrochemical deposition. This sensor was used to successfully measure the glucose in a human serum sample. It is a new strategy for detecting glucose through the nonenzymatic photoelectrochemical (PEC) method. Para-Cresol (PC) is a phenolic compound present in human urine samples. A higher concentration of PC can cause various diseases like kidney failure, cardiovascular diseases, liver damage, etc. Therefore, it is very necessary to detect the PC level in patients to monitor their body function. Barium tungstate ($BaWO_4$)–functionalized carbon nanofiber (BW-fCNF) composite–based electrochemical sensors were used to detect PC (Sundaresan et al. 2021). The BW-fCNF nanocomposites showed maximum performance toward PC determination with a very low detection limit of 0.006 μM and a wide linear range of 0.02–67.5 μM. So the BW-fCNF nanocomposites are potent to determine the PC present in the human urine samples.

TABLE 8.2
Relative Thermometric Sensitivities S_R (K^{-1}) of Dy^{3+}-activated Phosphors

Compounds	350 K	700 K	References
$GdVO_4{:}Dy^{3+}$	9×10^{-3}	4.0×10^{-3}	(Gavrilović et al. 2014)
$NaDy(MoO_4)_2$	7.5×10^{-3}	3.8×10^{-3}	(Perera and Rabuffetti 2019)
$Na_5La_{0.5}Dy_{0.5}(WO_4)_4$	1.8×10^{-2}	3.0×10^{-3}	(Perera and Rabuffetti 2019)
$SrWO_4{:}Eu^{3+}, Dy^{3+}$	1.7×10^{-2}	-	(Wang, Bu, et al. 2017)

8.4.1.1 Temperature Sensing Using Scheelites

Temperature measurement is often required for both industrial research and scientific applications. Conventional methods of temperature measurement can be classified into two categories: contact methods, including thermistors, thermocouples, and resistance temperature detectors; and noncontact methods such as estimation of emitted infrared light (Bentley 1998). Nowadays, many kinds of devices are available for temperature measurements like thermometers (solid, gas, manometric, optical fiber, semiconductor, quartz, ultrasonic, etc.), pyrometers, or thermocouples, etc. In recent years luminescence thermometry has been a new emerging field for scientists (Childs et al. 2000). Luminescent thermometry has several advantages over the existing methods because it can be adopted for living biological systems (cells or tissues), contact-sensitive objects, objects present in a hazardous environment, and objects with nanosized dimensions. In these conditions the existing methods are not applicable. Various luminescence materials like polymers, semiconductors, and organic-inorganic materials are used in luminescence thermometry, but mainly lanthanide-doped nanoparticles have garnered great interest for use in luminescence thermometry because of their narrow emission lines and relatively longer emission lifetime in different spectral regions. The $AgGd(MoO_4)_2{:}Er^{3+}/Yb^{3+}$ (AGM: Er^{3+}/Yb^{3+}) showed great temperature sensing properties in tumors due to the characteristic nature of the excited state of Er^{3+} ions (Pavitra et al. 2022). Moreover, scheelite materials $NaLa_{1-x}Dy_x(MO_4)_2$ and $Na_5La_{1-x}Dy_x(MO_4)_4$ (M = Mo, W) were also investigated as thermosensitive phosphors in the 300–700 K temperature range (Perera and Rabuffetti 2019). Inspection of relative sensitivity (S_R) values (given in Table 8.2) confirm that the thermometric performance of scheelite-type materials at 350 and 700 K is comparable to that reported for other Dy^{3+}-activated thermosensitive phosphors. This comparison highlights the potential of scheelite-type materials as optical sensors for intermediate temperatures (<1000 K).

8.4.1.2 Detection of Drugs

In modern days, the usage of medicine/drugs is increasing day by day to get rid of various deadly diseases. But sometimes, the doses of the drug make a crucial impact on the human body. Overdose or minimal dose of medicine may cause many deadliest diseases like cancer. Some psychiatric diseases like schizophrenia are a kind of neural disease (mainly affecting the central nervous system) that is treated with antipsychotic drugs like CPZ, pimozide, haloperidol, etc. But overdose of these drugs can create insomnia, endocrine disorder, and autonomic disease and some-times causes pseudo-Parkinson disease. Till now, the detection of drug dose in the human body is maintained using various techniques like gas chromatography, high-pressure liquid chromatography, spectrophotometry, capillary zone electrophoresis, etc. Though those techniques are good and sensitive with less error but incur high costs and require skilled technicians and huge solvent consumptions. So scientists are trying to make new detecting tools for drug detection in the human system. Recently metal tungstates have evolved as good detecting tools because they have good elec-trical conductivity, good chemical stability, and mechanical stability. Sundaresan et al. have developed ytterbium tungstate (Yb_2WO_6) to detect the antipsychotic drug chlorpromazine. This material has excellent selectivity and good stability toward CPZ (Sundaresan et al. 2019).

8.4.2 SCHEELITES WITH ANTIBACTERIAL ACTIVITY

In recent days antibiotics dependency has reduced because of the modification of bacterial genomes and resistance to antibiotics. This phenomenon is called multidrug resistance (MDR). Nowadays, many groups of researchers have developed new nanodrugs (metal nanoparticles, organic nanoparticles) for multidrug-resistant bac-teria. Scheelite nanomaterials are also a good candidate for killing bacterial cells without making them resistant. In some cases they also detect the bacterial cells and monitor their gene expression. The $NaTb(WO_4)_2$ nanogreen fluorescence materials are investigated for their antibacterial efficacy against both gram-positive bacteria (*Staphylococcus aureus*) and gram-negative bacteria (*Escherichia coli*) but this material shows slightly bacteriostatic in nature, i.e., the growth of the bacteria is not inhibited, but the rate of the bacterial growth is slow in nature (Munirathnappa et al. 2020). In some other cases, scheelite-type materials like $BiVO_4$ along with silver nanoparticles induced the generation of reactive oxygen species (ROS) in the presence of visible light (Booshehri et al. 2014). This activity is useful for the purification of water from various bacteria-contaminated water. The bare $BiVO_4$ nanoparticles also have some effects on bacterial cells. The antibacterial mechanism of $Ag/BiVO_4$ nanocomposite is represented in Figure 8.7 schematically.

Also, the scheelite material $PbMoO_4$ modulates some antibiotic, sometimes correspondences agonistic as well as an antagonistic behavior. With gentamicin against *S. aureus* the $PbMoO_4$ nanocrystals modulate the effect of antibiotics and kill bacteria more efficiently than with only gentamicin, but these nanocrystals also showed an antagonist effect with norfloxacin and imipenem (Moura et al. 2018). From this out-standing result, one can say that in the future, scheelite materials may be evolved as an alternative to antibiotics and bacterial monitoring systems. The $BiVO_4$, $Ag-BiVO_4$, and $Ag-BiVO_4$-rGO nanocomposites showed antibacterial effects (Nethravathi et al. 2022). The zones of inhibition (mm) for different bacterial strains, namely *Klebsiella pneumonia* (gram −ve), *Pseudomonas aeruginosa* (gram −ve), *S. aureus* (gram +ve), and *Enterococcus faecalis* (gram +ve) are showed in the Table 8.3.

Various studies on silver nanoparticles exhibited significant antibacterial activity against gram +ve as well as gram −ve bacteria, including multidrug-resistant bac-teria (MDR) such as *P. aeruginosa* (Sadeghzadeh-Attar 2018), *Salmonella serovars*

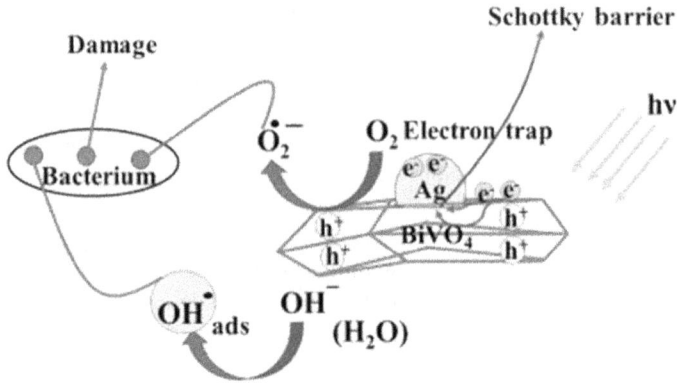

FIGURE 8.7 Photocatalytic inactivation of bacteria through Ag/BiVO$_4$ composites under visible light.

Source: Booshehri et al. 2014.

(Sadeghzadeh-Attar 2019), *E. coli*, *S. aureus* (Riaz et al. 2021), etc. However, Ag nanoparticles have shown some significant antibacterial properties; there are some side effects also, such as cellular toxicity and tissue toxicity, especially in human cells. Incorporating a very little amount of Ag into the BiVO$_4$ crystal can overcome these limitations (Nethravathi et al. 2022). The erythrocytes lysis analysis test was suggested as support for their explanation and proved that these materials are non-toxic to human cells. In the future this type of material may be considered one of the most efficient nanomaterials for antibacterial study.

8.4.3 SCHEELITES WITH ANTICANCER ACTIVITY

In the case of cancer cell treatment scheelite nanoparticles and their derivatives open a new window for researchers. Compared to other nanoparticles, luminescent scheelite materials have good optical stability and narrow band gap emission, and under the NIR excitation, they also have good tissue penetration without damaging tissue; thus they are being widely used for several biological purposes. Scheelite nanoparticles exhibit photothermal (PTT) or photodynamic therapeutic (PDT) activities; thus they are also used to diagnose various diseases. Calcium tungstate (CaWO$_4$), a radio luminescent particle, has had a good impact on treating cancer cells (P53 mutated human head-neck HN31 cancer cells) (Jo et al. 2018). Radioluminescent (RLP) nanoparticles mainly kill cancer cells through radiation therapy with better efficacy. Using uncoated CaWO$_4$ microparticles, complete suppression of the tumor growth in 3 months was observed, while using PEG–PLA-coated CaWO$_4$ NPs, the effects were significantly lower. Figure 8.8 represents (schematically) the behavior of a radioluminescent particle (RLP) toward a cancer cell.

The levels of cytotoxicity for a nanocomposite vary for different cells. Some particles are cytotoxic for cancer cells having no or very little effect on normal cells. The nanocomposite Fe$_3$O$_4$–CaMoO$_4$:Eu^{3+} has shown a cytotoxic effect in liver cancer

TABLE 8.3
Antibacterial Activity of $BiVO_4$, $Ag-BiVO_4$, and $Ag-BiVO_4-rGO$ Nanocomposite on Various Bacterial Strains

Treatment	Concentration	Staphylococcus aureus	Enterococcus faecalis	Klebsiella pneumoniae	Pseudomonas aeruginosa
$BiVO_4$	(100 µg/µL)	5.50 ± 0.50	6.75 ± 0.25*	5.83±0.17**	9.50± 0.50**
	(200 µg/µL)	7.83 ± 0.29*	8.17± 0.17**	7.50 ± 0.17*	10.00± 0.58*
$Ag-BiVO_4$	(100 µg/µL)	6.67 ± 0.33	7.43 ± 0.17	6.67± 0.33**	10.67± 0.33*
	(200 µg/µL)	8.50 ± 0.29*	9.83 ± 0.17*	8.52 ± 0.33*	12.13± 0.33*
$Ag-BiVO_4-rGO$	(100 µg/µL)	8.33 ± 0.33	7.67 ± 0.33	7.35± 0.33**	11.67± 0.33*
	(200 µg/µL)	11.33 ± 0.33**	12.83 ± 0.44	9.16 ± 0.17*	12.33 ± 0.33
Standard	(5 µg/µL)	10.50 ± 0.29	12.83 ± 0.44	10.83 ± 0.17	14.33± 0.333

Source   Nethravathi et al. 2022.

Note: *Symbols represent statistical significance: $*P < 0.05$, $**P < 0.01$ as compared with the control group.

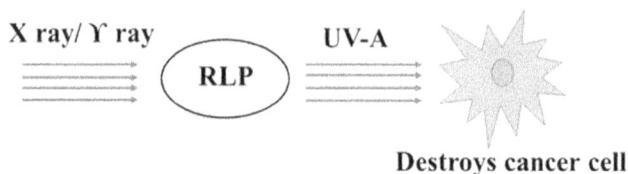

FIGURE 8.8   Schematic presentation of anticancer activity of RLP.

cell using the HepG2 cell line as well as in normal hTERT cells (human mesen-chymal stem cells), with less toxicity and good cell viability (Parchur et al. 2014). It was observed that on incubation of the HepG2 cells with the Fe_3O_4–$CaMoO_4$:Eu^{3+} hybrid nanocomposite for 24 h at various concentrations (5, 10, 20, 40, 80, 160, 320, and 640 mg mL^{-1}) showed a prominent dose-dependent decrease in their cell viability. The IC_{50} value was found to be approximately 193.26 mg mL^{-1} for liver cells, whereas in hTERT cells the hybrid nanocomposite showed no significant dose-dependent cytotoxic effects even at higher concentrations. These results confirmed that the scheelite nanocomposite shows toxic results against cancer cells but very little toxicity in normal cells. These facts also conclude that further research on scheelite nanoparticles will make a huge impact on cancer theranostics.

8.4.4   SCHEELITES IN WATER TREATMENT

In recent years water pollution has become a big crisis all over the world. Increasing wastes in industry and agriculture make water contaminated and scientists are trying to solve this problem in various ways. Many scheelite-type materials are employed to purify water. Some scheelite ABO_4 compounds (A = Ca, Sr, and Ba; B = Mo and W) with a wide band gap have been used as a potential photocatalyst for water treatment (Kowalkińska et al. 2021). The AWO_4^{2-} and $AMoO_4^{2-}$ type scheelite materials (A = Ca, Sr, and Ba) show photocatalytic activity under simulated solar light irradiation for phenol degradation. The catalytic behaviors of different tungstates and molybdates are shown in Figure 8.9.

Other metal molybdates (such as $CoMoO_4$ and $NiMoO_4$) have been demonstrated as promising catalytic electrode materials owing to their low cost, environmental friendliness, and rich reserves (Haetge et al. 2012, Yu et al. 2015). The $CaMoO_4$ nanosheet also offers a cost-effective 3D catalyst material for use in water-splitting devices for efficient and durable water oxidation under alkaline conditions (Gou et al. 2018). Some examples of scheelite-type materials with their photocatalytic degradations are shown in Table 8.4.

So, a detailed study on scheelite-type materials may be turned into one of the most efficient multifunctional nanomaterials for environmental challenges like water contamination by using the photocatalyst for massive water treatment.

8.4.5   SCHEELITES IN THE FOOD INDUSTRY

A report from WHO showed that about 420,000 deaths are caused for food-borne diseases every year. So food safety is an utmost concern for scientists all over the world.

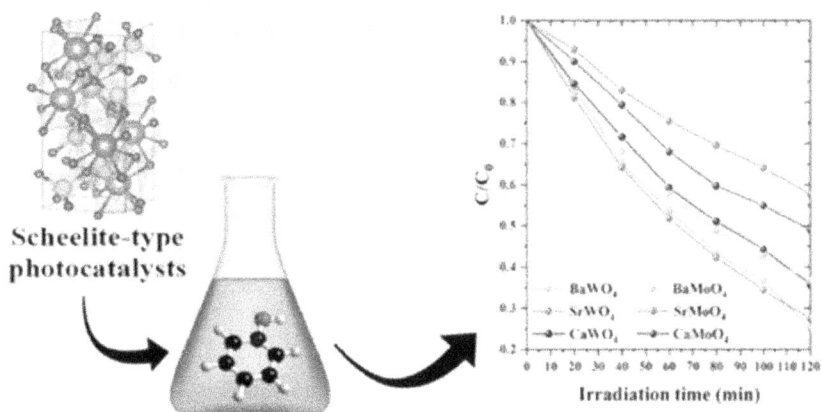

FIGURE 8.9   Schematic diagram of photocatalysis for water treatment.

Source: Kowalkińska et al. 2021.

TABLE 8.4
Photocatalytic Performance of Different Composites of BiVO$_4$ Nanomaterials for Dye Degradation

Name of the sample	Light source	Degradation (%)	Time (min)	References
BiVO$_4$	Visible light	4	180	[92](Wu et al. 2018)
Au-BiVO$_4$	Visible light	63	180	(Wu et al. 2018)
rGO/BiVO$_4$	Visible light	94	120	(Wang et al. 2014)
Ag-BiVO$_4$	Visible light	60	68.36	(Nethravathi et al. 2022)
BiVO$_4$	Visible light	60	52.72	(Nethravathi et al. 2022)
In$_2$S$_3$-BiVO$_4$	Visible light	36	180	(Wu et al. 2018)

There are so many techniques that are present for detecting food-borne pathogens like enzyme-linked immunosorbent assay (ELISA), mass spectroscopy, chromatography, etc., or reducing food-borne diseases. For diagnostic purposes various antibiotics and antifungal drugs are used to reduce the effect of food-borne diseases. Nowadays, scheelite materials (nanoparticles, macro particles, etc.) have attracted significant attention in food technology, because some of the previous techniques are time consuming (e.g. LC/GC-MS and ICP-MS methods), error prone (e.g. ELISA), and costly. To detect food contaminations (including toxins), biosensors are handy tools in day-to-day life. Recently, lateral flow assay (LFA) biosensors (Abdul Hakeem et al. 2021), also called paper biosensors, have been used to detect target molecules, proteins, or even whole cells. Various fungicides are commonly used to protect the postharvest decay of crops from fungal infections. Diphenylamine (DPA) is used

TABLE 8.5
Determination of DPA in Apple Juice Sample at BaWO$_4$/GCN/GCE

Sample	Added (µM)	Found (µM)	Recovery (%)	Relative standard deviation
Apple juice	5.00	5.12	102.40	2.67
	10.00	9.98	99.80	2.15
	15.00	15.10	100.60	2.67

Source: Karuppusamy et al. 2021.

worldwide as an antiscald agent. Mainly at the time of storage, the fruit's skin undergoes some kind of decay or distortion called scald. DPA is used as an inhibitor in fruits and vegetables to allow for storing them for a prolonged time in an air-cooled and controlled atmosphere. Consuming fruits or vegetables with DPA is dangerous to health and it has severe risks like eczema, hypertension, and bladder diseases as well as damages the red blood cells (RBCs). A scheelite-type material, barium tungstate decorated on graphitic carbon nitride nanocomposite (BaWO$_4$/GCN), was employed to detect DPA in apple juice (Karuppusamy et al. 2021). This composite showed outstanding results, as shown in Table 8.5.

8.4.6 Effect on Cells: Cytotoxicity, Cellular Uptake, and Drug Loading

The biocompatibility of scheelite materials is one of the important features of diagnostic and therapeutic approaches. The cytotoxicity of these materials is measured via various techniques like alamarBlue assay, (3-[4,5-dimethylthiazol-2-yl]-2,5 diphenyl tetrazolium bromide) (MTT) assay, XTT (methoxynitrosulfophenyl-tetrazolium carboxanilide) (XTT) assay, etc. The cytotoxicity measurement of NaTb(WO$_4$)$_2$ nanomaterials was carried out using alamarBlue assay in the HeLa cell line and it showed very little cytotoxicity even at the higher concentrations of 500 µg/mL (>90% viability is observed at 72 h compared to control cells). This result suggests that NaTb(WO$_4$)$_2$ is a good candidate for in vivo cell imaging purposes (Munirathnappa et al. 2020). A scheelite-like material NaLa(MoO$_4$)$_2$ also shows good cell viability in the case of human retinal pigment epithelium (ARPE-19) cells (Yang et al. 2015). Different morphologies of nanomaterials can affect cellular viability also. It was found that these nanomaterials with controllable morphologies, from nanorods to microflowers of various sizes in the range of 10 nm to 3 µm, showed their effect on cellular viability. All the abovementioned samples (differently sized) showed good biocompatibility with ARPE-19 cells even after a prolonged incubation time (72 h). When the size of NaLa(MoO$_4$)$_2$ was at the microscale, increasing concentrations of nanoparticles didn't cause changes in cell viability, while with decreasing size to nanoscale, the cell viability increased significantly even with increasing nanoparticle concentrations. The improved cell viability may be caused due to the enhanced surface area of the samples when the size is decreased to the nanoscale, which provides a more suitable environment for the cell adhesion and growth.

The drug loading capacity (e.g. curcumin) and sustained release of the drug also make AGM:Er^{3+}/Yb^{3+}@mSi core-shell nanoparticle a good choice for cancer therapy and photodynamic therapy (PDT) (Pavitra et al. 2022). A silica-coated $LaVO_4$:Eu^{3+} core-shell nanoparticles surprisingly exhibit good biocompatibility and cell viability in H522 and PBM cells even at higher concentrations (Ansari et al. 2011). Further studies will need to be conducted to know the effect on genes or with other proteins. So, it can be said that scheelite nanomaterials have a huge scope in cell biology to explain various unsolved mysteries. The effect of Fe_3O_4–$CaMoO_4$:Eu hybrid magnetic nanoparticles in the HepG2 cancer cell line was investigated (concentration-dependent way) using an MTT assay to estimate the cytotoxicity on cancer cells (Parchur et al. 2014). The IC_{50} value was found to be 193.26 mg/mL. It was evaluated that the biocompatibility of Fe_3O_4–$CaMoO_4$:Eu hybrid magnetic nanoparticles was excellent in the in vitro model. To detect the cell viability, $NaEu(WO_4)_2$ nanomaterials were used in HeLa cells (Munirathnappa et al. 2018). This material showed a good cell viability even with increasing concentration (shown in Figure 8.10, where NaEuW I and NaEuW II are mainly $NaEu(WO_4)_2$ nanomaterials prepared with water and ethylene glycol, respectively). Further studies are needed to understand the importance of this nanoparticle, in cell interactions and bioaccumulation/distribution for in vitro and in vivo applications.

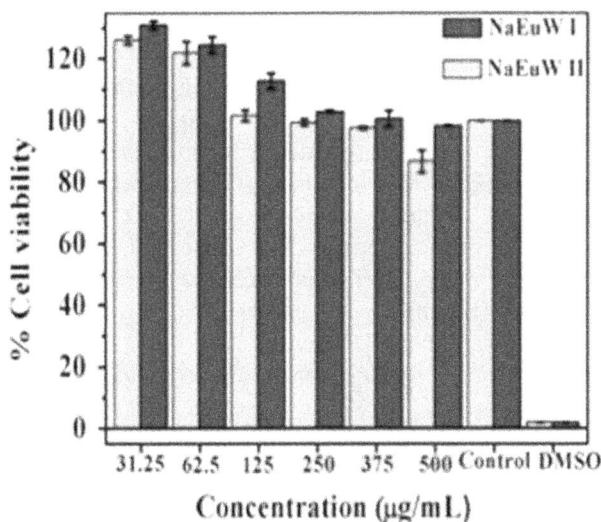

FIGURE 8.10 A plot of in vitro cell viability of HeLa cells incubated with the various concentrations of NaEuW I and II for a period of 48 hours.

Source: Munirathnappa et al. 2018.

8.4.7 Scheelites for Cell Imaging

Over a few decades, imaging techniques such as X-ray computed tomography, positron emission tomography, MRI, and fluorescence imaging have gained enormous interest in diagnostic purposes. Fluorescent imaging using various nanoparticles has been widely employed in clinical/biological applications including gene therapy, drug delivery, and biosensing (Grunert et al. 2018). Moreover, various quantum dots (QDs) have also been used for imaging biomolecules, yet they have some limitations like toxicity, chemical instability, low penetration depth, and low biocompatibility (Jin et al. 2015). In this context, researchers are trying to make some new materials with good biocompatibility and a high penetration rate. Apart from therapeutic applications, scheelite nanomaterials play an important role in diagnostic purposes also mainly in imaging technology due to their lower toxicity, higher chemical stability, optical stability, and fluorescence property. A highly uniform Eu^{3+} doped $NaGd(MoO_4)$ nanomaterial functionalized with poly(L-lysine) is a potent agent for optical and magnetic imagining purposes in vivo (Laguna et al. 2016). Again, the green fluorescence $NaTb(WO_4)_2$ nanomaterials are also promising candidates for cell imagining (Munirathnappa et al. 2020). It was observed that these fluorescent nanomaterials showed excellent bright green emissions in two different microorganisms (*E. coli* and *S. aureus*) and HeLa cells with negligible cytotoxicity. Both $NaTb(WO_4)_2$ bulk and $NaTb(WO_4)_2$ nano showed bright green fluorescence upon 378 nm excitation with remarkable variation in emission intensity. This result indicates that the size of the luminescent materials has an important role in the imaging property as well as biocompatibility. The size-dependent fluorescence emission properties confirmed the bright green fluorescence upon 378 nm excitation for both $NaTb(WO_4)_2$ bulk and $NaTb(WO_4)_2$ nano samples with significant variation in emission intensity. A mesoporous silica-coated $AgGd(MoO_4)_2{:}Er^{3+}/Yb^{3+}$ (AGM:Er^{3+}/Yb^{3+}@mSi) upconversion nanoparticle has potential in cell imaging with good biocompatibility and low cytotoxicity (Pavitra et al. 2022). The viability of HeLa cells was assessed using this nanomaterial, confirming its suitability for drug loading applications. Core-shell nanoplates of AGM:Er^{3+}/Yb^{3+}@mSi displayed bright field and dark field images of HeLa cell lines under 980 nm laser excitation. AGM:Er^{3+}/Yb^{3+}@mSi core-shell nanoplates showed the bright field and dark field images of HeLa cell lines at 980 nm laser excitation. So it can be concluded that scheelite nanomaterials show a great potential to create new imagining devices and achieve the goal with minimum effort. Figure 8.11 represents schematically the usages of scheelite nanomaterials with modification (e.g. polypeptide, protein or antibody) for cellular imaging purposes.

8.5 CONCLUSION

Over the past few decades, scheelite nanomaterials have made remarkable advances in the treatment of critical diseases, greatly promoting the application of modern precision medicine in the life system due to its enhanced therapeutic effect, antibacterial activities in multidrug-resistant bacterial strain and in bacterial cell killing, detection of various pathogens. Also in the future, these materials may be used in industrial purposes (food industry, water plant, etc.). Despite the remarkable achievements,

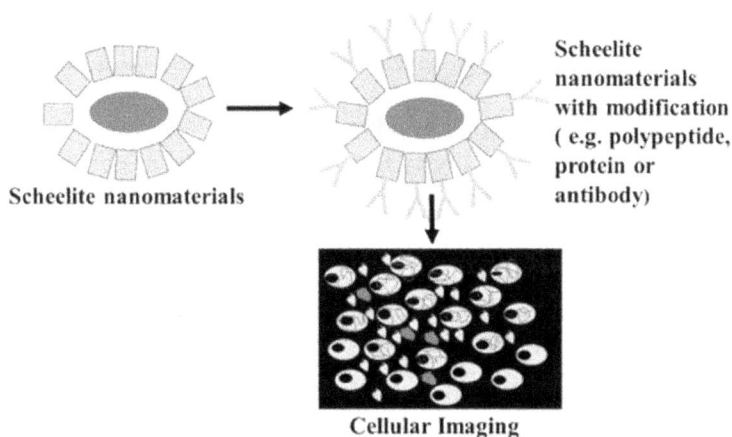

FIGURE 8.11 Schematic diagram of scheelite nanomaterials for cellular imaging.

there are still some challenges to using scheelite nanomaterials in the biomedical field. Some luminescent scheelite materials destroy cancer cells as well as normal cells also through radiation therapy. This is a big challenge for researchers to make them more nontoxic and biocompatible for normal cells. In summary, scheelite materials offer an excellent opportunity to practice precision medicine. It is expected that stable surface modification, low toxicity, and clinical trials will make them more competitive in the biological field.

REFERENCES

Abdul Hakeem, D., Su, S., Mo, Z., and Wen, H., 2021. Upconversion luminescent nanomaterials: A promising new platform for food safety analysis. *Critical Reviews in Food Science and Nutrition*, 62 (32), 8866–8907.

Aneesh, K., vusa, C.S.R., and Berchmans, S., 2017. Enhanced peroxidase-like activity of $CuWO4$ nanoparticles for the detection of NADH and hydrogen peroxide. *Sensors and Actuators, B: Chemical*, 253, 723–730.

Ansari, A.A., Alam, M., Labis, J.P., Alrokayan, S.A., Shafi, G., Hasan, T.N., Syed, N.A., and Alshatwi, A.A., 2011. Luminescent mesoporous $LaVO_4:Eu^{3+}$ core-shell nanoparticles: synthesis, characterization, biocompatibility and their cytotoxicity. *Journal of Materials Chemistry*, 21 (48), 19310.

Balaji, D., Durairajan, A., Rasu, K.K., and Babu, S.M., 2014. Sol-gel synthesis and luminescent properties of Eu^{3+}: $CsGd(WO_4)_2$ red emitting phosphors. *Journal of Luminescence*, 146, 458–463.

Bentley, R. E., 1998. Handbook of Temperature Measurement: Temperature and Humidity Measurement. *Springer Science & Business Media*, 1, 1–223.

Booshehri, A.Y., Chun-Kiat Goh, S., Hong, J., Jiang, R., and Xu, R., 2014. Effect of depositing silver nanoparticles on $BiVO_4$ in enhancing visible light photocatalytic inactivation of bacteria in water. *Journal of Materials Chemistry A*, 2 (17), 6209–6217.

Bu, W., Chen, Z., Chen, F., and Shi, J., 2009. Oleic acid/oleylamine cooperative-controlled crystallization mechanism for monodisperse tetragonal bipyramid nala(moO4)2nanocrystals. *Journal of Physical Chemistry C*, 113 (28), 12176–12185.

Cao, Y., Ding, X., and Wang, Y., 2016. A single-phase phosphor $NaLa_9 (GeO_4)_6O_2$: Tm^{3+}, Dy^{3+} for near ultraviolet-white led and field-emission display. *Journal of the American Ceramic Society*, 99 (11), 3696–3704.

Childs, P.R.N., Greenwood, J.R., and Long, C.A., 2000. Review of temperature measurement. *Review of Scientific Instruments*, 71 (8), 2959–2978.

de La Presa, P., Multigner, M., de La Venta, J., García, M.A., and Ruiz-González, M.L., 2006. Structural and magnetic characterization of oleic acid and oleylamine-capped gold nanoparticles. *Journal of Applied Physics*, 100 (12), 123915.

Deng, D., 2015. Li-ion batteries: Basics, progress, and challenges. *Energy Science and Engineering*, 3 (5), 385–418.

Dirany, N., Arab, M., Moreau, A., Valmalette, J.Ch., and Gavarri, J.R., 2016. Hierarchical design and control of $NaCe(WO_4)_2$ crystals: structural and optical properties. *CrystEngComm*, 18 (35), 6579–6593.

Durairajan, A., Thangaraju, D., Balaji, D., and Moorthy Babu, S., 2013. Sol-gel synthesis and characterizations of crystalline $NaGd(WO_4)_2$ powder for anisotropic transparent ceramic laser application. *Optical Materials*, 35 (4), 740–743.

Gai, S., Li, C., Yang, P., and Lin, J., 2014. Recent progress in rare earth micro/nanocrystals: Soft chemical synthesis, luminescent properties, and biomedical applications. *Chemical Reviews*, 114 (4), 2343–2389.

Gao, C.H., Hou, G.F., Zuo, D.F., Jiang, W.H., Yu, Y.H., Ma, D.S., and Yan, P.F., 2017. Syntheses, crystal structures, magnetisms and luminescences of two series of lanthanide coordination polymers based on tricarboxylic ligand. *ChemistrySelect*, 2 (3), 1111–1116.

Gavrilović, T. v., Jovanović, D.J., Lojpur, V.M., Dordević, V., and Dramićanin, M.D., 2014. Enhancement of luminescence emission from $GdVO_4$:Er^{3+}/Yb^{3+} phosphor by Li^+ co-doping. *Journal of Solid State Chemistry*, 217, 92–98.

Gou, Y., Liu, Q., Shi, X., Asiri, A.M., Hu, J., and Sun, X., 2018. $CaMoO_4$ nanosheet arrays for efficient and durable water oxidation electrocatalysis under alkaline conditions. *Chemical Communications*, 54 (40), 5066–5069.

Grunert, B., Saatz, J., Hoffmann, K., Appler, F., Lubjuhn, D., Jakubowski, N., Resch-Genger, U., Emmerling, F., and Briel, A., 2018. Multifunctional rare-earth element nanocrystals for cell labeling and multimodal imaging. *ACS Biomaterials Science and Engineering*, 4 (10), 3578–3587.

Guo, Y.-G., Hu, J.-S., and Wan, L.-J., 2008. Nanostructured materials for electrochemical energy conversion and storage devices. *Advanced Materials*, 20 (15), 2878–2887.

Haetge, J., Djerdj, I., and Brezesinski, T., 2012. Nanocrystalline $NiMoO_4$ with an ordered mesoporous morphology as potential material for rechargeable thin film lithium batteries. *Chemical Communications*, 48 (53), 6726.

Han, L., Xie, X., Lian, J., Wang, Y., and Wang, C., 2016. $K_2Y(WO_4)(PO_4)$: Tm^{3+}, Dy^{3+}: A potential tunable single-phased white-emitting phosphor under UV light excitation. *Journal of Luminescence*, 176, 71–76.

He, X., Zhou, J., Lian, N., Sun, J., and Guan, M., 2010. Sm^{3+}-activated gadolinium molybdate: an intense red-emitting phosphor for solid-state lighting based on InGaN LEDs. *Journal of Luminescence*, 130 (5), 743–747.

Hou, Z., Cheng, Z., Li, G., Wang, W., Peng, C., Li, C., Ma, P., Yang, D., Kang, X., and Lin, J., 2011. Electrospinning-derived $Tb_2(WO_4)_3$:Eu^{3+} nanowires: Energy transfer and tunable luminescence properties. *Nanoscale*, 3 (4),1568–1574.

Huang, J., Xu, J., Luo, H., Yu, X., and Li, Y., 2011. Effect of alkali-metal ions on the local structure and luminescence for double tungstate compounds $AEu(WO_4)_2$ (A = Li, Na, K). *Inorganic Chemistry*, 50 (22), 11487–11492.

Huang, S., Wang, D., Li, C., Wang, L., Zhang, X., Wan, Y., and Yang, P., 2012. Controllable synthesis, morphology evolution and luminescence properties of $NaLa(WO_4)_2$ microcrystals. *CrystEngComm*, 14 (6), 2235–2244.

Huang, S., Zhang, X., Wang, L., Bai, L., Xu, J., Li, C., and Yang, P., 2012. Controllable synthesis and tunable luminescence properties of $Y_2(WO_4)_3:Ln^{3+}$ (Ln = Eu, Yb/Er, Yb/Tm and Yb/Ho) 3D hierarchical architectures. *Dalton Transactions*, 41 (18), 5634–5642.

Huang, X., Li, B., and Guo, H., 2017. Highly efficient Eu^{3+}-activated K2Gd(WO4)(PO4) red-emitting phosphors with superior thermal stability for solid-state lighting. *Ceramics International*, 43 (13), 10566–10571.

Jin, G., Jiang, L.M., Yi, D.M., Sun, H.Z., and Sun, H.C., 2015. The influence of surface modification on the photoluminescence of CdTe quantum dots: Realization of bio-imaging via cost-effective polymer. *ChemPhysChem*, 16 (17), 3687–3694.

Jo, S.D., Lee, J., Joo, M.K., Pizzuti, V.J., Sherck, N.J., Choi, S., Lee, B.S., Yeom, S.H., Kim, S.Y., Kim, S.H., Kwon, I.C., and Won, Y.Y., 2018. PEG-PLA- coated and uncoated radio-luminescent CaWO4 micro- and nanoparticles for concomitant radiation and UV-A/radio-enhancement cancer treatments. *ACS Biomaterials Science and Engineering*, 4 (4), 1445–1462.

Kaczmarek, A.M., Liu, Y.Y., van der Voort, P., and van Deun, R., 2013. Tuning the architecture and properties of microstructured yttrium tungstate oxide hydroxide and lanthanum tungstate. *Dalton Transactions*, 42 (15), 5471–5479.

Karuppusamy, N., Mariyappan, V., Chen, T.W., Chen, S.M., Sundaresan, R., Rwei, S.P., Liu, X., and Yu, J., 2021. Scheelite type barium tungstate nanoparticles decorated on graphitic carbon nitride nanocomposite for the detection of diphenylamine in apple juice. *International Journal of Electrochemical Science*, 16, 210830.

Kowalkińska, M., Głuchowski, P., Swebocki, T., Ossowski, T., Ostrowski, A., Bednarski, W., Karczewski, J., and Zielińska-Jurek, A., 2021. Scheelite-type wide-bandgap ABO_4 compounds (A = Ca, Sr, and Ba; B = Mo and W) as potential photocatalysts for water treatment. *Journal of Physical Chemistry C*, 125 (46), 25497–25513.

Kumar, R.G.A., Hata, S., and Gopchandran, K.G., 2013. Diethylene glycol mediated synthesis of Gd2O3:Eu3+ nanophosphor and its Judd-Ofelt analysis. *Ceramics International*, 39 (8), 9125–9136.

Laguna, M., Nuñez, N.O., Rodríguez, V., Cantelar, E., Stepien, G., García, M.L., de la Fuente, J.M., and Ocaña, M., 2016. Multifunctional Eu-doped $NaGd(MoO_4)_2$ nanoparticles functionalized with poly(L-lysine) for optical and MRI imaging. *Dalton Transactions*, 45 (41), 16354–16365.

Lei, F., Yan, B., and Chen, H.H., 2008. Solid-state synthesis, characterization and luminescent properties of Eu^{3+}-doped gadolinium tungstate and molybdate phosphors: $Gd_{(2-x)}MO_6:Eu_x^{3+}$ (M = W, Mo). *Journal of Solid State Chemistry*, 181 (10), 2845–2851.

Lei, F., Yan, B., Chen, H.H., and Zhao, J.T., 2009. Molten salt synthesis, characterization, and luminescence properties of $Gd_2MO_6:Eu^{3+}$ (M=W, Mo) phosphors. *Journal of the American Ceramic Society*, 92 (6), 1262–1267.

Li, L., Liu, Y., Li, R., Leng, Z., and Gan, S., 2015. Tunable luminescence properties of the novel Tm^{3+} – and Dy^{3+} -codoped $LiLa(MoO_4)_x(WO_4)_{2-x}$ phosphors for white light-emitting diodes. *RSC Advances*, 5 (10), 7049–7057.

Li, Y., Wang, G., Pan, K., Qu, Y., Liu, S., and Feng, L., 2013. Formation and down/up conversion luminescence of Ln^{3+} doped $NaY(MoO_4)_2$ microcrystals. *Dalton Trans.*, 42 (10), 3366–3372.

Liu, J., Wu, Q., and Ding, Y., 2005. Controlled synthesis of different morphologies of $BaWO_4$ crystals through biomembrane/organic-addition supramolecule templates. *Crystal Growth and Design*, 5 (2), 445–449.

Liu, J., Xu, B., Song, C., Luo, H., Zou, X., Han, L., and Yu, X., 2012. Shape-controlled synthesis of monodispersed nano-/micro-NaY(MoO$_4$)$_2$ (doped with Eu$_3$$^+$) without capping agents via a hydrothermal process. *CrystEngComm*, 14 (8), 2936–2943.

Liu, S.-S., Yang, D.-P., Ma, D.-K., Wang, S., Tang, T.-D., and Huang, S.-M., 2011. Single-crystal NaY(MoO$_4$)$_2$ thin plates with dominant {001} facets for efficient photocatalytic degradation of dyes under visible light irradiation. *Chemical Communications*, 47 (28), 8013.

Liu, X., Hou, W., Yang, X., and Liang, J., 2014. Morphology controllable synthesis of NaLa(WO$_4$)$_2$: The morphology dependent photoluminescent properties and single-phased white light emission of NaLa(WO$_4$)$_2$: Eu^{3+}/Tb^{3+}/Tm^{3+}. *CrystEngComm*, 16 (7), 1268.

Liu, Y., Wang, Y., Wang, L., Gu, Y.Y., Yu, S.H., Lu, Z.G., and Sun, R., 2014. General synthesis of LiLn(MO$_4$)$_2$:Eu^{3+} (Ln = La, Eu, Gd, Y; M = W, Mo) nanophosphors for near UV-type LEDs. *RSC Advances*, 4 (9), 4754–4762.

Lu, L., Han, X., Li, J., Hua, J., and Ouyang, M., 2013. A review on the key issues for lithium-ion battery management in electric vehicles. *Journal of Power Sources*, 226, 272–288.

Macalik, L., Hanuza, J., and Kaminskii, A.A., 2000. Polarized Raman spectra of the oriented NaY(WO$_4$)$_2$ and KY(WO$_4$)$_2$ single crystals. *Journal of Molecular Structure*, 555 (1–3), 289–297.

Meng, J., Guo, H., Niu, C., Zhao, Y., Xu, L., Li, Q., and Mai, L., 2017. Advances in Structure and Property Optimizations of Battery Electrode Materials. *Joule*, 1 (3), 522–547.

Mikhailova, D., Sarapulova, A., Voss, A., Thomas, A., Oswald, S., Gruner, W., Trots, D.M., Bramnik, N.N., and Ehrenberg, H., 2010. Li$_3$V(MoO$_4$)$_3$: A new material for both Li extraction and insertion. *Chemistry of Materials*, 22 (10), 3165–3173.

Moura, J.V.B., Freitas, T.S., Silva, A.R.P., Santos, A.T.L., da Silva, J.H., Cruz, R.P., Pereira, R.L.S., Freire, P.T.C., Luz-Lima, C., Pinheiro, G.S., and Coutinho, H.D.M., 2018. Synthesis, characterizations, and antibacterial properties of PbMoO$_4$ nanocrystals. *Arabian Journal of Chemistry*, 11 (6), 739–746.

Munirathnappa, A.K., Ananda, K., Sinha, A.K., and Sundaram, N.G., 2018. Effect of solvent on the red luminescence of novel lanthanide NaEu(WO$_4$)$_2$ nanophosphor for theranostic applications. *Crystal Growth and Design*, 18 (1), 253–263.

Munirathnappa, A.K., Dwibedi, D., Hester, J., Barpanda, P., Swain, D., Narayana, C., and Sundaram, N.G., 2019. In situ neutron diffraction studies of LiCe(WO$_4$)$_2$ polymorphs: Phase transition and structure–property correlation. *The Journal of Physical Chemistry C*, 123 (2), 1041–1049.

Munirathnappa, A.K., Maurya, S.K., Kumar, K., Navada, K.K., Kulal, A., and Sundaram, N.G., 2020. Scheelite like NaTb(WO$_4$)$_2$ nanoparticles: Green fluorescence and in vitro cell imaging applications. *Materials Science and Engineering C*, 106, 110182.

Nethravathi, P.C., Manjula, M. v., Devaraja, S., and Suresh, D., 2022. Ag and BiVO4 decorated reduced graphene oxide: A potential nano hybrid material for photocatalytic, sensing and biomedical applications. *Inorganic Chemistry Communications*, 139, 109327.

Okubo, M., Hosono, E., Kim, J., Enomoto, M., Kojima, N., Kudo, T., Zhou, H., and Honma, I., 2007. Nanosize effect on high-rate Li-ion intercalation in LiCoO$_2$ electrode. *Journal of the American Chemical Society*, 129 (23), 7444–7452.

Ozdogru, B. and Capraz, O.O.Ö., 2021. Rate- dependent potential and electrochemical strain hysteresis in lithium iron phosphate cathodes for Li-Ion batteries. In *Electrochemical Society Meeting Abstracts 239* (No. 2, pp. 141–141). The Electrochemical Society, Inc.

Pan, Y., Zhang, Q., Zhao, C., and Jiang, Z., 2007. Luminescent properties of novel Ho^{3+} and Tm^{3+} doped gadolinium molybdate nanocrystals synthesized by the Pechini method. *Solid State Communications*, 142 (1–2), 24–27.

Parchur, A.K., Ansari, A.A., Singh, B.P., Hasan, T.N., Syed, N.A., Rai, S.B., and Ningthoujam, R.S., 2014. Enhanced luminescence of CaMoO4:Eu by core@shell formation and its hyperthermia study after hybrid formation with Fe_3O_4: Cytotoxicity assessment on human liver cancer cells and mesenchymal stem cells. *Integrative Biology (United Kingdom)*, 6 (1), 53–64.

Park, M., Zhang, X., Chung, M., Less, G.B., and Sastry, A.M., 2010. A review of conduction phenomena in Li-ion batteries. *Journal of Power Sources*, 195 (24), 7904–7929.

Pavitra, E., Lee, H., Hwang, S.K., Park, J.Y., Varaprasad, G.L., Rao, M.V.B., Han, Y.K., Raju, G.S.R., and Huh, Y.S., 2022. Cooperative ligand fields enriched luminescence of $AgGd(MoO_4)_2$:Er^{3+}/Yb^{3+}@mSi core–shell upconversion nanoplates for optical thermometry and biomedical applications. *Applied Surface Science*, 579, 152166.

Perera, S.S. and Rabuffetti, F.A., 2019. Dysprosium-activated scheelite-type oxides as thermosensitive phosphors. *Journal of Materials Chemistry C*, 7 (25), 7601–7608.

Postema, J.M., Fu, W.T., and Ijdo, D.J.W., 2011. Crystal structure of LiLnW2O8 (Ln= lanthanides and Y): An X-ray powder diffraction study. *Journal of Solid State Chemistry*, 184 (8), 2004–2008.

Rasu, K.K., Balaji, D., and Babu, S.M., 2017. Comparative analysis of $LiGd(WO_4)_2$:Eu^{3+} phosphors derived by sol gel and hydrothermal methods. *Journal of Crystal Growth*, 468, 159–161.

Riaz, M., Mutreja, V., Sareen, S., Ahmad, B., Faheem, M., Zahid, N., Jabbour, G., and Park, J., 2021. Exceptional antibacterial and cytotoxic potency of monodisperse greener AgNPs prepared under optimized pH and temperature. *Scientific Reports*, 11 (1), 2866.

Sadeghzadeh-Attar, A., 2018. Efficient photocatalytic degradation of methylene blue dye by SnO_2 nanotubes synthesized at different calcination temperatures. *Solar Energy Materials and Solar Cells*, 183, 16–24.

Sadeghzadeh-Attar, A., 2019. Preparation and enhanced photocatalytic activity of Co/F codoped tin oxide nanotubes/nanowires: A wall thickness-dependence study. *Applied Physics A: Materials Science and Processing*, 125 (11), 768.

Shimemura, T., Sawaguchi, N., and Sasaki, M., 2017. Structure and light emission of scheelite-type $ACe(WO_4)_2$ (A= Li, Na, or K). *Journal of the Ceramic Society of Japan*, 125 (3), 150–154.

Shivakumara, C., Saraf, R., Behera, S., Dhananjaya, N., and Nagabhushana, H., 2015. Scheelite-type MWO_4 (M = Ca, Sr, and Ba) nanophosphors: Facile synthesis, structural characterization, photoluminescence, and photocatalytic properties. *Materials Research Bulletin*, 61, 422–432.

Su, L., Fan, X., Cai, G., and Jin, Z., 2016. Tunable luminescence properties and energy transfer of Tm^{3+}, Dy^{3+}, and Eu^{3+} co-activated $InNbO_4$ phosphors for warm-white-lighting. *Ceramics International*, 42 (14), 15994–16006.

Sun, D.X., 2015. Hydrothermal synthesis of $NaY(WO_4)_2$:Tb^{3+} powders with assistance of surfactant and luminescence properties. *Journal of Materials Science: Materials in Electronics*, 26 (9), 6892–6896.

Sundaresan, P., Krishnapandi, A., and Chen, S.M., 2019. Design and investigation of ytterbium tungstate nanoparticles: An efficient catalyst for the sensitive and selective electrochemical detection of antipsychotic drug chlorpromazine. *Journal of the Taiwan Institute of Chemical Engineers*, 96, 509–519.

Sundaresan, P., Lee, C.H., Fu, C.C., Liu, S.H., and Juang, R.S., 2021. Ultrasound-assisted synthesis of barium tungstate encapsulated carbon nanofiber composite for real-time sensing of p-cresol in human urine samples. *Microchemical Journal*, 166, 106239.

Tan, G., Zhang, L., Ren, H., Wei, S., Huang, J., and Xia, A., 2013. Effects of pH on the hierarchical structures and photocatalytic performance of $BiVO_4$ powders prepared via the microwave hydrothermal method. *ACS Applied Materials and Interfaces*, 5 (11) 5186–5193.

Tian, L., Shen, J., Xu, T., Wang, L., Zhang, L., Zhang, J., and Zhang, Q., 2016. Dy^{3+} doped thermally stable garnet-based phosphors: luminescence improvement by changing the host-lattice composition and co-doping Bi^{3+}. *RSC Advances*, 6 (38), 32381–32388.

Tian, Y., Chen, B., Hua, R., Yu, N., Liu, B., Sun, J., Cheng, L., Zhong, H., Li, X., Zhang, J., Tian, B., and Zhong, H., 2012. Self-assembled 3D flower-shaped $NaY(WO_4)_2$:Eu^{3+} microarchitectures: Microwave-assisted hydrothermal synthesis, growth mechanism and luminescent properties. *CrystEngComm*, 14 (5), 1760–1769.

Tian, Y., Hua, R., Chen, B., Yu, N., Zhang, W., and Na, L., 2012. Lanthanide dopant-induced phase transition and luminescent enhancement of EuF_3 nanocrystals. *CrystEngComm*, 14 (23), 8110–8116.

Wang, J., Bu, Y., Wang, X., and Seo, H.J., 2017. A novel optical thermometry based on the energy transfer from charge transfer band to Eu^{3+}-Dy^{3+} ions. *Scientific Reports*, 7 (1), 6023.

Wang, S., Li, S., Wang, W., Zhao, M., Liu, J., Feng, H., Chen, Y., Gu, Q., Du, Y., and Hao, W., 2019. A non-enzymatic photoelectrochemical glucose sensor based on $BiVO_4$ electrode under visible light. *Sensors and Actuators, B: Chemical*, 291, 34–41.

Wang, T., Li, C., Ji, J., Wei, Y., Zhang, P., Wang, S., Fan, X., and Gong, J., 2014. Reduced graphene oxide (rGO)/$BiVO_4$ composites with maximized interfacial coupling for visible light photocatalysis. *ACS Sustainable Chemistry and Engineering*, 2 (10), 2253–2258.

Wang, Z., Lou, S., Li, P., and Lian, Z., 2017. Single-phase tunable white-light-emitting $Sr_3La(PO_4)_3$:Eu^{2+}, Mn^{2+} phosphor for white LEDs. *Applied Optics*, 56 (4), 1167.

Wang, Z., Zeng, H., and Sun, L., 2015. Graphene quantum dots: Versatile photoluminescence for energy, biomedical, and environmental applications. *Journal of Materials Chemistry C*, 3 (6), 1157–1165.

Wang, Z.J., Zhang, Y.L., Zhong, J.P., Yao, H.H., Wang, J., Wu, M.M., and Meijerink, A., 2016. One-step synthesis and luminescence properties of tetragonal double tungstates nanocrystals. *Nanoscale*, 8 (34), 15486–15489.

Wu, D., Bao, S., Wang, Z., Zhang, Z., Tian, B., and Zhang, J., 2018. Au-mediated composite In_2S_3–Au–$BiVO_4$ with enhanced photocatalytic activity for organic pollutant degradation. *ChemistrySelect*, 3 (17), 4889–4896.

Wu, X., Liang, Y., Chen, R., Liu, M., and Cheng, Z., 2011. Photoluminescence properties of emission-tunable $Ca_9Y(PO_4)_7$: Tm^{3+}, Dy^{3+} phosphor for white light emitting diodes. *Materials Chemistry and Physics*, 129 (3), 1058–1062.

Xu, L., Shen, J., Lu, C., Chen, Y., and Hou, W., 2009. Self-assembled three-dimensional architectures of $Y_2(WO_4)_3$:Eu: Controlled synthesis, growth mechanism, and shape-dependent luminescence properties. *Crystal Growth and Design*, 9 (7), 3129–3136.

Xu, L., Yang, X., Zhai, Z., Chao, X., Zhang, Z., and Hou, W., 2011. EDTA-mediated hydrothermal synthesis of $NaEu(MoO_4)_2$ microrugbies with tunable size and enhanced luminescence properties. *CrystEngComm*, 13 (15), 4921–4929.

Xu, L., Yang, X., Zhai, Z., Gu, D., Pang, H., and Hou, W., 2012. Self-assembled 3D architectures of $NaCe(MoO_4)_2$ and their application as absorbents. *CrystEngComm*, 14 (21), 7330–7337.

Xu, Z., Li, C., Li, G., Chai, R., Peng, C., Yang, D., and Lin, J., 2010. Self-assembled 3D urchin-like $NaY(MoO_4)_2$: Eu^{3+}/Tb^{3+} microarchitectures: Hydrothermal synthesis and tunable emission colors. *Journal of Physical Chemistry C*, 114 (6), 2573–2582.

Yang, M., Liang, Y., Gui, Q., Zhao, B., Jin, D., Lin, M., Yan, L., You, H., Dai, L., and Liu, Y., 2015. Multifunctional luminescent nanomaterials from $NaLa(MoO_4)_2$:Eu^{3+}/Tb^{3+} with tunable decay lifetimes, emission colors and enhanced cell viability. *Scientific Reports*, 5 (1), 11844.

Yi, G., Sun, B., Yang, F., Chen, D., Zhou, Y., and Cheng, J., 2002. Synthesis and characterization of high-efficiency nanocrystal up-conversion phosphors: Ytterbium and erbium codoped lanthanum molybdate. *Chemistry of Materials*, 14 (7), 2910–2914.

Yu, M.Q., Jiang, L.X., and Yang, H.G., 2015. Ultrathin nanosheets constructed $CoMoO_4$ porous flowers with high activity for electrocatalytic oxygen evolution. *Chemical Communications*, 51 (76), 14361–14364.

Zeng, C., Hu, Y., Xia, Z., and Huang, H., 2015. A novel apatite-based warm white emitting phosphor $Ba_3 GdK(PO_4)_3$ F:Tb^{3+}, Eu^{3+} with efficient energy transfer for w-LEDs. *RSC Advances*, 5 (83), 68099–68108.

Zhai, Y., Wang, M., Zhao, Q., Yu, J., and Li, X., 2016. Fabrication and Luminescent properties of ZnWO4:Eu3+, Dy3+ white light-emitting phosphors. *Journal of Luminescence*, 172, 161–167.

Zhang, J., Wang, X., Zhang, X., Zhao, X., Liu, X., and Peng, L., 2011. Microwave synthesis of $NaLa(MoO_4)_2$ microcrystals and their near-infrared luminescent properties with lanthanide ion doping (Er^{3+}, Nd^{3+}, Yb^{3+}). *Inorganic Chemistry Communications*, 14 (11), 1723–1727.

Zhong, J.S., Gao, H.B., Yuan, Y.J., Chen, L.F., Chen, D.Q., and Ji, Z.G., 2018. Eu^{3+}-doped double perovskite-based phosphor-in-glass color converter for high-power warm w-LEDs. *Journal of Alloys and Compounds*, 735, 2303–2310.

Index

For Product Safety Concerns and Information please contact our EU
representative GPSR@taylorandfrancis.com
Taylor & Francis Verlag GmbH, Kaufingerstraße 24, 80331 München, Germany

www.ingramcontent.com/pod-product-compliance
Lightning Source LLC
Chambersburg PA
CBHW060407220326
41598CB00023B/3049